Grazer Philosophische Studien

INTERNATIONAL JOURNAL FOR ANALYTIC PHILOSOPHY

VOL 85 - 2012

Rodopi

Amsterdam - New York, NY 2012

Die Herausgabe der GPS erfolgt mit Unterstützung des *Instituts für Philosophie der Universität Graz*, der *Forschungsstelle für Österreichische Philosophie, Graz*, und wird von folgenden Institutionen gefördert:

Bundesministerium für Bildung, Wissenschaft und Kultur, Wien
Abteilung für Wissenschaft und Forschung des Amtes der Steiermärkischen Landesregierung, Graz
Kulturreferat der Stadt Graz

Stadt GRAZ Wissenschaft

The paper on which this book is printed meets the requirements of "ISO 9706:1994, Information and documentation - Paper for documents - Requirements for permanence".

Lay out: Thomas Binder, Graz
ISBN: 978-90-420-3558-4
E-book ISBN: 978-94-012-0833-8
ISSN: 0165-9227
E-ISSN: 1875-6735

Printed in The Netherlands

INHALTSVERZEICHNIS TABLE OF CONTENTS

Themenschwerpunkt / Special Topic
BOLZANO & KANT

Gastherausgeber / Guest Editor
Sandra LAPOINTE

Weitere Artikel *Further Articles*

Grazer Philosophische Studien
85 (2012), 1–10.

INTRODUCTION*

Sandra LAPOINTE
McMaster University

Bolzano's contribution to philosophy, and in particular his contribution to the epistemology of logic and mathematics, is more often than not positioned in stark opposition to Kant. While Bolzano's formidable philosophical output is far from being exhausted by the contribution he made as a critic of Kant, his relationship to critical philosophy runs deep and is more complex than generally assumed. To start with, Bolzano paid unrelenting attention to the theories of those who followed in Kant's stride, the "new logicians" as Bolzano called them and he discussed their views with at least as much energy as he did Kant's. Indeed, Bolzano sought to determine what is distinctive of the "new logic" and thus offered a historical reflexion that is still, even today, philosophically enlightening when it comes to understanding this aspect of the reception of Kant's first *Critique*. As SANDRA LAPOINTE argues in the essay included here, Bolzano thought, for instance, that one thing that distinguishes the Kantian school of logic is their jointly held conviction that what is distinctive of logic is its "formality". He discussed the idea in detail, ultimately rejecting the theoretical assumptions on which the Kantian notion of formality rests and proposing his own alternative theory of logical form.

Historically, Bolzano stands in close relationship to Kant. Bolzano's work evolved in important ways as a response to the *Critique of Pure Reason*, but this supposes precisely that Bolzano agreed with Kant at least on the nature of the problems that shape philosophy. In Bolzano's case, disagreements also veil substantial points of convergence that make it hard not to see Bolzano's philosophy as distinctly post-Kantian. Rather than its "distance" from Kant, what characterises Bolzano's philosophy is the distinctive place it occupies "within" the reception of Kant's critical work. Bolzano's place within the reception of Kant's *Critique* is distinctive in

* My thanks to Clinton Tolley and Michael Kremer for their careful reading and valuable input.

many respects. First, Bolzano's discussion of Kantian theories is systematic in the following sense: it is rooted in a fully worked out alternative philosophical system whose sophistication and significance had no match at the time. Second, Bolzano's main theoretical focus—the epistemology of logic and mathematics—is nearly unique among Kant-commentators. Bolzano is among the few logicians and mathematicians of the nineteenth century to have engaged with Kantian theories and to have asked whether critical philosophy is apt to provide a foundation to logical and mathematical knowledge. Both factors concur to deliver what counts as the single most original and extensive critical discussion of Kant's philosophy at the turn of the nineteenth century. In this respect, it is fair to say the prolonged neglect of Bolzano's contribution to Kant-studies has been detrimental to our understanding of the Kant-reception and that this is a situation that needs to be redressed.

Assuredly, some of the issues that occupy Bolzano stand out as particularly central and enlightening when it comes to understanding the gist of Bolzano's criticism of Kant's theories. The question of the possibility of synthetic a priori knowledge is one of them. Bolzano's discussion of Kant's views on synthetic a priori knowledge serves as an illustration of the manner in which Kant's thought shaped Bolzano's in crucial respects. Bolzano accepted the Kantian view that what characterises mathematical truths is the fact that, while they are not analytic (they are synthetic), they are nonetheless a priori. This implies that he agreed with Kant, contra their predecessors, that there is a distinction to be made between analytic and synthetic judgements, that some synthetic judgements are a priori and that truths of the latter sort are constitutive of disciplines such as arithmetic and geometry. Yet, Bolzano's criticism of Kant's views on synthetic a priori knowledge is radical. Part of the disagreement comes from Bolzano's conviction that Kant failed to provide a correct analysis of the notions of analyticity and syntheticity, apriority and aposteriority, intuition and concept, among others. But the core of the dispute rests in Bolzano's belief that Kant failed to recognise the purely "conceptual" nature of a priori disciplines.[1] Bolzano in fact defines the a priori in

1. Kant, of course, had reasons to believe that mathematics is not purely conceptual. On Kant's account, only analytic knowledge is (by definition) purely conceptual. But the truths of mathematics are not analytic in Kant's sense—since they are not such that the concept of the predicate is included in the concept of the subject (Kant's definition of analyticity). Kant drew two related conclusions from this. On the one hand, he concluded that the truths of mathematics could not be purely conceptual. On the other hand, he concluded that what makes a

terms of purely conceptual knowledge so that, for Bolzano, to provide an account of mathematical knowledge *is* to provide an account of conceptual knowledge. On Bolzano's view, what it means for a proposition to be a purely conceptual truth is to stand in a conceptual, i.e. a deductive order: what characterises conceptual truths of a certain kind, say the truths of geometry, is not simply the fact that they do not contain "intuitions" but the fact that they stand in determinate deductive relations within an axiomatic structure. This, on Bolzano's account, explains both why conceptual truths are necessary and why we can know a priori that they are. Conceptual truths derive their necessity from the consequence relation in which they stand. We can know conceptual truths a priori, on Bolzano's account, because we can know a priori that they stand in the consequence relation in which they stand.[2] There is very little that is mysterious about this. Assuming a good definition of consequence, those are properties we would grant any deductive order.

What this means, of course, is that for Bolzano the problem of a priori knowledge—the central problem of the *Critique*—is completely divorced from that of analyticity as Bolzano conceives of it. This may be somewhat puzzling. Assuredly, the Kantian doctrine that has, in recent history, attracted the most interest is that of analyticity. Throughout the first half of the twentieth century, analytical philosophers have sought to do justice to Kant's insight and while it is no longer uncontested the distinction between analytic and synthetic sentences remains a staple in philosophy of language, logic and knowledge. One would expect to find that it plays as fundamental a role for Bolzano's theory as it does in the theories of his most distinguished analytical successors. Some have thought that it does: Bolzano's criticism of Kant's views on analyticity leads to a redefinition that is often described as having anticipated the standard Quinean account—the notion of logical truth—and this is a fair story to tell. Nonetheless, the question whether one should prefer Bolzano's account to Kant's is one that remains eminently open. For one thing, as Nicholas F. Stang argues in his paper, it is unclear that Bolzano's criticisms of Kant are entirely justified. In particular, a closer reading of

judgement synthetic is precisely the fact that it involves an intuition. Kant's theory culminates in the supposition that what makes the judgements of mathematics synthetic, yet *a priori* is the fact that they contain a putatively "pure intuition". Bolzano could never find a way to dismiss the idea vehemently enough.

2. This raises of course the question of the warrant for "primitive" truths. For a discussion of the problem, see (Lapointe 2011, 111ff.)

Kant shows that the idea according to which Kantian analyticity does not extend to judgements of syntactic forms other than traditional universal affirmatives is incorrect, and that Bolzano's criticism is in this respect not as potent as one might have initially thought. Besides, there are "problems" with Bolzanian analyticity *per se*, first and foremost the fact that it was explicitly not designed to account for cognitions whose predominant feature is their apriority—and thus does not preserve the Kantian intuition regarding epistemic and ontological necessity. Bolzano apparently misunderstood the problem to which a theory of analyticity is meant to provide a solution to be that of generality, not apriority. While this is retrospectively an odd mistake to make, it is not incomprehensible.[3] Besides it was tremendously fruitful. The substitutional procedure Bolzano applies to define analyticity allows him to introduce an operator that binds a variable thus anticipating Frege's (and Peirce's) views on quantification by half a century.

Of course, one might want to push the question further: why did Bolzano not call deductive knowledge "analytic", as we do today? Why did he insist on calling deductive (conceptual) knowledge "synthetic a priori"? To a certain extent, Bolzano's definitional choices explain this. It follows from Bolzano's definitions of the notions of analyticity, apriority, consequence, etc. that the relata of the consequence relation that defines the structure of axiomatic orders do not need to be analytic propositions in Bolzano's sense. The relation that defines the structure of axiomatic orders, the relation of *Abfolge*, can hold between truths that do not present the features associated with Bolzanian analyticity. But this is neither the only reason, nor the most important one. Bolzano's definitions are rooted in profoundly Kantian insights. For one, Bolzano agreed with Kant when Kant seems to imply that analytic knowledge is comparatively trifling and unimportant, thus making the problem of synthetic knowledge the central problem in the epistemology of logic and mathematics.

> In my opinion not even one principle in logic, or in any other science, should be a merely analytic truth. For I look upon merely analytic propositions as much too unimportant to be laid down in any science as proper theorems of it. Who would want to fill up geometry, for example, with propositions like: an equilateral triangle is a triangle, or is an equilateral figure, etc.? (1837, §12, 51f.)

3. See Lapointe 2011, chapter 4 for a detailed discussion.

Consequently, Bolzano subscribed to the Kantian view according to which "in all the theoretical sciences of reason synthetic a priori judgements are involved as principles". In fact, Bolzano pushed the view further. According to him:

> [...] we find judgments of this sort not only in mathematics, in the pure natural sciences and in metaphysics, as Kant proves it incontestably, but they are also to be found in logic, namely not merely among the theorems that belong to this discipline if we understand it, with Bolzano, according to a wider concept, but in the very part of it which one calls analytic and which has been worked on since Aristotle. (Příhonský 1850, 42f.)

Bolzano's rejection of the view that logic is analytic (in his, Leibniz's or Kant's sense) was enormously fertile. What's important to see is that, while Bolzano does not follow Kant, Kant's theories were nonetheless instrumental in Bolzano's reaching the conclusions he reached. They opened the path for a criticism of logic at large, and one in which Bolzano was first to engage.

Bolzano consistently devoted entire passages and sections of his published work to his discussion of Kant. Detailed analyses are to be found, in particular, in the *Contributions to a Better Founded Exposition of Mathematics* (1810), the *Science of Religion* (1834) and the *Theory of Science* (1837). One could worry that an examination of Bolzano's criticism of Kant's philosophy requires tedious and time-consuming mining into Bolzano's monumental work. But the reality is that few of Kant's critics deliver an analysis which, in spite of its daunting scale, is more easily surveyed. Indeed, a thorough review of Bolzano's criticisms is systematically presented in a little book that was published under the title *New Anti-Kant or Examination of the* Critique of Pure Reason *Following the Concepts Laid Down in Bolzano's* Theory of Science, in 1850. The *New Anti-Kant* was recently reedited by Edgar Morscher in the series *Beiträge zur Bolzano Forschung* (2003) and translated into French (Lapointe 2006). It will soon be available in English (Lapointe and Tolley forthcoming). While the book itself was not signed by Bolzano, there is little doubt that Bolzano's contribution to it was extensive. Toward the end of the 1830s, Bolzano had become aware of a French literary contest the topic of which was the presentation and critical discussion of Kant's philosophical system and the systems that emerged from it. Bolzano, who initially wished to take part, secured for the purpose the collaboration of his former student Frantisek Příhonský. The project turned out to take more time than expected and eventually took a life of its own. The result, published under Příhonský's name two years after Bolzano's death is a point-by-point assessment of the *Critique of Pure Reason* that compares very well with the most important pieces of Kant-criticism of the last two centuries. For those who wish to acquaint themselves with Bolzano's criticism of Kant, it is also an excellent starting point.

Příhonský's Introduction is all by itself an interesting document. Along with *On the Mathematical Method*[4], a slightly longer text Bolzano intended as an introduction to his *Theory of Magnitudes* and which was published only long after his death, it constitutes one of the few places to look for an "authorized" synopsis of Bolzano's logical views.[5] A comparatively longer part of this Introduction is devoted to the justification of Bolzano's "semantic realism": the view according to which the object of logic is a domain of abstract, objective entities Bolzano calls 'propositions in themselves' (and the subpropositional parts of the latter, 'ideas in themselves'). Bolzano's semantic realism is the pivot of his antipsychologistic stance. Throughout most of their history, analytical philosophers of language and logic have subscribed to the view according to which psychologism in logic (and by extension: the position according to which semantic content is to be determined on the basis of an investigation of the mind) must be eliminated and, to this extent, have mostly disregarded Kantian theories. Bolzano however did not disregard Kant and this aspect of his criticism is the occasion of a reevaluation of the Kantian standpoint. In his paper CLINTON TOLLEY argues that Bolzano's semantic realism does not succeed in overcoming the problem to which Kant's theory was meant to provide a solution. Tolley offers a Kantian response to one of the central aspects of Bolzano's criticism thus opening the door to a new kind of dialogue between analytical philosophy and Kantian scholarship.

What's interesting about Bolzano's semantic realism is not the fact that it is unproblematic—both critics and proponents acknowledge that there are epistemological and metaphysical problems associated with Bolzanian-type ontologies. What's interesting about Bolzano's focus on propositions and ideas in themselves is that it allows him to establish, for the first time, a series of important semantic concepts. As the next few pages of the Introduction of the *New Anti-Kant* make clear, it allowed Bolzano to talk about truth and reference with a systematicity and a clarity that was

4. See Bolzano 1981; see also Rusnock's translation in (Bolzano 2004).

5. The *Wissenschaftslehre and Religionswissenschaft in einer Beurtheilenden Uebersicht,* (Bolzano 1841) is another one.

unprecedented. According to Bolzano, there is a strict distinction to be made between (i) the idea in itself, (ii) the word that expresses it, (iii) the subjective idea of which it is the content and (iv) its object, if it has one. An idea in itself may or may not have an object; in the latter case it is "objectless". The notions of objectuality and objectlessness are put to work in a variety of contexts. First, Bolzano resorts to the notion of objectuality when attempting to define the conditions under which a proposition is true. In order to be true, the proposition expressed by 'A has b' must be such that the ideas 'A' and 'b' designate must both be objectual—and the object A must have the property b. Second, 'objectual' and 'objectless' are also used as second order predicates in the paraphrase of existential statements. As Bolzano sees it, to say that there are no round squares—that rounds squares don't exist—amounts to saying that the expression 'round square' designates an objectless idea.[6] Third, the notions of objectuality (the property for an idea of having an extension—*Umfang*) and content (*Inhalt*) are used as parameters in the definition of such fundamental notions as those of intuition and concept. If an idea is objectual, it has an extension—which can be singular or general. If an idea has parts, it is complex, otherwise it is simple. Combining these together we get, according to Bolzano, the following orthogonal classification:

	singular	general
simple	**intuition (e.g. 'this')**	concept (e.g. 'something')
complex	concept (e.g. 'God')	concept (e.g. 'equilateral triangle')

A great deal of what Bolzano has to say about Kant's epistemology of a priori knowledge rests on this thoroughly semantic approach to the distinction between concept and intuition and Bolzano's further justifications are the stock for an extraordinarily fruitful comparison with Kant. In his paper, Timothy Rosenkoetter focuses on Bolzano's and Kant's shared insight that intuition must be singular and argues that their views are more similar than one might think at first glance and that, indeed, Bolzano's positions might help us better understand Kant's.

One of the most interesting and original features of Bolzano's theory of intuition and concept, and in particular his notion of "purely conceptual

6. As it happens, we find here another point of agreement with Kant, the idea that existence is not real predicate.

truth", is the fact that it ultimately underpins his account of modalities as well as his views on a priori knowledge. This Příhonský discusses in the last part of the Introduction. A purely conceptual truth is a true proposition that does not contain any intuition, that is, any idea that is both singular and simple. The propositions that make up purely deductive orders, such as arithmetic, geometry, logic and ethics, are invariably purely conceptual. On Bolzano's account, such propositions are also, by definition, a priori. A proposition describes a *necessary* fact if that proposition follows from a purely conceptual truth.[7] Note, and this is important, that what makes a fact necessary is that the proposition that describes it "follows" from a purely conceptual truth (in a determinate conceptual order). By 'follows' here, Příhonský means 'is deducible' (*ableitbar*) a notion Bolzano defines as follows (cf. Bolzano 1837 §155, 113ff.):

The set of propositions T, T′, T″, ... is deducible from the set of propositions S, S′, S″, ... with respect to some components i, j, iff:

(i) *i*, *j*, ... can be varied so as to yield at least one true variant of S, S′, S″, ... and T, T′, T″, ...
(ii) whenever S, S′, S″... is true, T, T′, T″,... is also true.

What this means is at least the following: for Bolzano *a priori* knowledge is always deductive and cannot be explained without the support of a theory of logical consequence. This is an intuition that will prove to be fruitful in a number of Bolzano's successors. In general, we may understand one of the core motivations of early twentieth-century analytic philosophy as consisting precisely in explaining what warrants us to hold as true *a priori* propositions without however appealing to the idea of evidence or other similar epistemic criteria such as certitude or conviction. If the condition for knowing that an *a priori* proposition is true is that one grasp (part of) the axiomatic structure in which it is embedded, then the truth of *a priori* cognitions cannot be a mere function of evidence or intuition.

Bolzano's discussion of Kant ranges over dozens of issues that are painstakingly inventoried over some 10 pages in the Table of Contents of the

7. A proposition describes a contingent fact if it cannot be inferred from a purely conceptual truth. Finally a proposition describes a possible state of affairs if it does not contradict any purely conceptual truth, that is, if the proposition describing the non-existence of the state of affairs cannot be inferred from a purely conceptual truth. On this account, 'All equilateral triangles are equiangular' is necessary, 'Light travels at 299 792 458 m/s' is contingent and 'There are more than 8 planets in the solar system' is possible.

New Anti-Kant. The approach is chronological: Příhonský raises questions in the order in which they surface in the *Critique*, presenting Kant's view for the most part faithfully, unrelentingly appending his "*Bemerkungen*". Among the themes Příhonský discusses we find: the distinction between a priori and a posteriori cognitions, the distinction between analytic and synthetic propositions, the distinction between concepts and intuitions, Kant's demonstration of the possibility of synthetic a priori knowledge, the Kantian doctrine of space and time, the theory of the categories, the origin of error, the theory of definition, the supreme moral principle and the distinction between philosophy and mathematics, to name only a very few. The scope of the investigation is broad enough and Příhonský's discussion sufficiently detailed to make it impossible to summarize in a few points. But each topic—there are some 140 individual headings—is clearly announced, either in the form of a question (e.g. "Can intuitions enter in contradiction with the categories?"), a description (e.g. "Nullity (*Nichtigkeit*) of the doctrine of the schematism of the pure concepts of the understanding") or an assertion (e.g. "Neither the magnitude of a duration nor a distance in space belong to intuitions"), and the reader will find it easy to orient herself. Each question raised by Příhonský could be the occasion of a comparative essay almost all of which remain to be written. That is to say: for the Kant scholar, the *New Anti-Kant* is as remarkably rich as it is unavoidable. Likewise, the historian of analytical philosophy is likely to rediscover, through Příhonský and Bolzano, the significance of Kant's theories for the analytical tradition. In his paper, for instance, Waldemar Rohloff compares Bolzano's views on ordinary language and definition to Kant's, thus documenting an important moment in the history of our discipline: the birth of the modern conception of analysis. On the whole, the various points of comparison between Bolzano and Kant are as many dialogues between two traditions that not infrequently assume they have little to learn from another. The aim of the present collection of essays is to give a foretaste of the wealth of scholarship that can be gained through a study of Bolzano's criticism of Kant. And to restore Bolzano to his place in the history of philosophy. For among those whom history has unjustly overlooked, no one comes close to having Bolzano's merit and dedication.

Bibliography

Bolzano, Bernard 1841: *Wissenschaftslehre and Religionswissenschaft in einer Beurtheilenden Uebersicht.* Sulzbach: Seidel.

— 1837: *Wissenschaftslehre.* Sulzbach: Seidel.

— 1981: *Von der mathematischen Lehrart.* Stuttgart: Frommann-Holzboog. [Translated by Paul Rusnock, Amsterdam, Rodopi, 2004]

Kant, Immanuel 1992: *Lectures on Logic.* Cambridge: Cambridge University Press.

— 1800: *Logik.* Königsberg: Nicolovius.

— 1781: *Kritik der reinen Vernunft.* Riga.

— *Gesammelte Schriften.* Königl. Berlin: Preussischen/Deutsche Akademie der Wissenschaft, 1900– [referred to as 'Ak']

Lapointe, Sandra 2011: *Bolzano's Theoretical Philosophy.* Houndsmill: Palgrave.

Příhonský, František 1850: *Neuer Anti-Kant: oder Prüfung der Kritik der reinen Vernunft nach den in Bolzanos Wissenschaftslehre niedergelegten Begriffen.* Bautzen: Hiecke.

Grazer Philosophische Studien
85 (2012), 11–32.

IS LOGIC FORMAL? BOLZANO, KANT AND THE KANTIAN LOGICIANS[1]

Sandra LAPOINTE
McMaster University

Summary

In the wake of Kant, logicians seemed to have adhered to the idea that what is distinctive of logic is its "formality". In the paper, I discuss the distinction Kant draws between formality and generality of logic and argue that he ultimately conflates the two notions. I argue further that Kant's views on the formality of logic rest on a series of non trivial assumptions concerning the nature of cognition. I document the way in which these assumptions were received in his successors. In the second part of the paper I focus on Bolzano's criticism of the Kantian position and his redefinition of the notion of form. I argue that while what contemporary, post-Tarskian philosophers generally understand as the formality of logic ought to be traced back to Bolzano there are also important differences between the two positions.

As we understand the term today, Kant was not a logician and his interest in traditional logic was comparatively superficial. Of course, Kant did not entirely avoid logical issues and even innovated—one might consider the criticism of traditional syllogistics in *The False Subtlety of the Four Syllogistic Figures* (1762), or the classification of judgements he put forward in the section of the *Critique of Pure Reason* (1781) entitled 'Of the Logical Function of the Understanding in Judgements'. But his treatment of logical questions per se is often seen as relatively inconsequential.[2] By

1. I would like to thank Nick Stang and Clinton Tolley for their comments on previous versions of this paper.

2. See for instance (Kneale and Kneale 1962, 354ff.). This is not entirely surprising given what seems to have been Kant's position regarding the possibility of progress in logic: Kant had claimed that logic was „closed" (geschlossen) and „complete" (vollendet) (Bviii), which suggests that he did not think that major improvement could be achieved, let alone a reform undertaken.

contrast, Kant's dealings with questions concerning the subject matter of logic, its scope and its place within a theory of rationality should be seen as pioneering. It is true that in the *Critique* Kant discussed logic only to distinguish it from "transcendental" logic (in the *Introduction* to the eponymous section), that is, what Kant conceived as the investigation of the conditions of possibility of our knowledge of objects. But this discussion—a discussion that is developed in the first few pages of the *Introduction* to the *Logic* (1800)—turns out to be substantial. What's more, it was considerably influential. In the years that followed the publication of the (second edition of the) *Critique*, many authors sought to work Kant's ideas into their presentations of logic. The result is an interesting collection of logical *Systeme*, *Anfangsgründe*, *Grundrisse*, etc. whose authors all have in common a more or less explicit commitment to Kantian doctrines. They include Ludwig Heinrich von Jakob (1759–1827), Wilhelm Traugott Krug (1770-1842), Johann Christoff Hoffbauer (1766–1827), Johann Gottfried Karl Christian Kiesewetter (1766–1819) and Gottlob Wilhelm Gerlach (1786–1864).

The notion of a Kantian school of logic is not entirely new. Most notably perhaps, in his *System of Logic and History of Logical Doctrines*, (1868, §29), Friedrich Ueberweg introduced the label to pick out a group of authors that includes those mentioned above as well as others: Johann Gebhard Maass (1766–1823), Ernst Christian Gottlieb Reinhold (1793–1855), Carl Friedrich Bachmann (1785–1855), Jakob Friedrich Fries (1773–1843), Johann Friedrich Herbart (1776–1841) and Wilhelm Drobisch (1802–1896), for instance. Ueberweg thought relatively little of the Kantian logicians and his comments are accordingly terse:

> The logical works that proceed from the Kantian school, or which essentially share its tendency, refrain from entering upon the deeper problems, and do not make up for this want by perfect accuracy, sufficiency and clearness in the problems to which they have limited themselves. (Ueberweg 1868, §29)

In addition to being uncharitable, Ueberweg's discussion is problematic in other respects. For one, the list he provides includes at least one philosopher who clearly did not share Kant's ideas, openly criticized him, and indeed should be seen as a detractor of Kant's theories: Maass (Cf. Maass 1793).[3] More importantly, Ueberweg does not tell us what makes a logic "Kantian". Of course, the historical and theoretical connections

3. See (Allison 1973, 42).

between Kant and (most of) the philosophers Ueberweg includes in his list are often conspicuous and generally unquestionable. Nonetheless at least one interesting philosophical question remains untouched: what, if any, theoretical commitments did the Kantian logicians share?

Interestingly, we find the beginning of an answer to this precise question in Bernard Bolzano's *Theory of Science*.[4] That Bolzano should have had views on this is unsurprising. Bolzano was a keen commentator of the logical achievements of his predecessors and by the time Bolzano was writing the *Theory of Science* in the 1820s Kant's theories had "gained a pervasive influence upon the history of logic" (Calker 1822, 189). In the *Theory of Science*, Bolzano suggests that one of the distinctive features of the Kantian School is the fact that it conceives of logic as a "formal" discipline. (Cf. 1837, §12, 46f.) The idea is interesting on many accounts. For one, the idea that what distinguishes logic from other disciplines is its "formality"—in a sense that remains to be determined—seems hardly controversial so that Bolzano is here emphasizing a feature of Kant's theory that anticipates, at least at face value, a point on which most philosophers today agree. What's more, as has been recently argued (see MacFarlane 2000)[5], Kant was effectively the first modern philosopher to argue that logic is formal, so that Bolzano is pinpointing a feature of Kant's theory that was also truly innovative. Assuredly, Bolzano was not first to group Kantians around the idea that logic is formal. In the *Science of Logic* (1812), Hegel had criticised the Kantian idea that logic is merely formal—a fact Bolzano acknowledged in a footnote.[6] (1837, §12, 49) But Bolzano is

4. Bolzano is well known as a critic of Kant's philosophy. But he was an equally thorough commentator of the theories of those who followed in Kant's stride. In addition to offering systematic criticism of their ideas in the various section of the *Theory of Science* devoted to the "views of others", Bolzano also dedicates sections of his work to the "new logicians" (see 1837, §116, for instance). His examination of their views is sustained and exhaustive: the number of references to Gerlach, Hoffbauer and Jakob is at least equivalent to the number of references made to Leibniz—who is widely supposed to have had a major influence on Bolzano. And the number of references to Kiesewetter and Krug averages (or even surpasses) the number of references made to Kant himself.

5. McFarlane's analysis of the notion of formality in Kant differs from mine in many respects. For one, McFarlane argues that what Kant describes as the "general" character of logic is just one other way of saying that it is formal.

6. As Hegel puts it: "But in the first place, it is quite inept to say that logic abstracts from all content, that it teaches only the rules of thinking without any reference to what is thought or without being able to consider its nature. For as thinking and the rules of thinking are supposed to be the subject matter of logic, these directly constitute its peculiar content; in them, logic has the second constituent, a matter, about the nature of which it is concerned." (Science

not simply taking over Hegel's point. Bolzano's criticism of the Kantian conception of logic as formal and his alternative theory of logical form are both original and philosophically interesting. In the first part of the paper, I take up Bolzano's (and Hegel's) proposal that what distinguishes the Kantian school of logic is the idea that logic is formal and develop the idea. After drawing the relevant distinction between "generality" and "formality", I argue that Kant conflates the two notions (section 1). I argue further that Kant's views on the formality of logic rest on a series of non trivial assumptions concerning the nature of cognition (section 2). I document the way in which these assumptions were received in the Kantian school and identify two trends (section 3). In the second part of the paper I focus on Bolzano's criticism of the Kantian position and his redefinition of the notion of form (section 4). I argue that while what contemporary, post-Tarskian philosophers generally understand as the formality of logic ought to be traced back to Bolzano there are also important differences between the two.

1. *Generality and formality*

In the introduction to the "Transcendental Logic", Kant makes two important claims about logic, where 'logic' is to be understood broadly to correspond to traditional Aristotelian syllogistic. He claims first that inasmuch as it is "pure" (as opposed to "applied") logic does not rest on psychology. Second, Kant maintains that inasmuch as it is "general" (as opposed to "special") logic is also "formal". The first claim is comparatively unproblematic—Kant is adopting an antipsychologistic standpoint—and I want to focus on the second claim. What does Kant mean with "general"? And what does it mean, on his account, for logic to be "formal"? Finally, how are the two notions related?

The task of logic is to investigate and describe the "rules of thinking" *überhaupt*.[7] "Thinking" (*Denken*) is a technical term for Kant. When Kant speaks of "thinking", what he has in mind is not the undifferentiated totality of our cognitive performances, but a determinate subset thereof.

---

of Logic, §36) Here Hegel's point seems to be that to say that logic studies the rules of thinking, that the latter are its subject matter, is to say that logic has a content. But Hegel comes back to this criticism at various places in the *Introduction to the Science of logic* and is ultimately tightly linked to his views on knowledge at large. For this reason it cannot be discussed here.

7. By contrast the aesthetic is concerned with the rules of sensibility. (1781, B76)

On Kant's account, we are equipped with two essentially different types of cognitive capacities, sensibility and understanding that fulfil distinct yet complementary functions. We "receive" objects through sensibility and we "think" them through the understanding. More specifically, "thinking" consists in bringing the representations that are given to us through sensibility "under certain rules" (1800, A2) or, what amounts to the same, under certain "concepts" the source of which is the understanding. What results, on Kant's account, are "judgements", that is, doxastic states of a kind that includes both beliefs and cognitions. Kant's identification of concepts with rules should be noted. I won't speculate on the significance of this association for the Kantian epistemological theory as a whole, but in what follows I will assume that the terms are, for all relevant purposes, interchangeable.

Kant's use of the term 'logic' is not unambiguous. As Kant sees it, there are different types of rules to be investigated and, accordingly, different kinds of logics. In particular, thinking in Kant proceeds according to two different types of rules (or concepts). There are, on the one hand, the rules that make it possible for us to think about certain kinds of objects, the objects of the "special sciences" such as physics or mathematics. There are, on the other hand, the rules that make it possible for us to think *at all*. Kant calls the first kind of rules "contingent", the second "necessary". (Cf. 1800, A3) The investigation of the contingent rules of the special sciences pertains to the "special logics". The rules described by a special logic are contingent, not in the sense that they could have been otherwise but in the sense that there are circumstances in which they need not be applied. I am not presently applying the rules of physics or mathematics since I am not thinking about the particular objects that fall in their domain. The discipline that corresponds to traditional Aristotelian syllogistic proper Kant calls "(pure) general logic". The rules that are the object of general logic are "necessary" in the sense that, no matter what the object I am considering, I need to be applying them. As Kant sees it, general logic derives its generality from the nature of its subject-matter: what makes logic general *is* the fact that it is concerned with the "necessary" rules of thought. What this means can be understood along two ways. One could maintain, on the one hand, that what Kant means when he says that the rules of logic are "necessary"—and that logic is therefore general—is that I need to be applying these rules, if I want to be thinking *correctly*. This would amount to saying that the rules logic describes are normative of thinking: they are general normative constraints and their application is a necessary (but not sufficient) condition of acquiring correct beliefs (effective cognitions)—as opposed to erroneous ones. But one could argue, on the other hand, that what Kant means is stronger. In some places Kant claims that the rules logic describes are "absolutely necessary [...] because without them we cannot think at all" (Kant 1800, 3); they are the rules "without which there can be *no use whatsoever* of the understanding". (1781, B76; my emphasis) What these passages suggest is that the rules of logic are not merely normative; they are "constitutive" of thinking. On this second interpretation, the rules or concepts logic describes are necessary in the sense that without these concepts or rules, not only do we not have cognitions, we don't even have the beginning of an erroneous belief. What we have are mere pre-doxastic states that do not amount to belief. Consider what Kant says about thinking at large: if "thinking" is the production of doxastic states, and if bringing the manifold of sensation under the concepts or rules logic describes is a necessary condition of thinking (see above; cf. 1800, A2), then it seems to follow that the rules of logic are an indispensable ingredient, and hence constitutively involved in the production of doxastic states.

Kant's theory upholds a conceptual distinction between the idea that logic is general (that it is concerned with the necessary rules of thought) and the idea that it is formal (that it abstracts from the matter of cognition), and this is important. But Kant does not have independent criteria for generality and formality. It follows from Kant's view that I can only establish that logic is general if I can establish that logic is formal. The reason for this is that the procedure through which we discover the necessary rules of thinking *is* the procedure that consists in "set[ting] aside all knowledge that we must merely borrow from objects" (1800, A3), that is, the procedure which ascertains the rules of logic as formal.[8] In order to establish that a rule is necessary (and hence, in order to establish the rules that make up the subject-matter of general logic) I have to establish that it concerns the mere form of thought—whatever this involves. But if Kant has no independent means to establish a rule as necessary, the distinction between generality and formality collapses. The collapse of the distinction between generality and formality is documented by passages such as the following, for instance:

8. Kant takes it that it is possible to „separate" understanding from sensibility—and thus obtain „pure understanding". (See 1785, §57 A172) and hence the pure form of thinking. (Cf. 1781, B322f.)

> As general logic [i] it abstracts from all content of the cognition of the understanding and [ii] of the difference of its objects, and [iii] has to do with nothing but the mere form of thinking. (Kant 1781, B78)

In the context of Kant's theory, (i)–(iii) are roughly equivalent: 'content' and 'object' are used indifferently throughout the *Critique* to refer to the matter of cognition; and to abstract from the matter is to deal with the form of thinking. The point is that ultimately, the notion that does the work in Kant's theory when it comes to determining the nature of *general* logic is that of *formality*.[9]

2. *The two faculties account of cognition and the complementarity thesis*

In Kant's theory, the idea that logic is formal, that it is concerned with the "form of thinking" is a substantial metaphysical thesis that rests on two assumptions. It supposes, on the one hand, that we are equipped with two essentially different types of cognitive capacities, sensibility and understanding, that play distinct roles when it comes to explaining how we come to have cognitions. (Let us call this thesis the "two faculties account of cognition"). It assumes, furthermore that the application of the traditional Aristotelian distinction between matter (to sensibility) and form (to understanding) is philosophically enlightening when it comes to explaining the way in which cognitions in general—and logical cognitions in particular—come about.[10] (Let us call this "cognitive hylomorphism"). In Kant's theory, the idea that logic is formal assumes cognitive hylomorphism. In turn, Kant's cognitive hylomorphism presupposes the two faculties account of cognition.

Kant came to a first, provisional version of the two faculties account of cognition in his *Inaugural Dissertation* (1770), about a decade before writ-

9. Note that the absence of a strict distinction between generality and formality is not in itself problematic. On the contrary, that they ultimately amount to the same thing is a substantial and interesting theoretical claim and one, as we will see, to which Bolzano would subscribe. In principle at least, it is possible to argue that generality and formality coincide to deliver an interesting account of what is distinctive of logic. The concern with Kant's theory, at least from the Bolzanian standpoint, is that the notion of formality on which his theory of logic ultimately rests does not do the kind of work we would want it to do and, in particular, that it does not allow us to define eminently formal notions such as, say, validity.

10. Kant notoriously also applied the form vs matter distinction to intuition. I will not be discussing this aspect of his theory.

ing the first *Critique*.[11] There is however an important difference between the account of cognition presented in the *Inaugural Dissertation* and the account of the same as presented in the *Critique*. The *Inaugural Dissertation* supposes that there are two effectively distinct ways to cognise objects—through sensibility or through intellect—and correlatively, that what is cognised is to be divided into two classes. In the *Critique*, however, Kant argues that neither sensibility nor intellect can deliver fully-fledged cognitions when taken on their own. We arrive at cognitions as a result of the complementary action of sensibility and intellect. Kant appeals to the "complementary thesis" at different places in the *Critique.* In the Introduction to Transcendental Logic, he writes:

> If we call the *receptivity* of our mind to receive representations insofar as it is affected in some way *sensibility*, then on the contrary the faculty for bringing forth representation itself, or the *spontaneity* of cognition is the *understanding*. It comes along with our nature that *intuition* can never be other than *sensible*, i.e. that it contains only the way in which we are affected by objects. The faculty for *thinking* of objects of sensible intuition, on the contrary, is the *understanding*. Neither of these two properties is to be preferred to the other. Without sensibility no object would be given to us, and without understanding none would be thought. Thoughts without content are empty, intuitions without concepts are blind (1781, B75)

11. Cf. Kant 1770, §3. According to Lorne Falkenstein (1991), Kant's motivation for introducing the distinction was the belief that the distinction could solve some of the major problems of philosophy and metaphysics. The idea that we are equipped with two essentially different types of cognitive capacities that yield two essentially different types of representations should strike one as particularly original in context. (See Falkenstein 1991, 168f).) For one thing, Kant's two faculties account of cognition sets him apart from his Lockean and Wolffian contemporaries. Both Locke and Wolff were committed to the prevalently Cartesian thesis according to which the mind is without parts and has but one faculty of knowledge. On Descartes' view, "imagination" and "intellection" are not modules of the mind: they are applications of the unique cognitive faculty to different types of objects. (Cf. Meditation VI, 167 [AT 8]) Locke's theory is largely similar. Both sensation and reflection are types of „perceptions" or "observations" or "experiences" and what distinguishes them is the nature of their corresponding objects. Sensation arises when we observe "external sensible things", reflection from the perception of the "operations of our mind". (Essay, Book II, ch. I 1–5) On the Leibnizo-Wolffian view, the distinction between sensibility and intellect is to be explained in terms of the distinction between what is confusedly known and what is distinctly known. (See Kant 1770, §7) While both Leibniz and Wolff put forward theories on which representations may differ in quality, they were also committed to an account of cognition (representation) that involved a unique capacity. That Kant's two faculties account of cognition was original is all the more interesting in that, as we will see below, it also became distinctive of the Kantian school.

The complementarity thesis is pivotal to Kant's epistemological project as a whole. According to Kant, the fact that sensibility and intellect are complementary explains how we achieve cognition: in order to provide us with cognitions the intellect must combine (synthesize) the manifold of sensation according to given general concepts: the categories. Note, in this respect, the close connection, in Kant, between the complementarity thesis and cognitive hylomorphism. On Kant's account, the matter signifies the "determinable", the form its "determination" and "the understanding [the form—SL] demands first that something be given [the object, the matter—SL] in order to be able to determine it in a certain way". (Cf. 1781, B322f). The sense in which sensibility and intellect are complementary is precisely the sense in which matter and form are complementary, an assumption Bolzano—and some of the Kantian logicians—will deny.

3. *The Kantian logicians*

In the first section, I've argued that because Kant does not have distinct criteria for generality and formality the distinction between the two notions, in his theory, collapses. There is nonetheless a conceptual distinction between Kant's claim that logic is general (that it describes the necessary rules of understanding) and his claim that logic is formal (that it abstracts from the matter of thinking). That the two notions are conceptually independent is important to keep in mind when portraying the way in which Kant's conception of logic was received by his successors. We can distinguish two trends among the Kantian logicians. What is common to the two trends are two things. On the one hand, the Kantian logicians were committed to the two faculties account of cognition. For instance, the following authors were explicitly committed to the complementarity thesis:

Author	Two faculties account of cognition
Jakob, Ludwig Heinrich von (1800) *Grundriss der allgemeinen Logik und kritische Anfangsgründe der allgemeinen Metaphysik*, Halle, Hemmerde und Schwetschke	A representation of a manifold in consciousness is still not thinking of the latter. To this belongs that it be connected or unified in consciousness, that is, that it be represented as object or in a unity of consciousness. (Jakob 1800, §80, 32)

Krug, Wilhelm Traugott (1806) *Logik oder Denklehre*, Königsberg, Goebbels und Unzer	Concepts (conceptus, notiones) are representations that are produced through the unification of other representation and thus by admitting a represented manifold in the unity of consciousness. (Krug 1806, §24, 91)
Hoffbauer, Johann Christoff (1810) *Anfangsgründe der Logik*, Hemmerde und Schwetschke	An intuition is the representation of an object that is produced by the effect of the latter on the representing subject, the faculty of intuition is sensibility. [...] To think is to produce new representations out of some that were already given, the faculty of thinking is the understanding in the broad sense and the representations that are produced by the understanding are thoughts. [...] Even when I intuit an object that is present to a particular sense as singular object, I do not merely intuit, but also think. (Hoffbauer 1810, §§8–9, 22f.)
Kiesewetter, Johann Gottfried Karl Christian (1806) *Grundriss einer allgemeinen Logik nach Kantischen Grundsätzen*, Berlin, Lagarde	To arrive at some empirical cognition we must perceive the object through our internal and external senses. But this perception is still not a cognition. To this belongs also that we relate to an object by mean of a concept. Thus me must, through the understanding, form empirical concepts on the basis of empirical intuitions. (Kiesewetter 1806, §52, 20: see also §53, 20)
Gerlach, Gottlob Wilhelm (1817) *Grundriss der Logik*, Halle, Gebauer	How the understanding proceeds when building concepts thus becomes clear. Intuition delivers only something singular and manifold; but as the understanding unifies itself with it, singles out what coincides in this manifold, or the attributes which are identical in many perceived objects, and unifies it into a representation, it rises from the particular to the general, that is, it achieves a concept under which things that possess theses properties belong. (Gerlach 1817, §23, 12)

On the other hand, they were also all committed to the idea that logic is general in the sense that it is concerned with the necessary rules of thinking:

Author	Logic is general
Jakob, Ludwig Heinrich von (1800) *Grundriss der allgemeinen Logik und kritische Anfangsgründe der allgemeinen Metaphysik*, Halle, Hemmerde und Schwetschke	Logic can therefore be construed as a science of the necessary and general rules of thinking. (Jakob 1800, §67)

Krug, Wilhelm Traugott (1806) *Logik oder Denklehre, Königsberg,* Goebbels und Unzer	General logic is [...] pure if and to the extent that it presents the rules of mere thinking in their original determination [...] (Krug 1806, §11)
Hoffbauer, Johann Christoff (1810) *Anfangsgründe der Logik*, Hemmerde und Schwetschke	The rules of thought are rules which pertain to thinking as thinking (Hoffbauer 1810, §12) The science of the rules of thinking in general is pure logic, or what amounts to the same, the Analytic (ibid. §14)
Kiesewetter, Johann Gottfried Karl Christian (1806) *Grundriss einer allgemeinen Logik nach Kantischen Grundsätzen*, Berlin, Lagarde	Since pure general logic is meant to present the general rules of thinking, it cannot look at the particular properties of the objects of thinking but must abstract from the latter (the matter of thinking) and consider solely what is necessarily required for thinking as thinking. (Kiesewetter 1806, §2)
Gerlach, Gottlob Wilhelm (1817) *Grundriss der Logik*, Halle, Gebauer	Thus, since we only establish in logic the universal and necessary of thinking, logic becomes a universally valid (*allgemeingültige*) science, and to the extent that the cognition it includes concern the necessary and the constitutive (*gesetzmässige*) of a mental function (*Tätigkeit*) [...] it is an a priori science. (Gerlach 1817, §4)

The difference between the two trends resides in their respective attitude toward the idea that the distinction between form and matter is theoretically fruitful when applied to cognition—what I've termed 'cognitive hylomorphism'. Some Kantians—and perhaps indeed most of them—subscribed to cognitive hylomorphism to derive the idea that logic is formal:

Author	logic is formal
Jakob, Ludwig Heinrich von (1800) *Grundriss der allgemeinen Logik und kritische Anfangsgründe der allgemeinen Metaphysik*, Halle, Hemmerdeund Schwetschke	Logic abstracts from all differences in the objects that are thought. [...] (Jakob 1800, §66) Thinking always merely concerns the form of cognition [...] Since logic deals solely with this formal business, it is therefore in relation to its content a merely formal science. It deals with the formal rules of thought. (*Ibid.* §68)
Krug, Wilhelm Traugott (1806) *Logik oder Denklehre*, Königsberg, Goebbels und Unzer	In pure logic (or rather in the pure part of logic) formal thinking is considered in itself, such as it is determined a priori through the nature of the faculty of thinking [...] (Krug1806, §11)

Hoffbauer, Johann Christoff (1810) *Anfangsgründe der Logik*, Hemmerde und Schwetschke	The matter of a thought is that which in it corresponds to what is thought (*dem Gedachten*); its form that which in the latter is produced through thinking. The matter of thinking are the representations from which thoughts can be produced, and the form of thinking the manner in which this occurs (Hoffbauer 1810, §11) Pure logic is the science of the form of thinking (ibid. §17).
Kiesewetter, Johann Gottfried Karl Christian (1806) *Grundriss einer allgemeinen Logik nach Kantischen Grundsätzen*, Berlin, Lagarde	Since pure general logic is meant to present the general rules of thinking, *it cannot look at the particular properties of the objects of thinking but must abstract from the latter (the matter of thinking)* and consider solely what is necessarily required for thinking as thinking. (Kiesewetter 1806, §2, my emphasis)

But not all of them did. Gerlach's understanding of the relation between sensibility and understanding, for instance, does not rest on the traditional form/matter dichotomy.

These brief passages—and indeed each individual theory—would deserve to be examined more closely. What a cursory reading shows and the conclusion I would like to draw for now is that, contrary to what Bolzano (and Hegel) suggest, it is strictly speaking inaccurate to claim that what characterizes the Kantian school is the fact that it conceived of logic as formal. In this respect, Bolzano (and Hegel) may have overgeneralized. Or they may have inadvertently taken generality and formality to amount to the same thing. In Kant's theory, the idea that logic is formal presupposes cognitive hylomorphism and this is an assumption about which there was no consensus among Kant's followers.

The Kantian logicians shared a commitment to the two faculties account of cognition and to the idea that logic is general. In addition, and this is a point that was stressed by Bolzano—they also all subscribe to the more general methodological principle according to which it is by "reflecting on the exercise of the understanding" that one can come to discover the rules of logic. Note that in Kant's theory, the latter does not amount to an endorsement of psychologism. At least as Kant conceived of it—some of his followers came to disagree—the investigation of the rules (or concepts) of the understanding is not an empirical business but a transcendental one. (Cf. 1781, B90)[12] We discover the rules of thinking *a priori*. In particular,

12. For that matter it is not a "logical" exercise either. As Kant puts it: "I understand by an analytic of concepts not their analysis, or the usual procedure of philosophical investigation, that of analyzing the content of concepts that present themselves and bringing them to distinctness,

as Kant puts it, abstracting from all content of cognition, i.e. "attending to the mere form of the understanding in it", whatever it consists in, does not merely amount to collecting the different concepts that come to the fore when one sets the faculty of cognition into play. (Cf. 1781, B91) In this respect, it is neither the result of an introspective procedure nor of experimentation. What is relevant is the following: as Bolzano sees it, whatever this procedure consists in, for Kant and the Kantian logicians, the discovery of the rules of logic requires that I consider the "mental functions" involved in thinking and derive some of their features, and this presupposes that logical (and metalogical) knowledge is the result of an investigation of the cognitive faculty.

4. *Bolzano and the formality of logic*

The latter assumption was the main target of Bolzano's criticism: the rules of logic are not the result of an investigation of cognitive performances, transcendental or empirical. It's not that Bolzano disagreed entirely with Kant's theory of cognition. Bolzano accepted the view that we are equipped with two faculties when it comes to explaining how we acquire cognition—the faculty of concepts (understanding) and the faculty of intuitions (sensibility) (Cf. 1837, §278, 22). But he did not think much of the application of the traditional form/matter distinction to the latter and he rejected cognitive hylomorphism. Sensibility and understanding do not relate as form and matter—correlatively they are not "complementary" cognitive capacities. While Bolzano does not discuss it in detail this is a point he is keen to emphasize. It is false, according to Bolzano, that concepts without intuitions do not provide cognitions. (Bolzano/Příhonský 1850, v) In particular, it is false that the concepts of a priori discipline such as arithmetic and geometry require "pure intuitions". (Bolzano/Příhonský 1850, 75)

Bolzano's rejection of cognitive hylomorphism does not amount to a rejection of the idea that logic is (at least in part) a question of form. The heart of Bolzano's argument against Kant is that, contrary to what Kant

but rather the much less frequently attempted *analysis of the faculty of understanding* itself, in order to research the possibility of a priori concepts by seeking them only in the understanding as their birthplace and analyzing its pure use in general; for this is the proper business of a transcendental philosophy; the rest is the logical treatment of concepts in philosophy in general." (1781, B90) and the Kantian logicians are assuming, the subject-matter of logic is not thinking—and in particular the putative "form" thereof—but, to put it in a nutshell, *what is thought*: the objective content of cognitions, propositions "in themselves". Bolzano writes:

> All the scholars that subscribe to one or the other of the above claims assume tacitly that the objects that constitute the objects of logic belong under the concept of a thought, that is, if nothing else, that they must be thoughts. But what if this [...] presupposition was not completely right and the object of logic were more general? [...]What, therefore, if logic did not merely have to establish the rules that hold for thought truths (true thoughts, as they are also called), but for truths *überhaupt*; what if the validity of logical rules had to extent not to merely to thought propositions (thoughts), but also to propositions in themselves, independently of whether they are or are not thought by someone. (1837, §12, 47)

Bolzanian propositions are abstract, objective entities akin to Fregean *Gedanken:* they are both the *Sinn* of sentences (1837, §28) and the objective content of propositional attitudes such as beliefs and cognitions (1837, §34). The task of logic, on Bolzano's account, is to systematically describe the various properties of and relations that hold among propositions as well as their subpropositional parts, ideas in themselves. The first two volumes of the *Theory of Science* offer an extensive theory of the latter. The theory in question presents itself in the form of a series of implicit and explicit definitions, which are thoroughly discussed and compared to the views of others. Bolzano stipulates, for instance, that ideas and propositions in themselves are not real (*wirklich*) (1837, §54, §122), that propositions are composed of ideas (1837, §123), that ideas have a content (*Inhalt*) (1837, §56) and, unless they are "objectless" (1837, §67), an extension (*Umfang*) (1837, §66), that ideas are either simple (1837, §61) or complex (1837, §56), that there are singular ideas as well as general ones (1837, §68), etc. Bolzano proceeds to define the relevant notions: content, complexity, simplicity, extension, objectuality, etc. and introduces a number of further distinctions: ideas can be redundant (1837, §69), real or imaginary (1837, §70), they can be intuitive or conceptual (1837, §72, §75), as can propositions (1837, §133). Bolzano's "descriptive" approach to logic is intimately linked to his semantic realism: the task of logic is to describe the properties of a domain of objects, the domain of the "an sich".

What then, on this account, does it mean for logic to be formal? Logic is formal, on Bolzano's account, in the sense that it does not describe the properties of "individual propositions" but the semantic regularities that can be found in entire classes thereof:

> Logic never considers individual and fully determined propositions, i.e. one such that its subject, predicate and copula are already fully fixed, but a whole genus of propositions at once, i.e. all propositions at the same time which are such that although some of their constituents are fixed, they may be such or otherwise with respect to the others. (1837, §12, 48)

For Bolzano, the form of an idea or proposition is its structure. We call 'form' he explains, the way in which the parts of something are connected with one another (cf. 1837, §81, 389):

> One can distinguish two kinds of properties in objects that are composed of several parts: those such that by stating them one specifies which are the parts of which it is composed without determining the manner of its composition; and those that deal with the latter. But since we usually call the parts of which an object is constituted, taken together, its matter and the manner of their connection its form, one can call a property-idea which only states the parts of an object *an idea of its matter*, by contrast, one that describes the manner of the connection between these parts an *idea of its form*. (1837, §81, 389f.)

Note that strictly speaking Bolzano's use of the notion of form as applied to propositions is ambiguous. On the one hand, in passages such as the preceding Bolzano takes the form of a proposition (idea, or inference) to be one of its properties: it is the property it shares with all the propositions (ideas, or inferences) whose parts are connected in the same manner. In this sense, what I describe when I describe the structure or form of a proposition is something it shares with an entire "genus" or "species" of propositions, namely the class of propositions that is generated if we consider some or all of its non-logical parts as arbitrarily variable. Bolzano explains:

> [...] we consider the form of propositions and ideas when we keep an eye only on what they have in common with many others, that is, when we speak of entire species or genera of the latter. [. . .] one calls a species or genus of proposition formal if in order to determine it one only needs to specify certain parts that appear in these ideas or propositions while the rest of the parts which one calls the stuff or matter remain arbitrary." (1837 §12, 51)

Elsewhere, however, Bolzano says that propositions, in virtue of belonging to a certain class, "fall under" a certain form (Cf. 1837, §81, 393). Let us introduce the square brackets '[' and ']' to form designations for propositions and ideas in themselves—inverted comas are used to designate sentences and words. To say that the proposition:

[All men are mortal]

falls under the form:

'All As are Bs'

is to say that 'All As are Bs' is a "determinate connection of words or signs" through which the genus to which [All men are mortal] belongs can be "represented". (See 1837, §81, 393) On this understanding, propositional and inferential forms are linguistic expressions that use schematic letters to represent sets of substitution instances. A schematic expression represents the set of all substitution instances we obtain by arbitrarily varying the relevant components in a given proposition or inference.

Two comments at this point. First, while it is not unequivocal Bolzano's use of the notion of form—to designate the structural property *per se* and to designate the expression that represents this property—is also unproblematic. A schematic expression (form in the second sense) always represents a structural property (form in the first sense) and the only way to have cognitive access to the property is through the schematic expression. When Bolzano says that the form of [All men are mortal] is 'All As are Bs' he means both that the expression 'All As are Bs' represents the set of all propositions that have the same structure as [All men are mortal] and that we can generate that set by considering the propositions that are expressed when we arbitrarily substitute the components designated by the schematic letters, in this case all the non logical components.

Second, a "formal" expression in Bolzano's sense need not contain only logical terms essentially; correlatively a propositional or inferential form may contain non logical parts that are not varied at all. Take:

'X, who is a bachelor, is unmarried'

This expression is formal in Bolzano's sense since it designates a class of propositions, namely the set of all the propositions expressed through the variants we get by arbitrary substitution of 'X'. On Bolzano's account, any interesting semantic regularity established on the basis of the substitutional method counts as a "formal" property, even if the propositions involved contain non logical terms essentially. In the latter example, the proposition expressed is analytic in Bolzano's sense (with respect to 'X'; more on this in what follows), which explains why Bolzano makes it a point to stress that analyticity would fall within the scope of a logical investigation, even if the latter were limited to an investigation of "form". (Cf. 1837, §12, 50)

Despite the somewhat ontologically weighty underlying talk of species and genera of propositions, the advantage of Bolzano's definition of form over the Kantian one should be obvious. Kant's notion of form as applied to the cognitive faculty tells us little about the type of semantic regularities logicians are typically concerned with. In particular, as most logicians conceive of it, the subject matter of logic is logical consequence. What logic is concerned with is to clarify what it means for a conclusion to follow from a given set of premises or, what amounts to the same, what it means for an argument to be valid. Central to this endeavour is a notion of form that is very similar to Bolzano's. When contemporary logicians speak of the form of an argument what they have in mind is typically the result of a process of schematisation: in order to determine whether a given argument is valid, one replaces non-logical terms by schematic letters. An argument is valid if, however the schematic letters are reinterpreted, the result does not consists of true premises and a false conclusion. Notwithstanding the fact that Bolzano talks of propositions and not of sentences, it is hard not to notice the connection between standard definitions of validity and Bolzano's notion of *Ableitbarkeit*:

> I say that the propositions M, N, O, … are *ableitbar* from the propositions A, B, C, D … with respect to the variable parts i, j, … if every set of ideas that can be put in the place of i, j, … and that makes all of A, B, C, D, … true, also make all of M, N, O, … true. (1837, §155, 114)

Formal properties in Bolzano's sense are properties such as analyticity and *Ableitbarkeit* that are defined on the basis of Bolzano's substitutional method. But to say that *Ableitbarkeit* is a formal property in Bolzano's sense is not to say that the validity of the inference can be established by virtue of logical form alone. Something similar can be said about analyticity. A proposition is analytic in Bolzano's sense if it is the member of a set of equiveridical substitution instances:

> If however there is even only one single idea in a proposition that may be exchanged arbitrarily without altering its truth or falsity; i.e. if all proposi-

tions which turn up through the exchange of this idea with any other are all true or all false provided only that they have objectuality, then this property of the proposition is remarkable enough to be distinguished from all others for which this is not the case. (1837, §148, 83)

On Bolzano's account, establishing that a given proposition is analytic and true requires more than consideration of its form. This is clear in the case of proposition such as [X, who is a bachelor is unmarried], that is, propositions that are analytic in Bolzano's "wider" sense. In order to establish that [X, who is a bachelor is unmarried] is analytic (with respect to [X]) one needs to establish that all its objectual substitution instances are true. One needs to establish that the proposition expressed by:

'Caius, who is a bachelor, is unmarried'

is both objectual and true. That this holds for, say:

'Semprionus, who is a bachelor, is unmarried'

as well. That the propositions expressed by:

'Triangle, who is a bachelor, is unmarried'

is not objectual and therefore rejected from the range of acceptable variants, and so on. What must be noted is that establishing the objectuality/non-objectuality and truth-value of a proposition is a semantic concern in the pregnant sense: it requires a theory of reference and truth, which mandates that we pay attention to the "matter" of propositions and is therefore not a purely formal concern. One might be tempted to think that this is an idiosyncrasy of Bolzano's wider definition of analyticity. After all, propositions that are analytic in Bolzano's wider sense contain non-logical terms essentially, so we obviously cannot judge of their truth and analyticity solely on the basis of their form. But, technically, the same holds of propositions that are "logically analytic", i.e. analytic in Bolzano's narrower sense. Take the proposition expressed by:

'Caius, who is a man, is a man'

that is, what we usually understand to correspond to a logical truth. Bolzano's theory does not tell us that this proposition is true by virtue of its form, as we would expect.[13] Bolzano's theory tells us only that *if* all the objectual variants of this proposition are true, then it is analytic (with respect to all its non logical components). On Bolzano's account, we don't know a priori that logical truths are true. Propositions are analytically—and even logically analytically—true by virtue of what is the case, which makes it impossible for us to know a priori that they hold necessarily. But on the standard view, a theory of analyticity should at least account for our intuitions concerning the epistemic and modal necessity of logical truth. The conclusion to be drawn is that the kind of semantic regularity logical analyticity picks out is not, in Bolzano's theory, the kind of property we usually associate with the notion.[14]

The case is similar for *Ableitbarkeit*. While *Ableitbarkeit* is formal in Bolzano's sense, the property it denotes is not formal in the contemporary sense. The following inferences are both instances of *Ableitbarkeit*:

Formal *Ableitbarkeit*	Material *Ableibarkeit*
Caius is mortal i's *ableitbar* (with respect to 'Caius', 'men' and 'mortal') from All men are mortal Caius is a man	Caius is mortal is *ableitbar* (with respect to 'Caius') from Caius is a man

What this example shows is that *Ableitbarkeit* does not of itself warrant a distinction between instances of merely materially valid arguments and instances of formally valid ones. All inferences that preserve truth from premises to conclusion (for some arbitrary component) are instances of *Ableitbarkeit*. This, again, is problematic. Knowing that an inference is an instance of *Ableitbarkeit* does not amount to knowing a priori that *if* the premises are true, *then* the conclusion is as well. What I know, a posteriori, when I establish that an inference is an instance of *Ableitbarkeit* is that, for some arbitrary component(s), the premises and the conclusion are always true at the same time.

13. As Bolzano sees it, we can know that the proposition is analytic by virtue of its form. But this is different than saying that we can know that it is true by virtue of its form. (See Bolzano 1837, §148)

14. See Lapointe 2011, chapter 5 for a discussion of this point.

Conclusion

What is interesting about Bolzano's theory of analyticity and *Ableitbarkeit* is that it was first to deliver a substitutional account of (propositional, inferential) form and truth preservation that largely anticipated the contemporary Tarskian theory. What is disappointing about Bolzano's theories of analyticity and *Ableitbarkeit* is that they do not provide an account of what it means to be true or valid by virtue of form alone. One might be inclined to consider this an intractable flaw. But there are at least three reasons not to jump to that conclusion. Firstly, while Bolzano's theories of *analyticity* and *Ableitbarkeit* do not deliver an account of epistemic and modal necessity[15] they do point to interesting semantic regularities on the basis of a systematically defined substitutional procedure. More precisely, Bolzano's 'formal' properties systematically pick out *generalities*. This might seem like an unexciting remark but, in context, the systematic treatment of generality ought to have been one of the most important logical developments since Aristotle and anticipated Frege's account of quantification by almost half a century.[16]

Secondly, none of what I've said above implies that Bolzano's theory of logic is incapable of delivering an account of what it means for a truth to be a priori and necessary. The point rather is that these are not properties that derive from the "formality" of logic, at least not as Bolzano understands it. As Bolzano sees it, if by a "formal" discipline one means a discipline concerned with the description of the semantic regularities sets of substitution instances present, then the position according to which logic is formal is not absurd. Logic certainly is concerned with the definition of semantic regularities such as analyticity and *Ableitbarkeit*. The position is not absurd, but it is also not, according to Bolzano, adequate. This is an important point. Logic, while concerned with formal properties is not exclusively concerned with them. As Bolzano conceived of it and, he argued, as most of his contemporaries did as well, the scope of logic is not limited to the determination of logical relations such as analyticity and *Ableitbarkeit*. For one thing, it also includes the determination of properties such as apriority and aposteriority, for instance. But, on Bolzano's view, we cannot determine whether a proposition is a priori without considering certain features that pertain to its "matter". In order to establish that 'Equiangular

15. I have argued in (Lapointe 2011, chapter 7) that Bolzano's views on a priori knowledge rest elsewhere.

16. See (Lapointe 2011, chapter 4) for a discussion of Bolzano's views on generality).

triangles are equilateral' is a priori, for instance, we need to establish that it is composed solely of concepts and we can only do so by considering the nature of the components that effectively make up its content. (Cf. 1837, §12, 51) 'Equiangular triangles are equilateral' is a priori, on Bolzano's account, because none of its components: [equiangular], [triangle], [are] and [equilateral], is or contains an intuition in his sense, i.e. a deictic or indexical. (Cf. 1837, §133) What's more, and more importantly perhaps, as Bolzano sees it, formal properties, i.e. properties that are defined on the basis of the substitutional method, do not get to the heart of the matter when it comes to defining what is distinctive of logic as a theory of consequence. If we want to understand what makes a given inference *necessary*, we have to understand the relations of dependence between the propositions involved in this inference and this requires us to revert to a systematic understanding of the axiomatic structure in which they stand. Central to Bolzano's account of logic as a theory of axiomatic systems is his definition of the relation of *Abfolge*. But the relation of *Abfolge* is not defined on the basis of the substitutional method and it is not formal in Bolzano's or, as far as I can see, in any other sense.

Finally, the resulting view of logic, the idea that its formality really only amounts to generality (universality), is itself an interesting philosophical claim. In recent literature, such modally deflationary accounts of truth and logical consequence have been ascribed to Tarski himself and even vindicated. I am not claiming that the Bolzanian conception of logical truth and consequence is to be preferred. What I am claiming however is that it certainly deserves closer consideration.

References

Allison, Henry E 1973: *The Kant-Eberhard Controversy*. Baltimore: Johns Hopkins University Press.

Bachmann, Carl Friedrich 1828: *System der Logik*. Leipzig: Brockhaus.

Bolzano, Bernard 1837: *Wissenschaftslehre*. Sulzbach: Seidel.

Calker, Friedrich 1822: *Denklehre oder Logik und Dialektik*. Bonn: Eduard Weber.

Descartes, Rene 1641 (1996): *Meditations on First Philosophy*. Cambridge: Cambridge University Press.

Drobisch, Moritz Wilhelm 1836: *Neue Darstellung der Logik*. Leipzig: Voss.

Falkenstein, Lorne 1991: „Kant's Account of Intuition." *Canadian Journal of Philosophy* 21/2, 165–193.

Fries, Jakob Friedrich 1811: *System der Logik*. Heidelberg: Christian Friedrich Winter.
Gerlach, Gottlob Wilhelm 1817: *Grundriss der Logik*. Halle: Johann Jacob Gebauer und Sohn.
Hegel, Georg Wilhelm Friedrich 1812: *The Science of Logic*. Nürnberg: Schrag.
Herbart, Johann Friedrich 1834: *Lehrbuch zur Einleitung in die Philosophie*, (3rd edition). Königsberg: August Wilhelm Unzer.
Hoffbauer, Johann Christoph 1794: *Anfangsgründe der Logik*. Halle.
Jakob, Ludwig Heinrich 1791: *Grundriss der allgemeinen Logik*. Halle: Hemmerde und Schwetschke.
Kant, Immanuel 1800: *Logik*. Königsberg: Nicolovius.
— 1781: *Kritik der reinen Vernunft*. Riga.
— 1770: *De Mundi sensibilis atque intelligibilis forma et principiis* (*Kant's Inaugural Dissertation*). Königsberg.
— 1762: *Die falsche Spitzfindigkeit der vier syllogistischen Figuren*. Frankfurt und Leipzig, 1797.
Kiesewetter, Johann Gottfried Karl Christian 1806: *Grundriss einer allgemeinen Logik nach Kantischen Grundsätzen*. Berlin: Lagarde.
Kneale, William & Kneale, Martha 1962: *The Development of Logic*. Oxford: Clarendon.
Krug, Wilhelm Traugott 1806: *Logik oder Denklehre*. Königsberg: Goebbels und Unzer.
Lapointe, Sandra 2011: *Bolzano's Theoretical Philosophy*. Houndsmill: Palgrave.
Locke, John 1690: *An Essay Concerning Human Understanding*. Oxford: Oxford University Press, 1975.
MacFarlane, John 2000: Dissertation: *What Does it Mean to Say that Logic is Formal?* Pittsburgh University.
Maass, Johann Gebhard Ehrenreich 1793: *Grundriss der Logik*. Halle.
Příhonský, František 1850: *Neuer Anti-Kant: oder Prüfung der Kritik der reinen Vernunft nach den in Bolzanos Wissenschaftslehre niedergelegten Begriffen*. Bautzen: Hiecke.
Reinhold, Ernst 1827: *Die Logik oder allgemeine Denkformenlehre*. Jena: Cröker.
Ueberweg, Friedrich 1868: *System der Logik und Geschichte der logischen Lehren*. Bonn: Aldolf Marcus.

Grazer Philosophische Studien
85 (2012), 33–61.

A KANTIAN REPLY TO BOLZANO'S CRITIQUE OF KANT'S ANALYTIC-SYNTHETIC DISTINCTION

Nicholas F. STANG
University of Miami

Summary

One of Bolzano's objections to Kant's way of drawing the analytic-synthetic distinction is that it only applies to judgments within a narrow range of syntactic forms, namely, universal affirmative judgments. According to Bolzano, Kant cannot account for judgments of other syntactic forms that, intuitively, are analytic. A recent paper by Ian Proops also attributes to Kant the view that analytic judgments beyond a limited range of syntactic forms are impossible. I argue that, correctly understood, Kant's conception of analyticity allows for analytic judgments of a wider range of syntactic forms.

Introduction

Although he praised Kant's (re)discovery of the analytic-synthetic distinction as having great significance in the history of philosophy,[1] Bolzano was sharply critical of how Kant drew that distinction. His criticisms include:

(1) The distinction, as drawn by Kant, is unclear. (*NAK*, 34)

1. In *Neuer Anti-Kant* Příhonský writes: "Wir bemerken, daß, obgleich wir die Eintheilung der Urtheile in analytische und synthetische für eine der glücklichsten und einflußreichsten Entdeckungen halten, die je auf dem Gebiete der philosophischen Forschung sind gemacht worden, es uns doch scheinen wolle, als wenn sie von Kant nicht mit dem nöthigen Grade von Deutlichkeit aufgefaßt worden sei" (*NAK*, 34; cf. *NAK*, xxii). I say rediscovery because Bolzano recognizes that the distinction is inchoately present in Aristotle and Locke, and is made by Crusius in much the same way as Kant (*WL*, 87). I refer to the *Neuer Anti-Kant* (Příhonský 1850) by '*NAK*' followed by page number. I refer to Bolzano's *Wissenschaftslehre* by '*WL*,' followed by volume number, paragraph number and the page numbers in (Bolzano 1837). I refer to the *Kritik der reinen Vernunft* (*KrV*) in the usual fashion by giving the page number in the 1781 edition (A) followed by the page number in the 1787 edition (B). I refer to the Akademie edition of Kant's *Gesammelte Schriften*, (Kant 1900–) by '*Ak.*,' followed by volume and page number. Where applicable, I indicate which translation I have used. '*JL*' refers to the *Jäsche Logik*.

(2) The distinction, as drawn by Kant, inappropriately relies on the metaphor of the 'containment,' or 'inclusion' [*Enthaltensein*] of one concept in another. (*NAK*, 34)

(3) The distinction, as drawn by Kant, only applies to judgments within a narrow range of syntactic forms, namely, universal affirmative judgments. Kant cannot account for judgments of other syntactic forms that, intuitively, are analytic. (*WL* II, 88)

(4) Kant ignores the semantically interesting class of sentences, those whose truth-value is unaffected by substitution of one or more of their terms. (*NAK*, 35-36)

(5) Kant fails to prove there are synthetic judgments in mathematics. (*NAK*, 36)

(6) Kant's 'containment' criterion for analytic judgments is too broad; it includes judgments that, intuitively, are not analytic. E.g. [king of Macedonia] is contained in the concept [father of Alexander, King of Macedonia] but, intuitively, the judgment <The father of Alexander, King of Macedonia, is King of Macedonia> is not analytic.[2] (*NAK*, 35)

(7) Kant is wrong to identify what is 'contained' in a concept with what is 'thought' in the concept; the latter is a purely psychological matter of which other ideas I contingently associate with the concept. (*WL* II, 89)

The list above is not exhaustive and there are other criticisms. This paper is part of a larger project of offering a comprehensive interpretation of the Kantian analytic-synthetic distinction that answers all of them. Here however, I will focus on (1)–(7), and especially on (3). I will argue that, at least on this point, Bolzano is mistaken. Although Kant does not emphasize this point, his theory of analytic judgment has the resources to account for judgments of a wide variety of syntactic forms other than universal affirmative categorical judgments, that is, judgments of the form <All As are B>. I will propose a more charitable reading of Kant's texts that supports attributing to him the view that there can be analytic judgments of other syntactic forms. In doing so, I will indirectly address points (1) and (2), and to a lesser extent (4), (6), and (7). Point (5) I will not address at all

2. I use square brackets [...] to refer to concepts and angle brackets <...> to refer to judgments. Thus, '[A]' refers to the concept expressed by A, and <p> refers to the judgment expressed by the sentence p. In the context of Bolzano's philosophy, square brackets refer to *Vorstellungen an sich* and angle brackets refer to *Sätze an sich*.

other than to point out that, correctly understood, Kant's theory of analytic judgments is especially vulnerable on that point. By arguing that Kant can consistently classify as analytic a wider class of judgments than Bolzano, and many contemporary readers of Kant, assume, I am arguing that it is in principle harder for Kant to show that mathematical and natural scientific judgments do not fall within that wider class of analytic judgments.

In section one I begin my interpretation of Kant's analytic-synthetic distinction by offering an interpretation of the passage in the Introduction to *KrV* (A6/B10) where Kant introduces the distinction and appears to give a definition of what it is for a judgment to be analytic. I then go on to explain why it would be a problem for Kant's epistemological project in the *Kritik* if his definition of analyticity did not apply to judgments outside a narrow range of syntactic forms. I then examine several of Bolzano's examples of judgments that, intuitively, should count as analytic, but which do not pass the relatively narrow definition given in the Introduction to *KrV*. In section two I turn to the other main discussion of analyticity within the *KrV*, the section titled "On the supreme principle of all analytic judgments" (A150-3/B189-3). I argue that the meaning of that passage depends upon how Kant conceives of the principle of contradiction, and the relation between that principle and the principle of identity. In the third section, I argue that Kant thinks of these as two principles that immediately entail one another, and, on that basis, in the fourth section I offer an interpretation of the "supreme principle" section on which Kant, contra Bolzano, has a conception of analytic judgments that includes judgments of a much wider range of syntactic forms.

Bolzano's claim—that Kant's notion of analyticity is restricted to judgments within a narrow range of syntactic forms—is also put forth by Ian Proops in an influential recent paper (Proops 2005). Despite significant differences in their readings of Kant, Bolzano and Proops both attribute to Kant the view that analytic judgments only have a restricted range of syntactic forms; while Bolzano appears to read Kant as restricting analyticity to judgments of the form <All (A+B)s are B>, Proops correctly recognizes that Kant recognizes analytic judgments of the form <No (A+B)s are ~B>, but, on his reading, all analytic judgments must be either of these two forms. Whereas Bolzano takes this to be a criticism of Kant's view, Proops is neutral on whether Kant is wrong on this point. The Bolzano/Proops thesis will be my stalking horse throughout this paper.[3]

3. Two recent papers by Lanier Anderson (Anderson 2004, 2005) vindicate the idea that

1. *The problem of analytic judgments with alternate syntactic forms*

In the *Kritik der reinen Vernunft* Kant introduces the analytic-synthetic distinction in the following famous passage:

> In all judgments in which the relation of a subject to the predicate is thought, (if I consider only affirmative judgments, since afterwards the application to negative ones is easy,) this relation is possible in two different ways. Either *the predicate B belongs to the subject A as something that is (covertly) contained in this concept A*; or *B* lies entirely outside the concept *A*, thought to be sure it stands in connection with it. In the first case I call the judgment **analytic**, in the second **synthetic**. Analytic judgments (affirmative ones) are thus those in which the connection of the predicate is *thought through identity*, but those in which this connection is thought without identity are to be called synthetic judgments.[4] (*KrV*, A6–7/B10–11; my italics)

This passage has been the focus of a great deal of critical scrutiny, so before continuing I want to explain how I interpret it.

It is a basic article of Kantian lore that Kant gives several apparently different criteria for analyticity.[5] The passage quoted above contains two of them; another we will see later in the next section.[6] They are:

Kantian analytic judgment include judgments of a wider range of syntactic forms (e.g. disjunctive judgments). However, I do not have space in this paper to fully explain why I reject Anderson's view. For a telling, but not decisive, objection to Anderson's interpretation, see Proops (2005, 600).

4. Translations from *KrV* are taken, with occasional minor modifications, from Kant (1997).

5. See Van Cleve (1999, 17–21) for a helpful comparison of the different criteria.

6. There is also Kant's statement, immediately after the passage quoted, that analytic judgments are judgments of "clarification" [*Erläuterungurtheile*] while synthetic judgments are "judgments of amplification" [*Erweiterungsurtheile*] (A7/B11). Proops (2005) reads this as a distinct criterion on analyticity, in contrast with the containment-or-identity criterion of A6/B10. On this point I think Proops goes astray. He assumes that Kant's characterization of analytic judgments as judgments of clarification rather than amplification entails that analysis of a concept never reveals judgments not previously known to me. Admittedly, this is suggested by Kant's claim that "nun ist hieraus klar: daß durch analytische Urtheile unsere Erkenntnis nicht erweitert werde" (A7). But this sentence is removed in the B edition, and even in the A edition it continues: "sondern der Begriff, den ich schon habe, aus einander gesetzt, und mir selbst verständlich gemacht wurde." But, presumably, if a concept is made 'verständlich' to me, I come to know judgments I did not previously know, and the first part of the sentence only supports Proops' reading if 'erweitern' means provide us with knowledge of judgments we did not previously know. But '*erweitern*' is precisely the word Kant uses to characterize synthetic judgments in the first place [*Erweiterungsurtheile*], so the sentence from the A edition provides no support to Proops' interpretation not already found in the characterization of synthetic judgments as '*Erweiterungsurtheile*.' A more plausible reading, and one that does not conflict with

Containment criterion: A judgment <All As are B> is analytic if and only if the predicate concept [B] is contained in the subject concept [A].

Identity criterion: A judgment <All As are B> is analytic if and only if the connection of the subject [A] and the predicate [B] is 'thought through identity.'

I refer to these as criteria because I want to prescind for now from assumptions about which, if either, is definitional of analyticity for Kant, and which, if either, is simply a necessarily true bi-conditional. Both criteria have their problems—it is at least as unclear how one concept can be contained in another as it is how their connection can be 'thought through identity'—but first I want to focus on the containment criterion.

As we saw earlier, Bolzano objects to the Containment criterion that it inappropriately relies on a metaphor (*NAK*, 34f.). I think this is mistaken. Kant is not using 'containment' metaphorically; he means quite literally that in an analytic judgment the predicate is a part of the subject.[7] In other words, I think the Containment criterion should be understood—at least provisionally—as follows:

Containment criterion: A judgment <All As are B> is analytic if and only if the predicate [B] is a part of the subject [A.]

This criterion faces the immediate problem—Bolzano's point (6) from above—that it is too wide: it counts as analytic some judgments that, intuitively, are not analytic. For instance, since [bachelor] is part of [son of a bachelor] it follows that <Every son of a bachelor is a bachelor> is analytic, which it surely is not.[8] What this shows is that the notion of 'part' in the Parthood-criterion needs to be restricted; only certain parts [B] of a concept [A] are such that <All As are B> is analytic. This would be a serious problem

---

the Containment-or-identity criterion given a few sentences earlier, is that analytic judgments clarify concepts by bringing to our awareness marks that were contained in them (and thus were contained in our thoughts involving those concepts all along), and thereby allow us to know new judgments, but do not 'amplify' our knowledge because they only call to our attention the logical consequences of judgments we already knew. Furthermore, the idea that an analytic judgments cannot constitute a new piece of knowledge cannot be squared with Kant's insistence that, for all we know, there may always be more analytic marks to be discovered in an empirical concept (*JL* §103, *Ak.* 9:142); cf. *Ak.* 24:272, 915, 757.

7. Lapointe (2011, 20ff) makes a similar point.

8. This is my replacement for Bolzano's 'King of Macedonia' example.

if the Containment Criterion were Kant's ultimate view about what it is for a judgment to be analytic. I will argue in the final section of this paper that this is not the case, so the problem Bolzano points to—not all of the parts of a concept are analytic marks of the concept —does not cut very deeply.

One might object to the Containment criterion that talk of one concept being a part of another concept can only be metaphorical because parthood is only a relation among spatiotemporal particulars. The exact ontological status of concepts for Kant is somewhat unclear, but, I take it, it is quite clear they are not spatiotemporal particulars, so if parthood were only a relation between spatiotemporal particulars, then this objection would be sound. However, Kant himself does not restrict parthood to spatiotemporal objects. In various passages of the *KrV* and elsewhere Kant discusses the mereological structure of non-spatiotemporal things in themselves, and contrasts it with the mereological structure of spatiotemporal appearances.[9] He also consistently denies that things in themselves are parts of appearances. In none of these passages does Kant deny that things in themselves can stand in parthood relations with other things in themselves (*Ak.* 8:201–10). So, while it may be true that parthood is a relation that can only hold among spatiotemporal objects (although I have my doubts), this is not Kant's view.

To return to the Identity criterion, the meaning of that criterion depends upon the meaning of the phrase "thought through identity." If "thought through identity" just means *identity*, then the Identity criterion should be read as follows:

> *Identity criterion*: A judgment <All As are B> is analytic if and only if [A]=[B].

However, the Identity criterion, so understood, stands in obvious tension with the Containment criterion, because the Containment criterion entails

9. In the *KrV*, the Second Antinomy (A434–443/B462–471); *Prolegomena* §13 (*Ak.* 4:286). Some readers might object to my attributing views about the mereological structure of things in themselves to Kant on the basis of the Second Antinomy. Isn't the point of the point of the Antinomies to show that each of these opposing arguments is fallacious? While I don't have the space to fully defend my reading of the Antinomies, I will note that Kant's view is that the Antithesis of the Second Antinomy is fully correct *if we limit it to objects of sensible intuition*, and the Thesis of the Second Antinomy is fully correct *if it expresses how we would most rationally think about things in themselves*, although we cannot know its conclusion to be true. All I am claiming in the body of the text is that Kant does not regard mereological relations (composition, simplicity, etc.) as *ipso facto* restricted to spatial objects.

that judgments like <All gold is metal> are analytic while the Identity criterion entails that they are not. Thus, it might seem that Kant gives two incompatible criteria on analyticity in the course of the two sentences from the Introduction to *KrV* quoted earlier.

But, as the following passage from Kant's metaphysics lectures makes clear, Kant understands the Containment and Identity criteria as equivalent:

> All analysis is nothing other than the consciousness that a certain mark is contained in the concept of the thing. The mark that is contained in the concept of the thing is partially identical with the [concept] of the thing. Every analytic mark is identical, not with the whole concept, but rather only with a part. For instance, 'every body is extended,' is really judged through identity. (*Metaphysik Mrongovius*, *Ak.* 29:789)

When Kant says in the Introduction to *KrV* that in an analytic judgment the relation of the predicate to the subject is thought 'through identity' he means that the predicate is identical to a (perhaps improper) part of the subject-concept. There is no conflict, therefore, between the Containment and Identity criteria.[10] However, there does appear to be a conflict between the passage just quoted from the Mrongovius lectures and Kant's Critical view. In that passage, Kant denies that judgments in which subject and predicate are identical—which elsewhere calls 'identical' judgments and 'tautologies'—are analytic, but in a number of published writings in the Critical period he seems to accept that identical judgments are analytic (*KrV* B17; *Prolegomena* §2, *Ak.* 4:269). Since the Mrongovius passage occurs in an unpublished lecture transcript, while the published passages support the claim that identical judgments are analytic, I will assume that the view expressed by Kant in *KrV* is that identical judgments are analytic.

We have now seen how to combine the Containment and Identity criteria into a single criterion:

> *Containment Criterion*: A judgment <All As are B> is analytic iff the predicate [B] is identical to a proper or improper part of [A].

I have formulated these two criteria as criteria on the analyticity of judgment having the form <All As are B>. This raises the natural question,

10. Similarly, Bolzano is mistaken when he complains that Kant's definition of analytic judgments in *JL* as judgments "whose certainty rests on identity of concepts" applies only to identical judgments like <Man is man> (*WL* II, 87).

(how) does the analytic-synthetic distinction apply to judgments of other syntactic forms? I did this because in the crucial passage from the Introduction to *KrV* Kant is careful to restrict the range of this criterion: to judgments with one subject and one predicate (categorical judgments), and affirmative ones at that. He claims that the "application to negative ones [judgments] is easy" (A6/B10). Bolzano correctly points out that the Containment criteria, as formulated at A6/B10, only applies to judgments within a limited range of syntactic forms, but fails to note that Kant himself explicitly states that, as formulated, it is not supposed to apply more widely. Bolzano concludes from this that Kant cannot allow for analytic judgments of different forms. In the remainder of this section I want to focus on whether Kant is right about that, whether the 'application' of these criteria to negative, and other kinds of judgments, is as easy as Kant says it is.

Bolzano offers the following as examples of forms of judgments that, intuitively, are analytic, but fail the Kantian criterion because they are not of the right form:

(a) <A, which is B, is A.>
(b) <Every A is either B or ~B.> (*NAK* 35; *WL* II, 88)

Correctly understood, none of these is a clear counter-example to the Containment criterion, although the first does pose some problems. First I will explain why judgments of these forms are not problematic for Kant, and then I will explore other judgmental forms that do pose very serious problems for Kant's theory of analytic judgments.

Since Kant clearly intends the Containment criterion given above to apply in the first place to universal judgments—let us hope that the "application" to particular and singular judgments is "easy"—I will read (a) as:

(a) <All As, which are B, are A.>

This judgments can be read in either of two ways: either the relative clause 'which is B' is a component of the subject-concept [A], or it is not and expresses merely a separate judgment. Therefore (a) can be read either as:

(a*) <All (As which are B) are A.>
(a†) <All As are B, and are A.>

But (a*) clearly satisfies the Containment criterion on analyticity given above: the predicate [A] is a part of the subject [As which are B]. On this reading, (a*) is analytic. However, (a†) does not satisfy the Containment criterion because it does not have only one predicate: it is what we would now call the conjunction of the judgment <All As are B> and the analytic judgment <All As are A.> Whether this is analytic will depend upon whether conjunctions of judgments can be analytic; as any students of Kant's logic knows, Kant does not have a logical function of conjunction of judgments, so there may be serious problems in his giving a criterion for whether (a†) is analytic or not. (Intuitively, (a†) is analytic if and only if its first conjunct, <All As are B>, is analytic.) But this shows that the original judgment (a) is problematic for Kant, not in virtue of the relative clause 'which is B,' but in virtue of the fact that Kant does not recognize conjunctions of judgments as a separate logical form. Bolzano may ultimately be right that Kant cannot account for the analyticity of judgments of the form of (a), but not for the reasons he thinks. At the end of the paper, I will return to (a†) and argue that, charitably understood, Kant does have an account of analyticity of judgments of that form.[11]

This brings us to the second judgment form that, according to Bolzano, could not count as analytic on Kant's criterion:

(b) <All As are B or not-B.>

Bolzano is right that this does not satisfy Kant's criterion on analyticity, but this is not a weakness of Kant's view because Kant has his own philosophical reasons for thinking that (b) is not analytic. If all judgments of the form (b) were analytic, then it would be analytic that every object is fully determinate with respect to every contradictorily opposed pair of predicates [B] and [~B]. In the opening of the Transcendental Ideal section of *KrV* (A571–573/B599–601) Kant explicitly distinguishes that principle, which he calls the 'principle of complete determination,' from the principle that "of every two contradictorily opposed predicates only one can apply to" an object, which "rests on the principle of contradiction." This principle is:

11. See the discussion of this example in Siebel (2011, 110). I share Siebel's puzzlement at Bolzano's distinction between <A, which is B, is B> and <A, which is B, is A>. In the paper I've tried to go a little further in understanding the barriers Bolzano might have seen to the analyticity of the latter judgment.

(b*) <No A is B and not-B.>

It might be thought that (b) and (b*) are logically equivalent, but for Kant they are not. I do not have the space here to fully explain why they are not equivalent for Kant, but it suffices for our purposes to point out that, while (b*) is a logical principle for Kant, (b) is not. On Kant's view, logic does not determine whether objects are fully determinate with respect to every pair of contradictorily opposed predicates. Kant wants to leave room for the *logical* possibility that there is some object that falls *neither* under [B] *nor* under [~B].[12]

Abstracting for a moment from the particular examples Bolzano uses, his point is that the Containment criterion only applies to universal affirmative judgments. It would appear that Kant's criterion on analyticity does not even apply to judgments of any of the following forms:

(c) <No (A+B) is ~B>, e.g. <No gold is non-metallic>.[13]
(d) <No ~B is (A+B)>, e.g. <No non-metallic thing is gold.>
(e) <All ~B are ~(A+B)>, e.g. <All non-metallic things are not gold>
(f) < No A is (B + ~B)>, e.g. <No metallic thing is gold and not-gold.>

The Containment criterion states that a judgment is analytic if and only if the predicate concept is a proper or improper part of the subject concept. But in (c) through (f) the predicate is neither a proper nor improper part of the subject-concept. The prospect for allowing analytic judgments of different-syntactic forms looks hopeless, given the Containment criterion.

But recall that Kant explicitly restricts the Containment criterion to affirmative subject-predicate judgments, and (implicitly) to universal ones at that. But he also says that the "application" to—by which I take him to mean the extension of that criterion, since, as we have seen, the criterion itself is not applicable—to negative judgments is "easy". In the other main text on the analytic-synthetic distinction in *KrV*, "On the supreme prin-

12. I explore this issue in much greater detail in Stang (forthcoming). Kant does describe the principle of excluded middle ((b) in the main text) as "springing" from the principle of contradiction and the principle of identity (MM, *Ak.* 29:791) but I don't think that he means by this that it is analytic (as those principles are).

13. I follow Kant in writing negated particular judgments using 'No.' So <No A is B> is the negated particular judgment <~(Some A is B)>.

ciple of all analytic judgments" Kant writes: "if the judgment is analytic, whether it be negative or affirmative, its truth must always be able to be cognized sufficiently in accordance with the principle of contradiction" (A151/B190). This would appear to be Kant's application of the definition of analyticity to negative judgments.[14] In the next section I offer a detailed look at that text. Proops (2005) argues that the "supreme principle" section extends Kant's conception of analyticity to judgments of form (c). I will argue in the next section that, on Proops' reading of that text, judgments of form (d)–(f) do not count as analytic on Kant's conception. In the remainder of this section I want to explain why it would be a serious problem for Kant if judgments of forms (c)–(f) were not analytic.

Note that judgments of form (f) are not categorical judgments; they are disjunctive judgments. Since the Containment criterion, as given at A6/B10, applies only to judgments with one predicate and one subject, it is not at all clear that it can apply to disjunctive judgments, which have multiple predicates, or hypothetical judgments, which have multiple subjects and multiple predicates. On Proops' reading, the "supreme principle" section does not alter the picture; the expanded criterion Kant gives there still applies only to categorical judgments. This is problematic for several reasons. First of all, there is significant textual evidence that Kant held that there are analytic judgments with non-categorical form, although that evidence is mainly confined to unpublished student transcripts of his lectures on metaphysics. The clearest evidence occurs in the von Schön transcripts of Kant's lectures from the late 1780s:

> According to its form the principle of contradiction is a categorical judgment, and is therefore the judgment of all analytic judgments only insofar as they are categorical [*nur in so fern sie kategorisch sind*] and expresses in complete generality that a predicate may not contradict the subject; it determines merely the relation of the predicate to the subject. 'To any subject, [at most] one out of [a pair of] contradictory predicates applies' [*Cuilibet subjecto pradicatorum contradictoriorum competet unum*] is the principle of disjunctive judgments and the principle of hypothetical judgments is: every judgment has a ground. Therefore we have, as highest principles of all analytic judgments, the principle of contradiction, a principle of division and one of the connection of the conditioned with the condition. We cannot have more, because there are no

14. This is confirmed by the fact that in the A edition Kant writes "denn auf die verneinende ist die Anwendung leicht" (A6) while in the B edition he adds 'nachher' (B10); I take 'nachher' to be a reference to the "Supreme Principle" section, where Kant shows how to apply his Criterion to negative judgments.

> further relations of cognition [of subject and predicate] in judgments. These judgments are all analytic. (*Metaphysik von Schön, Ak.* 28:478)[15]

Aside from textual considerations, though, there are overwhelming systematic reasons why Kant needs to accept analytic judgments of a wide range of syntactic forms, including, (c)–(f). If, for example, a judgment of the form of (f) is not analytic, then it is either *a priori* or *a posteriori*. If it is *a posteriori*, this means it can only be known through experience. Intuitively, though, that is absurd and would constitute a one line *reductio* of Kant's epistemology. It must, therefore, be *a priori*.

If it is synthetic *a priori*, though, according to Kant's transcendental epistemology, it is made true by the forms of experience: space, time and the categories. Not only is this highly counter-intuitive, it also leads to problems for Kant's system. If it is *a priori* cognizable by us, then the *KrV* provides an account of our ability to cognize it, since Kant is quite clear that he believes that the *KrV* provides a systematic account of all our synthetic *a priori* knowledge. At the beginning of the 'System of all principle of pure understanding' Kant writes:

> Now our task is to exhibit in systematic combination the judgments that the understanding actually brings about *a priori* subject to this critical warning, for which our table of the categories must doubtless give us natural and secure guidance. For it is precisely these whose relation to possible experience must constitute all pure cognitions of the understanding *a priori*, and whose relation to sensibility in general will, on that very account, display all transcendental principles of the understanding completely and in a system. *A priori* principles bear this name not merely because they contain in themselves the grounds of other judgments, but because they are not themselves grounded in higher and more general cognitions. (*KrV*, A148–9/B178–9)

The *Kritik* does not contain an inventory of every *a priori* cognition of which we are capable—for that, we must wait for the 'system' for which the *Kritik* is merely a necessary propadeutic (*KrV*, A13/B27). But the *Kritik* must present systematically and completely all principles of *a priori* cognition, that is, all of the fundamental *a priori* cognitions from which other *a priori* cognitions are derived and which are not themselves grounded in 'higher and more general cognitions.' If judgments of the form (c)-(f) are not *analytic* then they are not grounded in the 'supreme principle' of analytic judgments. But then where in the *KrV* does Kant

15. Cf. *Ak.* 28: 522-3, and 28: 544.

account for our knowledge of such judgments? That the system of the *KrV* might not successfully ground all of our synthetic *a priori* knowledge is not in itself implausible; what is implausible is that Kant could have been unaware that something so basic as our *a priori* knowledge of <No A is B and not-B> could not be accounted for within his system. The same points apply to judgments of forms (c)–(e). In other words, the cost of attributing to Kant a conception of analytic judgment that precludes judgments of these forms from being analytic is attributing to Kant either an absurd philosophical position—that judgments of these forms are *a posteriori*—or one that violates one of the basic systematic aims of the *KrV*, namely, to systematically account for the sources of all *a priori* cognition. Minimal interpretive charity favors the contrary view, that Kant had a conception of analyticity wide enough to embrace judgments of syntactic forms other than universal-affirmative-categorical and universal-negative-categorical.[16]

The other systematic reason why Kant must accept analytic judgments of a wide range of syntactic forms, other than (c)-(f), is the anti-Leibnizian thrust of the analytic-synthetic distinction. One of Kant's main points in drawing the analytic-synthetic distinction is to argue that there are synthetic *a priori* truths, i.e. necessary truths that are not analytic. This is supposed to rebut the view—held in common by Leibniz, Wolff and Baumgarten—that the negation of a necessary truth entails a contradiction, i.e. that all necessary truths are logically necessary (Baumgarten 1757, §7, §101; Wolff 1962, I–2, §36).[17] Kant holds, on the contrary, that there are non-contradictory judgments that are nonetheless not 'really possible,'

16. Thus it is surprising to read Ian Proops claim: "I have argued that Kant settled for a non-exhaustive classification of judgments [into analytic and synthetic]. Some commentators have found this idea hard to accept. Their desire to find an exhaustive classification in the first *Critique* has inclined them toward reading Kant's discussion of the principle of contradiction at A151/B190–1 as containing a further characterization of analyticity in epistemic terms. Interestingly, however, such a tradition of interpretation does not appear to have originated in serious Kant scholarship" (Proops 2005, 610). While it may not have originated in "serious Kant scholarship," it does originate both in an appreciation of the systematic aims of the *KrV* and, more straightforwardly, in the text itself. Recall how Kant originally introduces the distinction: "In allen Urteilen, worinnen das Verhältnis eines Subjekts zum Prädikat gedacht wird [...] ist dieses Verhältnis auf *zweierlei Art* möglich. *Entweder* [...]; *oder* [...]" (A6/B10).

17. What I mean here by 'necessary' corresponds to Leibniz's 'strict' or 'metaphysical' necessity; among texts available to Kant in 1781, see the Fifth Letter to Clarke, *New Essays* (IV.ii.1, IV.vii.11, IV.xxix.9), *Theodicy* §174, *Monadology* §33, as well as the unpublished texts 'Primary truths' (Leibniz 1989, 98-101), 'On Freedom' (Leibniz 1989, 96), and 'First Truths' (Leibniz 1969, 267-271).

or, equivalently, that not all necessary truths are logically necessary. His analytic-synthetic distinction needs to allow him to formulate this anti-Leibnizian point; it needs to allow him to formulate, in his own technical terminology of 'analytic' and 'synthetic' judgments, the claim that some judgments that are not logically necessary are nonetheless 'really necessary.' To do this, he needs, as much as possible, his notion of 'analytic' judgments to line up with what Leibniz would call 'truths of reason': judgments whose negations logically entail a contradiction. In other words, if Leibnizian 'truths of reason' with alternate syntactic forms (e.g. (c)–(f) from above) do not count as Kantian analytic judgments, then in claiming that there are really necessary truths that are not analytic Kant would be making a claim Leibniz would accept.[18] Drawing the analytic-synthetic distinction in such a way would neuter the anti-Leibnizian thrust of the *Kritik* from its opening pages. There are, therefore, overwhelming reasons, both textual and systematic, for thinking that Kant must have a conception of analytic judgments that includes judgments with alternate syntactic forms. In the rest of the paper I will attempt to determine precisely what that conception is.

2. *The non-contradiction criterion*

Kant's discussion of the non-contradiction criterion in the "On the Supreme Principle of all Analytic Judgments" is sufficiently important to merit quoting in its entirety:

> Now the proposition that no predicate pertains to a thing that contradicts it is called the principle of contradiction, and is a general though merely negative

18. Some readers might reply that, since Leibniz took all judgments to be ultimately reducible to categorical judgments, Kant would not be blunting his critique of Leibniz by ignoring analytic judgments of alternate syntactic forms. Some evidence that Leibniz contemplated the reduction of hypothetical to categorical propositions is offered in (Proops 2005, 595); see also (Mates 1986, 58). However, Proops offers no evidence that Leibniz took disjunctive propositions to be reducible to categorical ones. Even if Leibniz did in fact envisage reducing all propositions to those with categorical form, Kant's objection to Leibniz would still be significantly philosophically weaker if it ignored the possibility of analytic judgments with non-categorical form: it would then no longer be an objection to the Leibnizian view that all necessary truths are (what Kant would call) analytic, but only to the conjunction of that view and the reductionist view just described. Kant would then not be targeting the heart of Leibniz's modal theory, but a dispensable commitment of that theory (i.e. that all propositions are ultimately categorical in character).

> criterion of all truth, but on that account it also belongs merely to logic, since it holds of cognitions merely as cognitions in general, without regard to their content, and says that contradiction entirely annihilates and cancels them.
>
> But one can also make a positive use of it, i.e., not merely to ban falsehood and error (insofar as it rests on contradiction), but also to cognize truth. [α] *For, **if the judgment is analytic**, whether it be negative or affirmative, its truth must always be able to be cognized sufficiently in accordance with the principle of contradiction.* [β] *For the contrary of that which as a concept already lies and is thought in the cognition of the object is always correctly denied, while the concept itself must necessarily be affirmed of it, since its opposite would contradict the object.*
>
> Hence we must allow that the **principle of contradiction** to count as the universal and completely sufficient **principle of all analytic cognition**; but its authority and usefulness does not extend beyond this as a sufficient criterion of truth. For that no cognition can be opposed to it without annihilating itself certainly makes this principle into a *conditio sine qua non*, but not into a determining ground of the truth of our cognition. Since we now really have to do only with the synthetic part of our cognition, we will, to be sure, always be careful not to act contrary to this inviolable principle, but we cannot expect any advice from it in regard to the truth of this sort of cognition. (*KrV*, A150-2/B189-191; italics and bracketed letters by author)[19]

In the first paragraph, Kant points out that the principle of contradiction (PC) is only a negative criterion on truth: any judgment that violates it is false, but some non-contradictory judgments are false or merely groundless. He then goes on to claim that some judgments can be cognized through the PC alone. Specifically, if a judgment is analytic, whether it is affirmative or negative, it can be cognized through the PC alone. Fully understanding what this means requires understanding how Kant understands the PC. The closest that Kant comes to explaining his conception of the principle is given in sentence [β].

Kant here describes the PC as governing the case of affirmative analytic judgments of the form <All (A+B) is B> as well as negative analytic judgments of the form <No (A+B) is ~B>. This suggests that the Non-Contradiction Criterion just consists in the following:

> *Non-Contradiction Criterion*: A judgment is analytic if and only if (i) it is affirmative and has the form <All (A + B)s are B>, or (ii) it is nega-

19. Cf. the discussion of this passage by Bolzano/Příhonský (*NAK*, 91).

tive and has the form <Not (A+B) is not-B>, where [A+B] stands for a concept in which the concept [B] is contained.

If this is all that Kant has in mind in this passage, then the Non-Contradiction Criterion adds little to the Containment Criterion. What it adds is an 'extension' to the case of negative judgments; it can accommodate negative judgments of the form <No (A+B) is not-B>. But it cannot accommodate analytic judgments of forms (d)–(f) from section one. This is how Proops (2005) reads the "supreme principles" passage; if that is the correct reading then Proops is correct in holding that all that passage adds to A6/B10 is an extension of the definition of analyticity to judgments of the form <No (A+B) is not-B>.

But there is another way of reading this passage. The first italicized sentence [α] suggests the following:

Cognition through PC Criterion: A judgment <*p*> is analytic if and only if <*p*> can be cognized through the principle of non-contradiction alone.

Initially, Kant formulates this criterion by writing "if the judgment is analytic [...]" ; this suggests that where I have written 'if and only if' I should have written 'only if.' But he goes on to write that "we must allow the principle of contradiction to count as the *universal* and *completely sufficient* principle of all analytic cognition" (my emphasis). I take this to mean that only analytic judgments can be cognized through the PC alone, which means that cognizability through the PC alone is both a necessary and sufficient condition for being an analytic judgment.

How might the Cognition through PC Criterion differ from the Non-Contradiction Criterion? Assuming (as seems plausible) that all of the judgments that pass the Non-Contradiction Criterion are cognizable through the principle of contradiction (since they are substitution instances of it, either in its 'affirmative' guise or its 'negative' guise), the two criteria are distinct if and only if there are judgments that can be cognized through the PC alone but which do not have the form <All (A+B) is A> or <No (A+B) is not-B>. But, intuitively, that is precisely the case with judgments with alternate syntactic forms from the previous section ((d)–(f)). Intuitively, those judgments can be cognized through the principle of non-contradiction because the denial of any of them entails a contradiction, although none of them are mere substitution instances of the PC.

This seems like the right conception of analyticity, not merely 'intuitively,' but also because it seems like a promising start for reconstructing the class of our *a priori* cognitions that do not rely on the synthetic *a priori* principles of experience.[20] But is there any ground for thinking that Kant had a conception of what it is to 'cognize a judgment through the principle of contradiction' other than merely recognizing it as a substitution-instance of that principle, either in its positive or its negative guise? In the next section, I will argue that, in fact, careful attention to Kant's texts supports attributing this view to him.

3. *The unity of the principle of contradiction*

In the passage on the "Supreme Principle of all Analytic Judgments" quoted earlier, Kant claims that all analytic judgments can be sufficiently cognized through the principle of contradiction, and this accounts for analytic judgments of the form <All (A+B) are B> and <No (A+B) is not-B>. However, this is puzzling, because only the latter seems to be an instance of the principle of contradiction in the strictest sense, while the former is an instance of what Kant calls the 'principle of identity.' How can the idea that there is a *single* highest principle of all analytic judgments be squared with this apparent duplicity in their application?

In his pre-Critical work of 1755, *Principiorum primorum cognitionis metaphysicae nova dilucidatio*, Kant explicitly claims that there are *two* basic logical principles: the principle of contradiction and the principle of iden-

20. Siebel (2011) quotes with approval Künne's claim that "unlike 'true' and 'necessary,' the world 'analytic' is a philosopher's term of art. Memories of doctrines associated with this term (be they Kantian, Fregean, Carnapian or whatever) should not be mistaken for pre-theoretical 'intuitions' concerning analyticity. There simply are no such intuitions one could appeal to" (Künne 2006, 219). I think Künne and Siebel are mistaken on this point; as Grice and Strawson point out "there is no need to appeal only to tradition; for there is also present practice. We can appeal, that is, to the fact that those who use the terms 'analytic' and 'synthetic' do to a very considerable extent agree in the applications they make of them. They apply the term 'analytic' to more or less the same cases, withhold it from more or less the same cases, and hesitate over more ore less the same cases. This agreement extends not only to cases which they have been taught so to characterize, but to new cases. In short, 'analytic' and 'synthetic' have more ore less established philosophical use; and this seems to suggest that it is absurd, even senseless, to say that there is no such distinction. For, in general, if a pair of contrasting expressions are habitually and generally used in application to the same cases, where these cases do not form a closed list, this is a sufficient condition for saying that there are kinds of cases to which the expressions apply; and nothing more is needed for them to mark a distinction" (Grice and Strawson 1968, 142f.).

tity, and neither can be derived from the other. The principle of identity is the basic principle of all affirmative truths, Kant there contends, while the principle of contradiction is the basic principle of all negative truths. One possibility would be that Kant retains, within the Critical period, this dichotomy of basic logical principles, and the Non-Contradiction criterion should really read:

> *Non-Contradiction Criterion*: The judgment <p> is analytic if and only if it is a substitution instance of *either* the principle of contradiction *or* the principle of identity.

Aside from its displeasingly disjunctive character, this principle fails to do justice to Kant's ambivalence in the Critical period about whether the PC and the PI are one or two principles. In some texts he refers to 'the principle of contradiction or identity'—suggesting there is one principle, than can either be formulated in terms of contradiction or in terms of identity —while in others he refers to them as separate principles.[21]

This ambivalence strongly suggests that Kant's considered view is that there is some sense in which the PI and the PC are the same principle, and some sense in which they are not. He criticizes Baumgarten for attempting to derive the PC from the PI:

> In §11 the author seeks to prove the principle of identity through the principle of contradiction. One cannot, however, prove it in this fashion, because the principle of contradiction, which is supposed to prove the principle of identity, already exhibits [the principle of identity], and would thus be used to prove itself. (*Metaphysik Vigilantius, Ak.* 29:964)

Or, as Kant expresses the same point more tersely but perhaps more clearly: "The author wishes to prove the principle of identity from the principle of contradiction; but the ground [*Ursach*] is itself the principle of identity, which he does not observe (*Metaphysik Herder, Ak.* 28:11). The term Kant uses here is '*beweis*en.' But '*Beweis*' is a technical term for Kant; it does not refer to just any logically valid argument, but to one in which the premises are the grounds of the conclusion, that is, a logically valid argument in which the premises are more basic than, and therefore can

21. There are a number of texts in which Kant writes as though they are two principles (E.g. *Ak.* 24:477, 28:9, 24:279, 28:544; and *Refl.* 4655), and a comparable number of texts in which Kant writes as though they are one principle (E.g. *Refl.* 3741, 4275, 4477, 4872, 5176, 5176; and *Ak.* 28:523).

explain, the conclusion (*Wiener Logik*, *Ak.* 24: 892). But this is not the case with Baumgarten's putative '*Beweis*' of the PC from the PI because the PI is no more basic than the PC and therefore cannot explain why the PC holds. In fact, Kant holds that there can be no '*Beweis*' strictly speaking of either the PC or the PI; they are basic principles that cannot be grounded in anything more fundamental (*Logik Blomberg*, *Ak.* 24:278f.).

Kant denies that the PI can be '*bewiesen*' from the PC, but he admits that the PC follows from the PI and vice versa:

> The proposition or principle of identity is valid for affirmative judgments just as the principle of contradiction is valid for negative judgments. Fundamentally, one can regard both of these principles as one; for when I posit one, the other follows immediately [*so folgt aus diesem auch schon das andere.*] The principle of identity is already conceived in the principle of contradiction. (*Metaphysik Pölitz*, *Ak.* 28: 544)

The PC entails the PI and vice versa. Unordered materials from the K_1 manuscripts include an illuminating passage on the same point, which Erdmann prefaces with the comment, "Especially illuminating is the proof [*Begründung*], missing in the later work, that the principle of identity and the principle of contradiction do not need to be distinguished by transcendental reflection":

> [...] judgments, according to their forms, are divided into the affirmative and the negative, for affirmative [judgments] stand under the principle of identity and negative [judgments] stand under the principle of contradiction. Nonetheless, there is no distinction in content between the affirmative and the negative. Here we are not concerned with the form, as in logic, but with the content, so among the two principles [of identity and contradiction], we choose one and make it the general principle, and it makes no difference which one we choose, except that the principle of contradiction makes more of an impression when we show that the opposite [of a judgment] is false. It makes more of an effect when one shows that [the judgment] is an absurdity and that it contradicts itself, than when one merely proves an identity. (*Ak.* 28:1519)

The PC and the PI have different logical forms, because one is affirmative and one is negative, but they present the same content. Consequently, it makes no difference which one we choose as the most general principle of analytic judgments; the other can be derived from it. The only difference is presentational: the principle of contradiction makes a stronger impression.

This idea, of two principles entailing one another and thus presenting the same content under different logical forms, recalls the discussion of 'equivalent' judgments in the *Jäsche Logik*: "they are to be regarded as a mere substitution of words that signify one and the same concept, where the judgments themselves also remain unaltered as to form" (*JL*, *Ak.* 9:116). Kant's example of a pair of equivalent judgments is:

(1) Not all men are virtuous.
(2) Some men are not virtuous.

He says of these judgments that they "have the same form"—which cannot strictly be true because (1) is a negative universal judgment and (2) is a particular infinite judgment—and that they "say one and the same thing." Their equivalence consists in the fact that they can be 'immediately converted' into one another. Putting this together with the discussion of the PC and the PI—and discounting the sloppy claim in the *JL* that equivalent judgments have the same form—I conclude that Kant's view is that the PC and the PI are equivalent judgments, in the sense discussed in the *Jäsche Logik*.

Kant goes on to make clear that universal affirmative judgments can always be 'contraposed,' that is, immediately converted into equivalent negative judgments. Thus,

(1) <All bodies are extended.>

can be immediately converted to

(2) <No non-extended thing is a body.>

And, since negative universal judgments can themselves be 'converted'—their predicate and subject can be reversed *salva veritate*—(2) is equivalent to

(3) <No body is non-extended.>

Thus, (1), (2) and (3) are 'equivalent' judgments, judgments that "say the same thing" although they instantiate different logical forms. But note that (1) has the form <All (A+B) is B> and (3) has the form <No (A+B) is not-B> In other words, we have shown that, on Kant's theory of equivalent judgments, the PI in the former formulation is equivalent to the PC in the latter formulation. This allows us to eliminate the displeasingly disjunctive character of the Non-Contradiction Criterion, using the notion of equivalent judgments:

> *Non-Contradiction Criterion*: A judgment <p> is analytic if and only if <p> is equivalent to a substitution-instance of the PC.

Even if we formulate the PC strictly in the form <No (A+B) is ~B>, this will deliver the result, not only that judgments of the form <All (A+B) are B>, but also that judgments of form (d)–(e) from section one are analytic:

(d) <No ~B is (A+B)>, e.g. <No non-metallic thing is gold.>
(e) <All ~B are ~(A+B)>, e.g. <All non-metallic things are not gold.>

Given Kant's rules of immediate inference (outlined in *JL,* §44–55), these are all equivalent to <No (A+B) is ~B.> Consequently, this constitutes a significant advance over Proops (2005). Strictly speaking, judgments (d) and (e) have neither of the forms countenanced by Proops for analytic judgments, <All (A+B) is B> and <No (A+B) is ~B>. Thus, strictly speaking, no judgment of form (d) or (e) can be analytic, on Proops's reading. The way to accommodate judgments (d)–(e) in the Non-Contradiction Criterion is via the notion of 'equivalent judgments,' but, as I argue in the next section, this is the thin end of the wedge. Once we allow judgments equivalent to substitution instances of the PC to count as analytic, the door is open to include analytic judgments of a much wider range of syntactic forms.

4. *Cognition through the PC*

I have used Kant's discussion of judgments that are equivalent—those that can be immediately converted into one another—in the section on 'Immediate Inferences of the Understanding' in the *Jäsche Logik* to understand the relation between the PC and the PI. What started us down the path was the idea that there might be more to a judgment being 'cognizable through the' PC than its merely being a substitution instance of the PC. And we have seen that this is so: cognizing that a judgment is equivalent to the PC although not itself a substitution instance of that principle is,

I argued, a way of cognizing a judgment 'in accordance with' [*nach*] the PC, and judgments that can be so cognized are analytic. Consequently, as we saw in the previous section, the Non-Contradiction Criterion can be understood as wider than it might initially seem, and admits analytic judgments of a (slightly) wider range of syntactic forms.

But a natural continuation of this idea allows us to extend the range of analytic judgments even further. Recall the 'Cognition through PC' Criterion from earlier:

> *Cognition through PC Criterion*: A judgment <*p*> is analytic if and only if <*p*> can be cognized through the principle of contradiction alone.

In what sense is a judgment that satisfies the Non-Contradiction Criterion cognized through the PC? The natural answer is that the judgment is either a substitution instance of the PC—in which case it is cognized by being subsumed as a case under the general principle that is the PC—or it is equivalent to a substitution instance of the PC—in which case it is inferred from the PC. Judgments of the latter kind are cognized 'through the PC' in the sense that they are inferred from the PC. Consequently, judgments that are not equivalent to PC substitution-instances but which are entailed by such substitution-instances are 'cognized through the PC' in exactly the same sense, and should be counted as analytic.

What this means depends, however, upon what notions of entailment and inference are involved. Fortunately, though, Kant's texts contain a natural answer. The discussion of 'equivalent' propositions in the *Jäsche Logik* that I used in the previous section to motivate the idea that the PC and the PI are equivalent is embedded in a larger discussion of 'immediate inferences of the understanding.' Equivalent propositions are defined as judgments that immediately entail one another. Immediate inferences are 'immediate' in the sense that they have only one premise. The rules of immediate inference are the conversion rules by which we transform premises of syllogisms into premises that have the forms recognized by the traditional syllogistic figures. Kant describes the immediate inferences of the understanding as follows:

> §44. *Peculiar nature of inferences of the understanding*. The essential character of all immediate inferences and the principle of their possibility consists simply in an alteration of the *mere form* of judgments, while the *matter* of the judgments, the subject and predicate, remains *unaltered, the same*.

Note 1. By virtue of the fact that in immediate inferences only the form of judgments is altered and not in any way the matter, these inferences differ essentially from all mediate inferences, in which the judgments are distinct *as to matter* too, since here a new concept must be added as mediating judgment or as middle concept (terminus medius) in order to deduce the one judgment from the other. (*JL*, *Ak.* 9:115)

Here Kant points out, correctly, that the point of a conversion rule is to alter the form of the judgment—so that it can be used as a premise in a syllogism form—while not altering the 'matter' of the judgment, intuitively, 'what the judgment says.' Immediate inferences are to be distinguished from mediate inferences, syllogisms, in virtue of the fact that the conclusion of a syllogism always combines one concept from the major premise and one concept from the minor premise, via the middle term (which does not appear in the conclusion). So the conceptual 'matter' of a syllogistic conclusion is always different than its conceptual 'input.' This is not the case with immediate inferences.

I have suggested that if a judgment is cognized through the PC in virtue of being equivalent to a substitution instance of the PC, then it is cognized through the PC in virtue of being immediately inferred from the PC. Consequently, any judgment immediately inferred from a substitution instance of the PC is cognized through the PC, and thus, according to the "Supreme principle" section, should be counted as analytic. In other words,

Immediate Inference: A judgment <p> is cognized through the PC if it is either a substitution instance of PC or immediately inferred from a substitution instance of the PC.

This is a sufficient condition on cognition through the PC, and thus, by the Cognition through PC Criterion, on a judgment's being analytic. I will argue that it is not a necessary condition; there are other ways of cognizing a judgment through the PC than immediately inferring it from (a substitution instance of) the PC.

I noted earlier that Kant's definition of immediate inferences as inferences with only one premise admits of a crucial exception, which Kant explains in this passage:

With inferences of the understanding a *judicium intermedium* [intermediate judgment, or middle premise] may also be made, to be sure, but then this mediating judgment is merely *tautological*. As, for example, in the immediate

> inference, All men are mortal, *some men are men*, hence some men are mortal, the middle concept is a tautological proposition. (*JL*, *Ak.* 9:115)

Tautological judgments, as Kant here defines them, are just instances of the form, <A is A>. But notice that in this passage Kant claims that the judgment <All men are moral> *immediately entails* <Some men are mortal>, even though this inference requires as a premise the tautology <Some men are men.> But this inference is clearly a syllogism. But this means that the rules of syllogistic inference are themselves admissible as rules of immediate inference provided that at least one of the premises is a tautology. Consequently, the rules of syllogistic inference are themselves included among the rules of immediate inference, provided those syllogistic rules are applied to pairs of premises at least one of which is a tautology (e.g. <Gold is gold.>).

This means that we can amend the Immediate Inference Criterion to read:

> *Immediate Inference*: A judgment <p> is cognized through the PC if it follows by immediate inference from a substitution instance of the PC, or by syllogistic inference from a substitution instance of the PC and a tautology.

By definition, immediate entailment is not transitive: if <p> immediately entails <q> and <q> immediately entails <s> it does not follow that <p> immediately entails <s> (although it might). So if there were a judgment <p> that followed by immediate inference from <q>, where <q> followed by an immediate inference from a substitution instance of the PC, <p> would not necessarily follow immediately from that substitution instance. But clearly <p> would be cognizable 'according to' [*nach*] the PC alone, for the only starting points for our cognition of <p> is a substitution instance of the PC; everything else is done through immediate inferences (including the special case of syllogistic inference with a tautological premise). Thus, the 'Immediate Inference' principle is too restrictive, and needs to be reformulated as:

> *Inference*: <p> is cognized through the PC iff <p> follows by a series of inferences from a subsitution instance of the PC, where each inference is either an immediate inference in sensu stricto (i.e. a one premise transformation rule), or is a syllogistic inference with one tautological premise.

But we can go further than this. If some judgment follows, not from a single substitution instance of the PC but from several substitution instances thereof, by a series of immediate inferences and syllogistic inferences, then that judgment itself is cognized according to the PC alone. After all, our cognition of the judgment is based solely on instantiating the PC and then performing immediate inferences on it, and deriving consequences using syllogistic reasoning. The only basis for objecting to this is that, at this stage, the syllogistic rules may be applied in cases where neither of the premises is a tautology, for instance:

(1) All gold is metal. [equivalent to a substitution instance of the PC]
(2) All metal is conductive. [equivalent to a substitution instance of the PC]
(3) ∴ All gold is conductive. [By (1) and (2) by syllogism]

Although neither of these premises are tautologies in the strict sense of *JL* §44, I see no reason to deny that if we cognize (3) by deriving it from (1) and (2) we have cognized (3) according to the PC alone. Thus, I propose the following as the final and definitive criterion for 'cognition through the PC alone' in the sense of A151/B190:

> *Inference*: A judgment <p> is cognized through the PC iff <p> is derived by a series of immediate inferences and syllogisms from a set of premises, each of which is a substitution instance of the PC.

But, plugging this into the 'Cognition through PC' Criterion on analyticity, we get:

> *Cognition through PC Criterion*: A judgment <p> is analytic iff there is a series of immediate inferences and syllogisms from a set of premises, each of which is a substitution instance of the PC, whose conclusion is <p>.

Because tautologies (<A is A>) and simple containment truths (<(A+B) is B>) are immediately equivalent to substitution instances of the PC, this is equivalent to allowing the premises to be either substitution instances of the PC or tautologies or simple containment truths. From the perspective of this criterion on analyticity, the Containment-or-identity criterion of A6/B10 applies and *was only intended to apply* to the special case of judgments that are straight-forward instances of the PI, that is, those that have the form <All (A+B) are B.> Since most of Kant's examples of analytic judgments are of this form, the Containment-or-identity criterion is sufficient for most of his philosophical purposes, but by no means exhausts his conception of analyticity.

I began by talking about several of Bolzano's criticisms of how Kant draws the analytic-synthetic distinction; I have focused on his third criticism —that it applies only to judgments within a narrow range of syntactic forms—but along the way, I have indirectly touched on several of his other criticisms (e.g. that it rests on the metaphor of inclusion). However, my interpretation, if correct, makes one of Bolzano's critiques—that Kant never adequately proves that mathematical judgments are synthetic —harder to answer. If analytic judgments have only a limited range of syntactic forms, then Kant can prove that mathematical judgments are not analytic merely by proving they do not have these forms. However, if my Inference Criterion is the correct interpretation, Kant has to prove the more ambitious claim that mathematical judgments like 7+5=12 do not follow by logical inference from any number of Containment-truths and substitution instances of the PC. Whether, and how, Kant can prove this lies beyond the scope of this essay.[22]

For now, I would like to conclude by making two related points. First of all, to return to an observation I made earlier in the paper, Kant's analytic-synthetic distinction does not occur in a vacuum. Instead, it is made within a very specific argumentative context: he wants to deny the Leibnizian/Wolffian view that all necessary truths are derivable from definitions plus logical laws. Consequently, it is incumbent upon him to formulate his own view in such a way as to enable him to formulate precisely his disagreement with Leibniz. If Kant's analytic judgments are in fact a proper subset of what Leibniz calls "truths of reason" then in claiming that there are necessarily true non-analytic judgments (the synthetic *a priori*) Kant is making a claim Leibniz need not deny. Specifically, if Kant's analytic judgments are restricted to the limited range of syntactic

22. Thus, one of the advantages of interpretations like those of Anderson (2004, 2005) and Proops (2005) is that they make clear how Kant could have taken himself to have proven that arithmetic is synthetic, given how brief his actual arguments for this conclusion are. In future work I will argue that in this respect Anderson is a false friend to Kant; he equips Kant with a short, powerful argument for the synthetic nature of arithmetic at the cost of burdening him with an indefensible conception of analyticity.

forms that both Bolzano and Proops claim they are, then in claiming that there are necessary judgments that are not analytic Kant is not expressing a disagreement; of course Leibniz would accept there are necessary truths that do not belong in that limited class of judgments, because Leibniz has to accept that there are necessary truths of a vastly wider range of syntactic forms, e.g. <No A is B and not-B.> If Kant were to limit the analytic in this way, he would be drawing the analytic-synthetic distinction in such a way as to make one of his central anti-Leibnizian points moot.

If, on the other hand, Kant has the wider conception of analyticity that, I have argued, he does, then he is in a perfect position to express his disagreement with Leibniz. The Cognition through PC Criterion, in its final formulation above, fits remarkably well with Leibniz's definition of 'truths of reason' in *New Essays* and other texts—'propositions that follow from definitions and identities'[23]—since definitions (e.g. <(A+B) is B> and identities (e.g. <A is A>) will be what Kant would consider substitution-instances of the principle of identity (PI). Clearly, Kant and Leibniz will differ over the definition of certain concepts, hence about which truths have the form <(A+B) is B>, and they may differ over the rules of inference, but this will not be a difference over what it is to be an analytic judgment, or 'truth of reason.' Furthermore, the set of Kantian analytic judgments and Leibnizian truths of reason may not overlap completely, but this will be due to the differences in the logical rules that are built into these definitions. This leads me to my second point.

I started out by discussing Bolzano's claim that Kant's definition of analyticity applies only to judgments with a narrow range of syntactic forms, and with a contemporary reading of Kant that seems to vindicate this charge, Proops (2005). I have argued that, properly understood, Kant's

23. *New Essays* IV.xviii.9, 'The Source of Contingent Truths' (Leibniz 1989, 98), 'On Freedom' (Leibniz 1989, 69), Fifth Letter to Clarke (Leibniz 1969, 696), Monadology §31–33 (Leibniz 1969, 646). In some of these passages Leibniz says: p is 'strictly' necessary iff ~p entails a contradiction, in other places he says: p is 'strictly' necessary iff p can be demonstrated from identities and definitions. These are equivalent. It is relatively clear that the latter definition in terms of demonstration is equivalent to the Kantian Cognition through PI criterion. The former definition, in terms of contradiction-entailment, is equivalent because, if ~p entails a contradiction, then the negation of that contradiction, which is a substitution instance of the PI, entails p. Conversely, if p can be demonstrated from a finite set S of PI substitution instances, then from ~p we can demonstrate the negation of the conjunction of the elements of S. Assuming we are allowed to introduce PI substitution instances at any stage in a demonstration (as I argued in the text), this means we can demonstrate from the negation of the conjunction of the elements of S, using a disjunctive syllogism, the negation of a PI substitution-instance. This shows that the Kantian and the Leibnizian criteria are equivalent.

conception of analyticity includes analytic judgments of a wider range of syntactic forms. However, the definition of analyticity given may not actually allow for analytic judgments of all the syntactic forms Kant would want to recognize. Whether this is true or not depends upon the strength of the logical rules built into the definition of analyticity. The final point I want to make is that if Kant's definition of analyticity does not extend to judgments of some syntactic form that, intuitively, should be counted as analytic, the fault is not to be laid at the door of his definition of analyticity or his philosophical intentions in distinguishing analytic from synthetic judgments. The fault lies in Kant's logic. Kant simply does not possess a logic powerful enough to generate all of the analytic judgments from the simple containment truths, i.e judgment of the form <All (A+B) is B>. So, in the end, Bolzano and Proops may actually be right, but they will be right for different reasons than they thought. It isn't that Kant intentionally limited analyticity to judgments within a narrowly circumscribed range of syntactic forms; he simply did not possess a sufficiently powerful logic to derive all of the analytic judgments from substitution instances of the principle of identity.[24]

References

Anderson, R. Lanier 2004: "It Adds Up After All: Kant's Philosophy of Arithmetic in Light of the Traditional Logic". *Philosophy and Phenomenological Research* 69, 501–540.

— 2005: "The Wolffian Paradigm and its Discontents: Kant's Containment Definition of Analyticity in Historical Context". *Archiv für Geschichte der Philosophie* 87, 22–74.

Baumgarten, Alexander Gottlieb 1757: *Metaphysica.* 4th Ed. Halle: Karl Hermann Hemmerde. Republ. as *Metphysica/Metaphysik: Historisch-kritische Ausgabe.* Trans. and ed. by Günter Gawlick & Lothar Kreimendahl. Stuttgart-Bad Cannstatt: Frommann-holzboog, 2011.

24. Clinton Tolley, Robert Hanna, Paul Rusnock and Mark Siebel gave me helpful comments on an earlier version of this paper. I'd also like to thank Otávio Bueno, with whom I had a very helpful conversation about Kantian rules of inference. Special thanks go to Sandra Lapointe for encouraging me to write about Kant and Bolzano in the first place, shepherding this project through several years and several (thankfully) abandoned versions of the present paper, and for very helpful comments on the penultimate draft.

Bolzano, Bernard 1837: *Wissenschafteslehre*, 4 vols. Sulzbach: Seidl. Republ. by Wolfgang Schultz, Leipzig, 1929–31.
Grice, Herbert Paul & Strawson, Peter F. 1956: "In Defense of a Dogma". *The Philosophical Review* 65(2), 141–158.
Kant, Immanuel 1902ff: *Gesammelte Schriften*, 29 vols. Berlin: Königlich Preußische Akademie der Wissenschaften, later Berlin-Brandenburg Akademie der Wissenschaften.
— 1997: *Critique of Pure Reason*. Trans. and ed. by Paul Guyer & Allen Wood, Cambridge: Cambridge University Press.
Künne, Wolfgang 2006: "Analyticity and Logical Truth: from Bolzano to Quine." In: Mark Textor (ed.), *The Austrian Contribution to Analytic Philosophy*. London & New York: Routledge, 184–249.
Lapointe, Sandra 2011: *Bolzano's Theoretical Philosophy*. Houndsmill: Palgrave-Macmillan.
Leibniz, Gottfried W. F. 1969: *Philosophical Papers and Letters*. Trans. and ed. by Leroy Loemker, Dordrecht: Kluwer.
— 1989: *Philosophical Essays*. Trans. and ed. by Roger Ariew & Daniel Garber, Indianapolis: Hackett.
Příhonský, Franz 1850: *Neuer Anti-Kant*. Bautzen: A. Weller. Republ. by Edgar Morscher, St. Augustin: Academia, 2003.
Proops, Ian 2005: "Kant's Conception of Analytic Judgment". *Philosophy and Phenomenological Research* 70, 588–612.
Siebel, Mark 2011: "'It Falls Somewhere Short of Logical Precision.' Bolzano on Kant's Definition of Analyticity". *Grazer Philosophische Studien* 82, 91–127.
Stang, Nicholas Forthcoming: "Indeterminacy and Transcendental Idealism". *British Journal of the History of Philosophy*.
Wolff, Christian 1962f: *Gesammelte Werke*. Hildesheim: Georg Olms.

Grazer Philosophische Studien
85 (2012), 63–88.

BOLZANO AND KANT ON THE PLACE OF SUBJECTIVITY IN A *WISSENSCHAFTSLEHRE*

Clinton TOLLEY
University of California, San Diego

Summary

Throughout his career, Bolzano presents his account of knowledge and science as an alternative to 'the Critical philosophy' of Kant and his followers. The aim of this essay is to evaluate the success of Bolzano's own account—and especially, its heavy emphasis on the objectivity of cognitive *content*—in enabling him to escape what he takes to be the chief shortcomings of the 'subjective idealist philosophy'. I argue that, because Bolzano's own position can be seen to be beset by problems that are both recognizably similar to, and possibly even *worse* than, those that he takes to afflict Kant's account of the elements of our knowledge, Bolzano's attempt to fully overcome the alleged vices of Kant's idealism by 'extruding' semantic content from the mind must be judged to be less than satisfactory.

§1. *Introduction: the mind and the 'elements' of science*

In the first *Critique*, Kant famously argues that the objects of which we can have genuine knowledge are objects that 'conform' to our capacities for knowledge (Bxvi).[1] Kant thinks that, as a consequence, any account of science must be grounded on an analysis of the nature and limits of our cognitive capacities. Kant himself attempts just such a grounding of the sciences of mathematics and physics over the course of the first *Critique*, the *Prolegomena*, and the *Metaphysical Foundations of Natural Science*. Because Kant thinks that all of our knowledge arises from two basic capacities (two 'stems', cf., B29)—on the one hand, our capacity for sensing (intuiting,

1. I will cite Kant's first *Critique* according to the B-edition pagination, unless it is a text found only in the A-edition; I will cite Kant's other works according to the Akademie Ausgabe (Kant 1902–) volume and page number. All decisions on translation are my own, though I have consulted (and usually followed, especially in the case of the first *Critique*) the Cambridge Edition of Kant's works (Kant 1991–).

being given) objects, what Kant calls our 'receptivity' or 'sensibility', and on the other hand, our capacity for thinking, judging, and inferring about objects, what Kant calls our 'understanding' (B74)—these two capacities themselves must form the genuine 'elements' in any account of knowledge. The correspondingly elementary 'doctrines' for the account of science in general, therefore, will be the sciences of sensibility and understanding, or what Kant calls 'aesthetic' and 'logic', respectively (B76). Accordingly, it is this pair of sciences that comprises the 'doctrine of the elements [Elementarlehre]' in the analysis of our theoretical knowledge that Kant provides in the first *Critique* itself (B29).

As Bolzano makes clear in his 1837 masterwork, the *Wissenschaftslehre*,[2] Bolzano takes Kant's fixation upon mental capacities and activity to have gotten philosophy off on the wrong track. By restricting its focus in this way, Bolzano thinks that Kant entirely neglects the primary locus of the *truth* we are seeking in science. This is because, like Frege after him, Bolzano thinks it is evident that what is true, the genuine bearer of truth, is not a mental act or capacity at all, but instead is the 'matter [Stoff]' that is contained in these acts (WL §291, II.108), what Bolzano calls 'propositions [Sätze]' (WL §24, I.108). Since knowledge ('cognition [Erkenntnis]') consists in the grasping of truths (WL §§36–37, I.163ff.), and since 'science [Wissenschaft]' more generally is a 'collection [Inbegriff]' of truths (WL §1, I.4), it is hopeless to think—as Kant seems to—that we could come to know what knowledge and science are without first looking to the nature of truth itself and the nature of bearers of the property of being true (propositions).

What the pursuit of such analysis will ultimately show, Bolzano thinks, is that the distinction between act and content must be drawn, not just for knowledge and science as a whole, but for each of its component parts as well. That is, even in the mere act of 'representing [Vorstellen]' objects, Bolzano thinks we find a separate content or 'matter [Stoff]' in addition to the act itself, and in addition to the object represented (WL §49, I.218). Similarly, we will see that all acts of 'judgment [Urteil]' are such as to 'contain [enthalten]' a 'proposition [Satz]', something likewise distinct from both the subjective act of judging and the object and property that the judging is about (WL §34, I.154; cf., WL §290, III.108). Even in acts of

2. I will cite the *Wissenschaftslehre* as 'WL', according to the volume number and pagination of the original edition (Bolzano 1837), inserting the section number for ease of reference. Translations throughout are my own, though I have consulted, and at times followed, the partial translation (and additional paraphrases) contained in (Bolzano 1972).

'inference [Schluss]' Bolzano thinks we will recognize that we must separate out the mental event from the 'derivation [Ableitung]' it contains, which itself 'asserts a relation of genuine grounding or consequence [Abfolge] between given propositions' (WL §46, I.213; cf., WL §223, II.392). Once the necessity of drawing these distinctions has been recognized, Bolzano thinks we will then see straightaway that, at the very least, the genuine *Elementarlehre* in the doctrine of science must *also* include an analysis of the contents of such activity and not just an account of the nature of the activity itself (WL §16, I.61f).

Bolzano, however, presses for an even more radical revision of Kant's picture. This is because he is confident that once we think more carefully about the nature of these contents, we will come to see that such contents, their truth or falsity, and their 'connection with each other as grounds and consequences', are entirely 'objective', in the sense of being what they are 'entirely *independently* of our cognition' of them, and more generally, independent of the make-up of our mental capacities and activity (WL §16, I.61; my ital.). Hence, even though these contents can ultimately be grasped in mental acts—or, as Bolzano more frequently puts it, they can '*appear* [erscheinen] in the mind' (cf., WL §21, I.84; my ital.)[3]—the properties that these contents have in virtue of their relation to mental acts cannot be their most fundamental properties or the properties that these contents possess '*in themselves* [an sich]' (ibid.; my ital.). From this, Bolzano concludes that it is 'the doctrine of representations, propositions, true propositions, and inferences *an sich*', and this doctrine *alone*, that forms the proper subject-matter for the *truly* 'elementary' doctrine of science (WL §15, I.59).

The nature of Bolzano's counter-thesis is summarized nicely in the conclusion to the 1850 *New Anti-Kant*,[4] a work written out and published by one of Bolzano's collaborators, Franz Prihonsky, just after Bolzano's death, but which comprises the results of years of co-operative work together:

> One shouldn't first investigate and analyze the capacity for cognition [Erkenntnisvermögen] of humans or of thinking beings as such; it is necessary, rather,

3. Cf., WL §48, I.217, and WL §121, II.3. See also Bolzano's correspondence with Franz Exner in (Bolzano 1935; 'BW'), translated in (Bolzano 2004; 'CE'); here, CE 142; BW 64. I have silently amended this translation throughout.

4. I will cite this work according to the original pagination (Prihonsky 1850), which is included in the margins of both the recent reprint in (Prihonsky 2003) and the French translation in (Prihonsky 2006), and will also be included in a forthcoming English translation by Sandra and myself (to be published by Palgrave).

> to sound out [ergründen] the nature of truths *an sich*—or, to speak more generally, the nature of propositions *an sich*, their constituents and kinds, their connection among one another and similar such relations. (NAK, 229-30)

This way of ordering things also helps to bring out the deeper ambition of Bolzano's project. By shifting the focus at the outset away from the 'subjective constitution of our mind', as Kant puts it (B38), and onto something that is decidedly more objective (truths, their components, and their interconnections), Bolzano ultimately hopes to undercut the entire motivation for Kant's 'Copernican' thesis concerning the nature of our knowledge. Whereas Kant took his bearings from the alleged insight into the limitations of our capacities that he thought was yielded by the 'analytic' of our sensibility and understanding, Bolzano thinks we should begin instead with the insights we already possess into truth itself, insights like: the truth is what it is, independently of our cognition of it; and: we know that we know at least some truths. By emphasizing the fact that we are already in possession of at least some truths, and then zeroing on the nature of the property of being-true and the bearers of this property, Bolzano hopes to light the way to an escape from what he took to be the skeptical, ultimately anti-scientific conclusions of the 'subjective idealist philosophy' (WL Preface, I.vii).

Now, while Bolzano's unfailing commitment of what has come to be called 'semantic objectivism' has been seen by many of his readers to provide a much-needed antidote to Kant's apparent psychologizing of the conditions for science in general,[5] those who are impressed by Kant's humility about the scope of our knowledge will surely wonder whether Bolzano has gone too far in the opposite direction. Indeed, from the point of view of Kant's philosophy, the combination of (a) such a bold confidence in the appearance of our possession of truth, with (b) an eschewing of the need for any investigation into the conditions for the possibility of this apparent knowledge, will surely seem to return us to pre-'Critical', dogmatic philosophy.

Since the full comparative assessment of the two positions would require much more space than I have here, my main goal in what follows will be a more modest one. I will try to assess whether Bolzano's new semantical objectivist thesis is, in fact, successful in overcoming the pitfalls he takes to beset the theory of knowledge contained in Kant's 'subjective' idealism. To this end, I will try to confront Bolzano's position with certain replies

5. For a discussion of the significance of this phrase, see (Benoist 2006).

and counter-challenges that would seem natural from the point of view of Kant's philosophy.

I begin in §2 by presenting Bolzano's basic arguments for the existence of representational content *an sich*, emphasizing the extent to which Bolzano breaks from Kant by taking such content to be radically independent of mental activity, even while acknowledging that it makes its 'appearance' in such activity. I will then turn to several difficulties that Bolzano's proposed objectivism would seem to face, when viewed from the Kantian point of view. I turn in §3 to the challenges that Bolzano's position runs up against precisely because of Bolzano's acceptance of an appearance/*an sich* distinction. More specifically, I will turn to difficulties facing Bolzano's thesis that it is possible to have knowledge of such content as it is 'in itself', in light of the worry that the features Bolzano takes to belong to the content in itself are *merely* features that belong to its appearance. This will lead us in §4 to point up the parallels between the predicament in which the Bolzanian finds herself and that faced by Kant when pressed to say something more about *Dinge an sich*. I will then turn in §5 to the further question of whether or not Bolzano is able to avoid introducing a *Restriktionslehre*—or perhaps even two—of his own, once we couple the difficulties faced by the idea of knowledge of content 'an sich' with Bolzano's additional thesis that all of our knowledge of *objects* takes the form of an apprehension of *contents* (propositions) that represent them.

My conclusion will be that, even if Bolzano is right to insist on the recognition of what might be called a 'semantical *an sich*', until Bolzano can explain how it is possible for us to have *knowledge* of such contents *as* they are 'in themselves', Bolzano's position threatens to leave us in an *even more* precarious position than Kant does, with respect to our claims to knowledge of a mind-independent reality. This is because Bolzano's account threatens to limit our knowledge, now not just to an appearance of an object (as does Kant), but instead to an appearance of *a representation of* an object. For in the Bolzanian model, not only would we be unable to know how the objects of our representations are 'an sich', we do not even seem able to know how our representations are, in *themselves*, representing these objects *as* being.

§2. *The extrusion*[6] *of representations from the mind: content 'in itself'*

Let me begin by looking more closely at the nature of Bolzano's thesis concerning the removal of subjectivity from the elementary doctrine of science. Bolzano takes this to follow from a more careful analysis of the 'content' ('matter') of our knowledge. Bolzano accepts that this content is itself something that is representational in nature, in the sense that it possesses intentionality: it 'has' or 'is related to [sich bezieht auf]' an object (WL §49, I.218-19)—or at least purports to 'have' one, since it might ultimately be 'objectless [gegenstandslos]' (WL §67, I.304f). Crucially, though, Bolzano thinks that one and the same object can be represented in different manners by distinct representations. For example, Bolzano thinks that the two concepts, <2^4>and <4^2>, each pick out the same number, though they do so in different ways (cf., WL §96, I.445f.).[7] Bolzano takes this fact to show that, in addition to the mental acts of representing and their objects, we must keep track of this further feature of representations, and it is this feature that Bolzano calls the 'matter [Stoff]' of an act of representing (cf., WL §§48–49, I.217f. and §271, II.9). Insofar as mental acts are said to 'contain [enthalten]' the things that function as their 'matter [Stoff]' (cf., WL §34, I.154), I will refer to this matter as the *content* of the mental acts.[8]

---

6. I take this way of putting Bolzano's counter-thesis from (Dummett 1993, 22f. and 131).

7. I will use angle-brackets to refer to the content (i.e., to what (as we will see below) Bolzano calls the 'objective' representation (proposition), representation (proposition) 'in itself') expressed by the expression within the brackets.

8. It should be noted that this is not how Bolzano himself means to use the term 'content [Inhalt]', and moreover that he expressly distinguishes what he means to pick out by 'content' from what he here calls 'matter [Stoff]' (cf., WL §271 Anm, II.9f.). In Bolzano's hands, 'content' is used instead to pick out the elements ('parts') of a composite 'matter': the 'sum of composite parts out of which an [objective] representation consists' (cf., WL §56, I.244 and, for the particular case of the content of a proposition as 'the sum of all its … parts', §123, II.5; see also CE 93; BW 10). For Bolzano, then, it is only the matter of mental acts that has a 'content' in this strict sense of the term. For our purposes, however, what we need to keep track of is only the generic notion of something (whether simple or composite) being 'contained in' a mental act, and not what this matter itself contains, and so it is to the matter that I will refer to by the 'content' of a mental act. (In fact, this use of 'content [Inhalt]' is one that Bolzano himself concedes (to Exner) makes a certain degree of 'good sense', provided that we recognize the need to nevertheless keep track of the difference between (the whole of) what is contained in a mental act and what this content itself 'contains', as parts (cf. CE 95; BW 11). (Thanks to Sandra Lapointe for emphasizing to me the need to head off possible misunderstandings based on terminological imprecision here.)

Bolzano takes the need for a content/object distinction to be further confirmed, first, by the fact that the content and its object bear obviously distinct properties and, secondly, by the fact that there can be contents that 'are' something themselves, and yet that ultimately 'have' no object. As an example of the former, Bolzano refers us to the representation <Greek philosopher>, something which, as a content, is a single unified thing, even though it has many objects. As an example of the latter, Bolzano refers us to the representation <nothing>, which, as a thinkable content, is itself something (and even has component parts: it is composed out of <not> and <something>), but which has no object and is hence *gegenstandslos* (cf., WL §49, I.219f.).

Yet as important as the distinction between content and *object* is for Bolzano, he takes it to be equally important—and especially crucial for his criticism of Kant—that all contents, save that of intuitions, represent what they do in a way that can be common to, or identically present in, many different individual *acts* of representing.[9] For example, you and I can both judge that two and two makes four—that is, we can both be said to judge the *same* thing—despite there being a *different* mental event that transpires in each case, one in your mind and one in mine. For this reason, Bolzano insists that this single identical content itself (here, the proposition <two and two makes four>) must be distinguished from any one of the multiplicity of mental events in which it is, has been, will be, or even *may* be present. While the number of mental events can be 'multiplied' indefinitely, the content remains one and the same (WL §48, I.217f.).

Though Bolzano thinks that the failure to draw the content/object distinction is something that will 'entangle us in the grossest absurdities' (WL §19, I.79), Bolzano takes the failure to separate the mental act or event from its content—to separate, that is, what he also calls the 'subjective' representation, or representation in its 'subjective' sense, from the 'objective representation', or representation in an 'objective' sense, or simply

9. Bolzano's doctrine of intuition is complex. Bolzano is quite clear that he takes both the content/object distinction and the act/content distinction to apply even of the content of demonstrative and indexical representations, such as intuitions; cf., WL §72, I.326f. Bolzano also admits, however, that 'we are not able to bring about [hervorzubringen] an individual intuition that we have had once for a second time in ourselves', and also that, because of this, 'it is even less possible to make *another person* have a subjective intuition which has the same corresponding objective representation as an intuition present *in us*', which implies that 'if by *communication* [Mitteilung] we mean the eliciting of a subjective representation in another to which belongs the same objective representation as belongs to our own, then we must assert that it is not possible to communicate intuitions' (WL §75, I.334f.).

the 'representation in itself [an sich]' (WL §48, I.217)—to be even more pernicious.[10] In fact, Bolzano thinks that the failure to 'distinguish sharply enough [scharf genug unterscheiden]' between subjective representations (i.e., 'representations as *thought* [gedachte]') and representations *an sich* is 'the source of most of the current errors in logic'—Kant's included (WL §12, I.47).

Yet Bolzano does not think that it is enough to simply *distinguish* such objective content from the subjective acts in which it is contained. Rather, Bolzano takes the above reflections to show that such content is in *no way dependent* upon the acts themselves, not even conceptually. Bolzano thinks, first, that such content is not dependent on our activity for its coming into being, since it is in no way the effect of our mental activity. This is made especially clear in Bolzano's correspondence with Franz Exner, a professor of philosophy in Prague:

> By *representation*, when I take the word in its objective meaning, I by no means understand a certain *appearance* [Erscheinung] in the mind of a thinking being, nor an *activity* [Tätigkeit] of this being, nor something *produced* [etwas Erzeugtes] in this being through this activity. (CE 142; BW 64)

This independence follows from the fact that, on Bolzano's account, this content, considered strictly as it is 'in itself', cannot be ascribed *any* 'existence [Dasein] (*Existenz* or actuality [Wirklichkeit])' *at all* (WL §19, I.78). In Bolzano's words, it is 'not something existing [etwas Seyendes]' and is 'not to be found in the realm of the actual' (WL §48, I.217).[11]

Bolzano reaches this conclusion from several routes. On the one hand, Bolzano takes it for granted that everything existing or actual is singular, whereas (as we noted above) multiple different subjects can 'grasp [auffassen]' the very same content in multiple acts at multiple times, without the content itself being 'multiplied' (ibid.). The identity of the content across real multiplicity shows the content itself to be ideal. On the other hand, Bolzano thinks that there could be contents that are never actually

10. As Dummett has noted in (Dummett 1993), there is a considerable degree of parallel between Bolzano's language and several of Frege's remarks in his 1884 *Grundlagen der Arithmetik*. For the insistence on keeping psychology and logic 'sharply separated', see especially the end of Frege's introductory remarks; for Frege's own use of the terms 'subjective' and 'objective representation, see a footnote to *Grundlagen* §27.

11. Even so, Bolzano thinks that (like objective representations) propositions are objects (WL §25, I.115), and that the proposition expressed by 'there are [es gibt] propositions and truths' is true (WL §30, I.144). As we will see in a moment, he is also willing to say that they 'obtain or subsist [bestehen]'.

be thought of or grasped by anyone at all—save, perhaps, by God. Even so, he also thinks that each of those contents would still represent whatever it does nonetheless (cf. WL §48, I.217f.). For both of these reasons, Bolzano concludes that these contents do not 'presuppose a living being as the subject in which they proceed [vorgehen]' and 'do not require a subject to whom they represent'; instead, such contents simply 'obtain or subsist [bestehen]' (WL §48, I.217).

All of this contrasts with what is true of 'subjective' representations, which Bolzano *does* take to be 'something *actual*, which has an actual existence in a determinate time' (ibid.). More specifically, as he tells Exner, Bolzano takes these to be 'appearances or products *in* the mind of a thinking being' and also to require a certain 'act' on the part of the mind in order to come into being in the first place (CE 142; BW 64). In fact, with respect to both sensible and intellectual subjective representations, Bolzano's views are quite close to Kant's. Like Kant (cf., B33f.), Bolzano thinks that, for us to *intuit* an object, 'it is necessary that the object enter [trete] into an entirely special relationship with us, through which a simple [einfache] representation that refers only to one single object is produced [erzeugt] in our consciousness' (WL §74, I.332). Because of this, Bolzano agrees with Kant that:

> we can infer from the possession of a subjective intuition to the existence of an object corresponding to it that has brought about [hervorgebracht] the intuition through its effecting [Einwirkung] our capacity for representation [Vorestellungsvermögen]. (WL §77, I.347)

Bolzano also agrees with Kant (cf., B129f) that an additional 'act' is required for us to go on to *judge* about an object, or indeed to have any 'composite [zusammengesetzte]' representation whatsoever. As he tells Exner, 'the combining [Verbinden] of representations is an act [Act] separate from the bringing about [Hervorbringen] of them' (CE 115; BW 32). Bolzano elaborates on this in the *Wissenschaftslehre* as follows:

> To explain the arising [Entstehung] of a composite representation, it is not enough ... to suppose merely that the several simple representations that are its parts become active in us simultaneously. Rather, we must presume a peculiar activity [Tätigkeit] in our soul through which it transpires that the several representations become connected into a whole of the sort that constitutes a new representation. We call this *the forming* [Bilden] *of representations through connection* [Verknüpfung] *or synthesis*. (WL §285, III.85)

In the following section, Bolzano is even more straightforward on the generality of this point: '*each* of our composite representations *only* becomes composed [zusammengesetzt] through the activity of our own self [durch die selbsteigene Tätigkeit]' (WL §286, III.87; my ital.). For this reason, Bolzano claims that 'the most proximate cause [Ursache] of the arising of a composite [subjective] representation in our mind' is 'a mental activity [Geistestätigkeit]' (WL §77, I.347). And because Bolzano takes judgment itself to be a composite representation, judgment, too, will require such an act. As the 'appearance of a proposition', a judgment consists in a 'whole' that is 'composed [zusammengesetzt] out of parts', each of which is a subjective representation (WL §291, III.109).[12]

By contrast, Bolzano thinks that a connection to mental activity is *not* included in the very definition of representational content *an sich*. As we have already seen, Bolzano thinks that such content is what it is independently of its relation to any actual activity on the part of subjects—indeed, independently of the existence of subjects at all. The extent of this commitment is made evident by the fact that Bolzano even rejects that such dependence obtains (or would obtain) in the case of God. To be sure, Bolzano accepts that if God does exist, then all representations are actually thought, all propositions are actually judged, all truths are actually known (WL §19, I.78; WL §25, I.113). For this reason, Bolzano is ready to accept that 'thinkability [Denkbarkeit]' might very well be a '*property* [Beschaffenheit]' that, in fact, 'belongs to each proposition' (WL §23, I.92), and also that 'cognizability [Erkennbarkeit]' might very well be a property that, in fact, applies to each truth (WL §26, I.116). Bolzano insists, however, that this universal validity of such concepts does not require that either the concept of thinkability or the concept of cognizability be contained analytically in the very concept of a proposition or the very concept of a truth, as a '*component* [Bestandteil]' of their contents (cf. WL §23, I.92, and §26, I.117). Rather, Bolzano takes it to be evident, for example, that 'we can think of the concept of a proposition *an sich*

12. What the precise nature of this act is, Bolzano admits he cannot say exactly, noting only that the relevant act of 'combination' or 'connection' is itself of an 'entirely peculiar' sort, while at the same time confessing that he is unable 'to determine which property this effort [Einwirken] must have in order to produce [erzeugen] a judgment', rather than something else (WL §291, III.109f.). Bolzano is willing to say that its source is our 'capacity [Vermögen]' or 'power [Kraft]' to judge (WL §290, III.108), though he does not think that this adds much by way of explanation. In any case, the key point for us here is simply that there can be no judgment without some sort of mental activity; its connection to such activity is included in its very definition.

without recalling [erinnern] to ourselves its property of being something that can be thought by some understanding [Verstand]' (WL §23, I.92). A similar sort of conceptual disjointedness shows itself, thinks Bolzano, in our ability to think the *concept* of truth without thinking about its *objects* (the truths themselves) as being things that bear the property of being cognizable (WL §26, I.117).[13]

The upshot is that while Bolzano accepts that every representation or proposition is, in fact, thinkable, what he means by this is *only* that every representation or proposition *an sich*, now considered qua object in its own right, bears the property of being thinkable. But Bolzano does *not* think that the representation *of a representation an sich*—i.e., the concept <representation *an sich*>, and with it the concept <proposition *an sich*>—is *itself* unthinkable without including within itself, as one of the components of *its* content, a representation (the concept) <being thinkable>. Nor does Bolzano think that such an implication-relation obtains between the concept <truth> and the concept <being cognizable>, even if *all* the objects that fall under the former concept also fall under the latter and so also bear the property it represents—i.e, even if they are 'reciprocal concepts [Wechselbegriffe]' in this sense (WL §25, I.114).[14]

13. Here Bolzano is drawing upon one of his most fundamental distinctions, one that is rightly celebrated as directly anticipating contemporary discussions concerning the intensionality of meaning—namely, the distinction between (a) the *properties* of the *object* of a representation and (b) the *components* of the *content* of a representation of an object, or what is contained 'in' the representation *an sich* itself; cf., WL §25, I.113; see also WL §§63–64; NAK 12ff; and (Bolzano 2003, §3 #5). For example, though Shakespeare has the property of being the author of *Macbeth*, and hence can be represented through the concept <the author of *Macbeth*>, Shakespeare can also be represented through concepts like <the author of *Hamlet*>, and so on. Though the object of each of these concepts has the very same properties, the concepts themselves must be distinguished because they have different component parts.

14. For helpful discussion of this point (and others), see (Sebestik 1992, 115–33). These features of Bolzano's view speak against Joelle Proust's claim in (Proust 1989) that though representations *an sich* are 'de jure independent from our knowledge', they are not thereby '*removed* from human understanding', because they are the 'origin' of knowledge (52; my ital.). It is surely right that they do provide the origin of knowledge, yet it is also true that, unlike Leibniz, Bolzano thinks that the concept of these contents does not in any way require that we think of them as available to, let alone inhabiting, any understanding at all. In this way, 'in themselves' they *are* 'removed' from understanding altogether.

§3. *Kantian questions about the 'appearance' of content*

Recognition of just how absolute a separation Bolzano means to effect between representational content (representations *an sich* and propositions), on the one hand, and mental acts, on the other, helps to bring out the depth of the gulf that he means to place between himself and Kant. It helps also to bring into even sharper relief the precise reasons why Bolzano takes Kant's approach to the elements of science to be so fundamentally misguided. Importantly, Bolzano does not think that Kant's approach is vitiated by the fact that, in a discussion of science, Kant talks about the mental activity of subjects. As we have already seen, Bolzano himself admits that such 'appearances' (representations in the 'subjective' sense) can constitute their own object of inquiry. What is more, Bolzano himself actually *insists* that a complete doctrine of science must eventually incorporate the findings of such analyses of our capacities, as is shown from the fact that Parts III and IV of Bolzano's own *Wissenschaftslehre* are devoted to just such analyses, providing both a 'theory of knowledge [Erkenntnislehre]' as well as an 'art of discovery [Erfindungskunst]'.[15]

Rather, what ruins Kant's approach is that when Kant talks about the kinds of representations that are peculiar to understanding and so of interest in logic—namely, 'concepts, judgments, and inferences'—Kant wrongly treats such representations *only* as '*acts of thinking*' and '*ways of thinking*', and so 'treats them *only* as (actual or possible) appearances in the mind of a thinking being' (WL §16, I.60–61; my ital.). Bolzano thinks Kant thereby fails to recognize that the true roots of this sort of analysis must lie deeper still, in the doctrine of the contents of such acts. That is, even if Kant is right to think that certain portions of the doctrine of science depend on considerations of mental capacities—such that, for example, 'the rules of the art of discovery' are such as to 'depend [abhangen] on the laws to which the cognizability [Erkennbarkeit] of truth for us humans is bound [gebunden]'—Bolzano thinks that 'there is no doubt' that all of *these* sorts of rules 'depend *much more* on those properties which pertain

15. Bolzano even goes so far as to claim that the Wissenschaftslehre as a whole is, for this reason, 'dependent' on psychology (cf., WL §13, I.54). Of course, neither this claim, nor the fact that Bolzano thinks it is necessary to include inquiries into the nature of human mental capacities within the discipline that merits the title 'logic', need imply that Bolzano thereby 'commits the sin of psychologism', as Rolf George notes in (George 1997, 235). This is because Bolzano never claims that the truly *elementary* doctrine of logic (*Elementarlehre*) is itself dependent on psychology. Rather, as we have seen, these elements of Bolzano's logic are in no way dependent upon psychology at all.

to propositions and truths *an sich* themselves' (WL §15, I.58; my ital.). For this reason, despite it being true that Bolzano himself eventually moves on to discuss the nature of subjective representations in his own *Wissenschaftslehre*, the fact that Bolzano devotes his own *Elementarlehre* to 'undertaking the discussion of representations, propositions, and truths *an sich*' is something that is meant to mark a 'very essential [wesentliche] difference' between his position and the Critical philosophy (WL §16, I.61).

Yet even as the foregoing has helped to bring out such 'essential' doctrinal differences between Bolzano and Kant, it has also brought to light a striking overlap of a different sort. For, as it is impossible for any reader of Kant not to notice when reading Bolzano, the very terms that Bolzano uses to convey the two sides of the fundamental distinction he has in mind are none other than the terms that Kant uses to pick out a distinction that stands at the heart of Kant's own philosophy—namely, the distinction between the *appearance* of something and the way that something is *in itself*.

To be sure, Bolzano transposes this distinction from the more ontologically-directed context in which Kant's thought most frequently operates, into a new intensional-semantical register. That is, whereas for Kant, it is a *Ding*, something real, that both appears but also has a way of being *an sich*, for Bolzano, what 'appears' is instead an 'unreal' or ideal content, rather than that content's object (though this might *also* appear).[16]

Even so, several of the key features that Bolzano takes to distinguish appearances from the content 'an sich' find their straightforward parallels in features that Kant takes to distinguish appearances from the 'Dinge an sich'. This is already evident from our discussion in the previous section of Bolzano's account of subjective representations. Like Kant, Bolzano takes the appearances at issue to be things that are mind-dependent both for their existence (they exist 'in the mind', they presuppose a thinking subject in which they exist) and for their generation (they require an act upon, or by, the mind to be 'produced'). Also like Kant, Bolzano places the mind in a passive relationship with respect to what it is that is appearing by way of the appearance: just as, for Kant, the mind does not produce or make up the affecting thing, so too, for Bolzano, the mind does not 'make up' the content that 'appears' in its subjective representations; rather, the mind is only able to 'apprehend or grasp [auffassen]' it (cf., WL §122, II.4).

16. For the insistence on the difference between the uses of this distinction, see (Jakob 1902, 33); compare as well (Proust 1989, 51f) and (Sebestik 1992, 21).

But then, precisely because Bolzano places an appearance/an sich distinction at the heart of his own account of the elements of science, those approaching Bolzano's position from the perspective of Kant's philosophy will now be curious to hear more about how Bolzano means to construe the appearing-relation that is involved in this new semantical dimension, and will like to know, more generally, why Bolzano is confident that we are able to *know* such contents *as* they are *an sich*.

Concerning the first topic, it is clear that Bolzano cannot assign to his contents the role that Kant assigns to *Dinge*—namely, that of 'causing' or 'bringing about' their own appearance through something like affection. This would require that contents possess some degree of actuality or reality, a sufficient *Wirklichkeit*. As we saw above, however, Bolzano insists that contents *an sich* are not something 'existing' or 'actual [wirklich]' in any sense of the term. In fact, as we have also seen, on Bolzano's official account of the 'arising' of appearances in the mind, their true 'cause' is either the *object's* entering into a 'special' relationship with our minds, or the exercise of powers and capacities of our own *mind* itself. How 'effects' by these sorts of 'real' causes should somehow consist in the 'appearance' of some third, 'unreal' thing—that is, how this causal interaction enables us to 'grasp' contents—is never sufficiently addressed.[17]

To his credit, Bolzano himself is quite open about this gap in his account. This emerges from one of his exchanges with Exner, in which Exner poses the following series of questions about our cognitive relation to truths *an sich*: 'How can the non-existing be grasped [aufgefasst] by something existing? What can grasping mean here? [...] What is this objective truth? How do you know [wissen] about it? How do you come by it?' (CE 152–53; BW 74–75). In reply, Bolzano actually admits that he cannot 'define' what he means by 'grasp' in these contexts. Bolzano himself, however, does not think this is worrisome, since, as he tells Exner, he ultimately only means to use 'the word 'grasping [Auffassen]'' as 'a figurative [bildliche] expression', in the service of an 'attempt at making intelligible [Verständigung]' what he means by 'objective representation' (CE 162; BW 84–85).

Yet however admirable such honesty surely is, Bolzano's reply is nonetheless not very satisfying. Its insufficiency becomes even more pointed

17. Here again there are obvious parallels to Frege and his difficulties in accounting for what he, too, calls the 'grasping [Erfassen]' of the objective contents that he calls 'thoughts [Gedanke]', a 'mental process [seelische Vorgang]' that he finds to be 'the most mysterious of all', since it transpires right on the border of the realm of the subjective and actual (i.e., what is psychological), and the realm of the objective and non-actual (i.e., what is logical) (cf., Frege 1969, 157).

when viewed against the backdrop of the issues concerning of knowledge and givenness that are at the core of Kant's own analysis of appearances. This is because a gap at such a crucial moment in Bolzano's story threatens to undercut Bolzano's confidence *in general* that he is able to say anything at all about how such contents are 'in themselves', rather than simply restricting himself to claims about how they appear to him, or perhaps to everyone with minds like ours. For it is entirely unclear how an appeal to an indefinable transaction between ourselves and contents *an sich* is supposed to be sufficient to get Bolzano off the hook of a predicament parallel to the one that Kant takes us to be in with respect to '*Dinge* an sich'. This becomes an especially pressing issue because at the end of the day, Bolzano agrees with Kant that the only things that are 'actually' present 'in' any of our minds are *appearances*, and not what appears (for Bolzano, the content) as it is 'in itself'. Why, then, doesn't Bolzano, too, think he has to admit that, as Kant puts things, whether or not 'we believe ourselves to cognize things in themselves'—or now, *contents* in themselves—the truth of the matter is that 'we have to do with nothing except appearances anywhere (in the world of sense), even in the deepest researches into its objects'? That even in the semantical context, 'the transcendental object remains unknown to us' (cf., B62–63)?[18]

Now, as we have seen, Bolzano clearly thinks he can say a good deal about contents *an sich*. In particular, he thinks that we can know that certain straightforward correlations obtain between such contents and their appearances. For one thing, Bolzano thinks that there is a distinct *kind* of content for each of the basic distinct kinds of mental acts of representation (representing, judging, knowing, inferring). For another, Bolzano thinks that every *individual* subjective representation must contain an objective representation as its content, such that every representing contains an objective representation (cf., WL §271, III.9), every judging contains a proposition (cf., WL §291, III.108), every cognizing contains a true proposition (cf., WL §36, I.163), and so on. What is more, Bolzano thinks that the mental acts 'have' the very same *object* as the one that is represented by the objective representational content that is contained in the act (WL §271, III.9).

Finally, and perhaps most importantly, in the case of composite representations, Bolzano thinks that the subjective representations have the

18. A similar worry is raised by Melchior Palagyi in §7 of (Palagyi 1902, 34).

same number of 'parts' as the objective representation that they contain,[19] which also stand in the same kind of structure or organization as the objective representation as well. With respect to judgment, for example, Bolzano claims, first of all, that 'a judgment must contain just as many corresponding [entsprechende] subjective representations as there are objective representations that can be distinguished within the proposition that is the matter of the judgment' (WL §291, III.109). Yet Bolzano recognizes that such correspondence of parts is not sufficient, since 'different judgments can arise through different combinations [Verbindungen] of the same parts' (ibid.). Hence, Bolzano also insists that the *way* in which the parts are combined in judgment must likewise 'correspond' to the combination that is present in the proposition that appears (ibid.).

With this, then, Bolzano has gone all the way to identifying certain features as common to both the appearance of the content and the way that it is in itself. But at this point we must echo Exner and ask: what are Bolzano's *grounds* for any of these correlational claims? How can he claim to know *any* of the features that belong to the appearance of content *also* belong to the content *an sich*, when he has not yet explained how it is possible for content *an sich* to be given for inspection by us in the first place? Absent a more substantial story about either the 'grasping' or the 'appearing' relation, there is surely room to doubt even the more restricted claim that there is, in the content as it is in itself, something or other that 'corresponds' to any given feature of its appearance. For how can Bolzano rule out, for example, the possibility that in every case of our engagement with composite subjective representations—say, in a judgment—that in such cases, the compositeness or combination that is present in the appearance of some further content is *only* present in its appearance, and not in the content as it is *an sich*?

Questions surrounding the nature of judgment have a special bite, since by Bolzano's own lights, what is responsible in general for there being combination in the appearance of contents is *not*, in the first instance, anything about the *content* itself 'an sich'. Rather, the combination present in my subjective representations, and in my judgments in particular, is something brought about by that mental *activity* that we saw Bolzano himself taking to be 'the proximate cause' responsible for the 'arising' of such composite appearances in the first place (cf., above, §2). But if

19. This should not be confused with the claim that either the appearance of the content or the content itself has the same number of parts as the *object* of the content—a claim that Bolzano denies (cf., WL §§63–64).

the arising of the appearance swings free of the relevant content in this way—or at least, until its link to the content is more fully explained—isn't it possible that whatever content there may be to judgments (and more generally, of any composite appearances) is not composite 'in itself' at all, but instead that all such compositeness is present only in the realm of appearances?

For his part, Kant would surely pursue exactly this line even further and claim that, if we can know apriori that features like compositeness will be present in certain appearances, then the genuine explanation for this must lie, not at all in an appeal to the way that any *content* is 'in itself', but instead in further reflection on just the *activity* that Bolzano himself takes to be its cause. As Kant himself would have it, all the 'combination [Verbindung]' or synthesis that is 'in' our representations can be *completely* accounted for by *our* 'having combined [verbunden]' our representations in just this way (cf., B129-30). This places the reason for combination's presence in appearances ultimately in a fact about our own intellectual powers—namely, that our capacity for thinking and judging is 'discursive' rather than 'intuitive', and so '*requires* that [any] manifold must first be run through, taken up, and combined [verbunden], in order to make [machen] a cognition out of it' (B102, my ital.).

How, then, would Bolzano hope to rule out a more thorough-going subject- or mind-dependence of the constitution of the relevant appearances? For this sort of account to be blocked, and for any of these worries to be addressed decisively, Bolzano himself would have to demonstrate to us that he has some sort of access to the content as it is 'in itself'—e.g., access to whatever proposition it is that is allegedly making its appearance in a series of numerically distinct judgments, i.e., in subjective representations. The problem here is that, on Bolzano's own telling, the constitution of what is really 'in' our mind in any actual thinking or judging of such contents is something that has been 'brought about' by our own mental acts of combination. What Bolzano needs, however, is access to the contents through a means *other than* through these very judgments themselves, so as to be able to compare the objective and subjective representations.

§4. *Content as an 'Etwas = X'*

We saw earlier that Bolzano himself is officially reticent to try to say more on topics in this neighborhood. Is there anything more to say on his behalf? Here we might revisit the strategy that Bolzano uses to try to bring the idea of such content before our eyes in the first place. Bolzano asks us to think only about what is 'common' to multiple mental acts that appear to have one and the same content. From this angle, the access or grasping of contents *an sich* would seem to come about through a kind of generalization.

Those seeing the matter from Kant's point of view, however, would argue that such generalization will only move us to the consideration of a concrete relation that is *internal*, as it were, to the world of *appearances* of content. Crucially, this would not yet bring in any special entity that is both responsible for there being such a relation, and yet lies behind each of the members in the relation taken singly or collectively. Nor would it bring into view something that has a way of being 'in itself', in complete independence, not just from any actual acts, but even from any *possible* mental acts, not just from its actual appearances but even from its *possible* appearances.

Yet even if the Kantian were to grant that it is necessary to recognize *some* such utterly appearance-independent entity on the basis of such generalization, the challenge then posed is that of saying how else we could refer to something of this sort otherwise than by means of the completely indeterminate concepts expressed in phrases like the ones Kant himself makes use of when referring to a *Ding an sich*, such as 'the something = X' (cf., A250). Perhaps, then, we could say things like: 'the something = X that is common to this set of appearances: my judgment, here, now, that two and two makes four, my judgment, there, then ..., your judgment ..., S's judgment ... (etc.)'.[20] But even this way of thinking about 'the content' is obviously not a way of thinking about it *as* it is 'in itself', but rather as it is *in relation to* a set of appearances. The trick, it would seem, is to reach a state of mind where we can let go altogether of any such relational determination of our target of thought, where we don't have to 'recall' to ourselves that this content can (and does) appear. But how?

Besides the appeal to identity across difference, the second main strategy that Bolzano has for trying to bring us to a position to know what he means by terms like 'representation *an sich*' is the *via negativa*. That is, Bolzano hopes that by getting us to *exclude* certain features common to subjective representations (features like: existence, actuality, needing to be borne by

20. We have ourselves sought to represent such a proposition by '<two and two makes four>', but we might ask whether this expression is ultimately anything more than an abbreviation of the one above.

a subject, needing to be brought about by a subject) from our thoughts about the objective contents, we will be able to make the transition to the objective content itself. This itself quite fits well with Bolzano's own official explanation of the meaning of the phrase 'in itself [an sich]' in the *Wissenschaftslehre* itself: this addition to a word 'has the goal of *preventing* us from adding at will something in thought that should not be thought of in relation to the word, according to the proper determination of this word' (I.247; my ital.).

The problem that we run up against here, however, is that it is unclear what exactly *remains* to be thought, in relation to the words like 'proposition' and so on, once *all* connections and relations to concepts associated with mental activity have been pushed out of view. What is it that is left to be grasped by a term like 'content' in complete isolation from the concept of that which 'contains' this content (i.e., in isolation from the concept <subjective representation>)?

In any case, Bolzano himself is again quite upfront that any hopes for a more straightforward positive non-relational definition here will be disappointed. This follows from the striking (at this point, almost maddening) admission by Bolzano to Exner that, just as he is unable to define or explain what he means by 'grasping', so too is he ultimately *unable* to provide anything like a genuine 'definition [Erklärung]' of the terms he uses to pick out contents either—i.e., of 'representation *an sich*', 'proposition', 'truth', etc. (CE 99; BW 16). Rather, all Bolzano thinks he is able to offer with respect to these terms as well are 'attempts to make intelligible [Verständigungen]' such concepts (ibid. cf., WL §23, I.91; WL §52, I.228).

Here, however, we reach an impasse, since, as we ourselves have seen, most, if not all, of Bolzano's attempts to make intelligible the concepts associated with content *an sich* do so by nothing other than *linking them to mental activities*. More specifically, they seem to take two forms: the first involves the appeal to the mental activity of grasping or appearing, even while Bolzano admits that this is only a 'figurative' way of describing what actually takes place. The second takes the form of pointing to the pervasive function that the objects of these concepts (the contents themselves) have within our mental (and linguistic) lives, and then asking us to abstract away from this function, so as to still 'think' such concepts yet without 'recalling' these connections to mental activity. The worry here is that there simply might not *be* anything left to think after such an abstraction, or that, even if there is something, our grasp of it as it is 'an sich' will be very thin indeed.

Now, even if this is right, this would not *force* Bolzano to deny that there 'is' such content—any more than Kant's attempt at a similarly 'negative' exposition of what *he* means by 'thing in itself' requires him to deny that there are such things. Rather, what this line of questioning is intended to point up is, first, the real possibility of severe limitations on the knowledge that Bolzano is able to claim about such contents themselves, and hence the threat of a substantial restriction on what can be contained in Bolzano's alternative *Elementarlehre* for the theory of science. Here the parallel with Kant is again instructive, since Kant, too, could (and, in effect, does) claim that the embrace of the distinction between *Dinge an sich* and things as they appear is the beginning of all philosophical wisdom and that the task of winning through to this distinction must itself set the stage for any true 'doctrine of elements'. Yet, in Kant's hands, the concept of what lies on the other side of the appearances is something that functions within the doctrine of science itself only as a 'boundary concept [Grenzbegriff]' (B311), and his attempts at forcing the recognition of this distinction would seem to be officially intended only to serve a 'critical' function: to prepare the way for a later presentation of the elements of the doctrine of science.[21]

Of course, allotting this sort of role for a notion of the *an sich* is considerably different from claiming, as Bolzano does, that the doctrine of what falls under such a 'boundary concept'—i.e., the positive doctrine of how *things* are 'in themselves'—could *itself* function as the necessary *Elementarlehre*. And even if Bolzano were to revise his conception of the status and informativeness of the doctrine of content *an sich* along the lines suggested by Kant's treatment of the concept of the *Ding an sich*, it would be a significant further step indeed for Bolzano to then go on to model his claims about the *appearances* of such contents in a manner *also* parallel to Kant's distinctively *transcendental* idealism. This is because conceding the poverty of our knowledge of the semantical 'in itself' need not go hand in hand with any contention that we can know anything at all *apriori* about the world of the appearances of such contents—what Kant would call their universal and necessary *form* (whether the logical forms of judging or the mathematical forms of intuiting).

It is worth noting at this point that Bolzano is actually quite critical of Kant's appeal to apriori knowledge, himself attributing instead all of the

21. In fact, this is how Kant characterizes the whole first *Critique* in its entirety: 'it is a treatise on the method, not a system of science itself' (Bxxii).

'certainty' that we have, concerning things like logical and mathematical laws, to the fact that these laws are 'easily able to be repeatedly tested' and are 'always proved correct'—i.e., because they are confirmed in experience—rather than due to any alleged apriori insight (cf., WL §315, III.244f). Yet despite such anti-apriorist leanings, Bolzano is willing to say a good bit about the mind's role in the conditions for the possibility of the appearances of content—especially concerning judgments—and this might provide the groundwork for an attempt to sort out more clearly what it is that our minds contribute, rather than what is present in them due to some other factor, e.g., the objects affecting us. This would require an exercise close to what Kant might call 'transcendental reflection' (cf., B317).

§5. *The appearance of objects through the appearance of contents*

In the previous sections we wrestled with Bolzano's distinction between representational content considered 'in itself' and its 'appearance' in subjective representations, in order to tease out the obstacles that a Kantian would take to lie in way of Bolzano's alternative *Elementarlehre*, due to the obscurity of the possibility of knowledge of such content as it is 'an sich'. With this we were occupied with a problem inspired by Kant, but which arises due to the new distinctions introduced by the Bolzanian framework itself. In this final section, I want to turn our attention instead to the question of whether Bolzano's framework puts him in a better position to address a concern that is already manifest within the framework of the Critical philosophy itself. This is the question of what to say, on Bolzano's behalf, now not about the possibility of knowledge of the semantical 'in itself', but rather about the possibility of knowledge of the *ontological* 'in itself'—i.e., the question of the extent of our knowledge of *things*, *real* objects, as *they* are in themselves.

Kant famously limits our knowledge of things to knowledge of the appearances of things. Aside, perhaps, from the *existence* of such things, Kant thinks that all that we are presented with in our perceptions and experiences is how these things appear to us in our acts of intuition; as Kant puts the point concerning outer intuition in the *Prolegomena*, '*all* the properties that make up the intuition of a body belong merely to its appearance' (4:289; my ital.). The same is true of 'inner' intuition as well (B67). In other words, each of the properties that we are presented with in our intuitions 'is not to be encountered in the object [of our intuition] *in itself* at all, but is always to be encountered in its *relation* to the subject and is inseparable from the *representation* of the object' (B70n; my ital.). But since the only things I can *know* are things which can be 'present [gegenwärtig] and given to me' (cf., *Prolegomena* §9, 4:282), and since it is only the appearances of objects, rather than the objects as they are in themselves, that can be immediately 'present' to me in this way (i.e., in my intuitions), it follows, thinks Kant, that the only knowledge I can have concerning real things is knowledge of their appearances.

Since the basis for Kant's claims lies in his analysis of our mental capacities, rather than in an analysis of the concept of the *content* of knowledge (truth) itself, Bolzano thinks that Kant's account of the limits of our knowledge cannot be correct. As we have seen, Bolzano accepts that *every* truth is, in fact, knowable, even if he admits that not every truth is knowable by some human currently alive (cf., WL §314, III.232f). Bolzano actually goes further, claiming that we 'must avoid naming a limit [for human cognition], because there is none *present* [vorhanden]', and hence claims also that 'the sum of human knowledge [Wissen] is capable of increasing ad infinitum' (WL §314 Anm, III.238). What is more, Bolzano takes the very concept of an object about which *nothing* can be known to be a self-contradictory or 'absurd [ungereimt]' concept (cf., WL §314, III.235). For these reasons, Bolzano holds an 'opinion' directly contrary to Kant's: 'it is hardly possible to give a determination of the limit of our capacity for cognition' (WL §314 Anm, III.238).

Of course, we are now in possession of grounds for doubting Bolzano's confidence in our ability to know ('grasp') truths as they are *an sich*, insofar as he agrees with Kant that what is actually present in the mind is always only a truth *as thought* or *as judged*—in short, always only the *appearance* of the truth. Worries about our knowledge are compounded once we recall the fact that Bolzano takes *all* of our knowledge or cognition in general, and hence all of our knowledge of *objects*, to consist in the appearance or grasping (i.e., judging) of true propositions about them: 'the cognizability of an object is the possibility of making a judgment that is true about it' (WL §26, I.117; cf., WL §36, I.163).[22] In other words,

22. Bolzano's student Prihonsky emphasizes just this point in the conclusion to *Neuer Anti-Kant*: '[I]t should not be forgotten that the object of knowing [Kennen] is first of all *only truths* and not actual, existing objects. The latter are knowable [erkennbar] for us *only by means of* [mittels] *the truths* that we know. For if one of our representations pertained to an actual object or not, and to how this object is otherwise constituted, we could only ascertain [entnehmen]

all of our knowing of objects is mediated by the grasping of propositions about them.

Why is this problematic? Let us assume, for the moment, that true propositions are transparent with respect to their objects—i.e., that in (successfully? or adequately?) grasping a *true* proposition as it is *an sich*, we thereby come into possession of knowledge of *objects* as they 'actually' are. Now, as we have just emphasized, because, in the Bolzanian framework, the proposition itself, as the content of judgment, is something that is an additional 'thing' that has a way of being 'in itself', it is itself something for which the question can arise as to how we could know whether or not we are grasping it as *it* 'actually' is *in itself*, rather than simply as it 'appears' in its relation to our minds. As we have seen, this sort of question is one for which Bolzano has no straightforward answer. To the contrary, if the considerations from the previous section are on the right track, then all Bolzano can claim to know of propositions comes by way of abstract representations of them through the form: <something = X that is common to such and such set of judgments>. Yet even this is itself a representation of the proposition via its relation to mental activity, rather than a representation of it as it is 'an sich'. But then, on the reasonable assumption that it is the true proposition *as it is an sich*, rather than how it appears, that is the conveyor of genuine knowledge about how its object 'actually' is, such genuine knowledge threatens to lie one step beyond our reach.

Things get even worse if we now question the assumption introduced above about the transparency of true propositions with respect to the objects they represent. For how does Bolzano hope to justify the claim that true propositions can represent their objects *as the objects are 'in themselves'*? More generally, what could it even mean for the *objects* of propositions to be grasped as *they* (the objects) are 'in themselves', and yet *in* propositions? Is it somehow in the intrinsic nature of objects to appear 'in' propositions? And if not, then wouldn't the way an object is given in a proposition differ from the way it (the object) is in itself?

What is striking is that just such a distinction would seem to be implied by Bolzano's own analysis of the representationality of the content of propositions. This is due to the fact that he takes the core semantical element of the proposition itself—the copula—to be *without any correlate* in the realm of objects themselves.[23] This puts significant pressure on the

this *through certain truths alone* that assert something about it' (NAK, 230; my ital.).

23. Bolzano thinks that all propositions have the form: <A has b> (WL §126-27, II.9), and thinks that, in any true proposition, both the <A> and the <b> will have objective correlates (an

idea that there can, in principle, be a coincidence between content of the sort embodied by true propositions and the way its objects 'actually' are, since whenever objects 'appear' in propositions, they appear in a form that *cannot* be their own (e.g., *as* 'having' properties, *as* 'being' a certain way).[24] But then, until *this* gap is closed, even the (as of yet still mysterious) ability to grasp true propositions (contents) as *they* are in themselves would not yet be sufficient secure knowledge of the sort that goes beyond the way *objects* appear in such contents to the way they are 'an sich'.

If, however, the worries of the foregoing sections are well-founded, if the previous analysis is correct, and we do *not* possess even the latter ability, but only can grasp true propositions as they appear, then the presence of this gap in Bolzano's account would now place us at *two* steps remove from the way objects or things are 'in themselves'—which is to say, Bolzano's account would introduce one *more* intermediary step from our representations to objects than is present in Kant's 'idealist' philosophy itself. But then this would suggest that, rather than avoiding Kant's conclusions concerning the limits of our knowledge, Bolzano's position would seem to require an *even more* restricted conception of the 'objects' of possible knowledge, with the end result being that an *even more* modest conclusion needs to drawn by the Bolzanian concerning the scope of our knowledge than was drawn by Kant himself. This is because, at least from the Kantian point of view, on Bolzano's picture, it looks like we cannot even have knowledge of appearances of objects; instead, we only have knowledge of the appearances *of the appearances* of objects (i.e., the subjective representation of the objective representation of objects in propositions).

None of this, of course, would show Bolzano's views to be false. Perhaps Bolzano is exactly right to insist on the threefold distinction between act, content, and object. Perhaps, furthermore, Bolzano is also right to insist that we need not ultimately lose our grip on the concept of this content even if we peel away all of the natural associations the concept surely has with the concepts of mental activity and of a thinking subject—that something meaningful, something 'in itself' will still remain.

What we have shown, though, is that the thesis of such radical separation between the contents of acts and the acts themselves—between the

object and a property, respectively; WL §196, II.328f). The <has>, by contrast, does not 'have' any object whatsoever (WL §78, I.360).

24. For more on this, see my discussion of Bolzano in §3 of (Tolley forthcoming). See as well (Benoist 2006, 27f), for a related discussion concerning Bolzano's lack of any concept of distinctly propositional referents (such as a state of affairs or fact).

concepts <truth>, <proposition>, and so on, and concepts like <mind>, <subjectivity>, and others—threatens to plunge Bolzano toward the same epistemic predicament that beset Kant's allegedly more 'subjectivist' construal of the elements of our knowledge.

What this suggests, minimally, is that, however much is gained by taking on board Bolzano's more 'objectivist' construal of the content of cognition, there are reasons to be cautious about conferring on this content such a radical conceptual autonomy from the concepts of our mental activity and its capacities. Whether this should lead to a full-scale 'return to Kant', or should simply encourage the exploration of positions that occupy the middle-ground between Bolzano and Kant, is of course another matter.[25]

It does seem, however, that the considerations advanced in the preceding give encouragement to those, often taking their cue from Kant's works, who think that there is very little, and perhaps nothing at all, to say about the concepts of representational content, propositions, and truth, *outside* of reflection on their role in contexts that include 'minded' subjects like us—that truth is nothing if not the goal of inquiry, belief, and assertion, that even if the reason why certain things are true is objective (i.e., the facts), the concept of something's being true bears an internal link to possible knowledge and hence to the life of the mind, that the idea of representational content becomes an entirely empty shell once the idea of its being *for* a possible consciousness has been removed, and so on.[26] Perhaps embracing this sentiment is, as Bolzano seemed to think, the first step down a slippery slope that runs toward the embrace of a 'subjective idealist' philosophy. What I hope to have shown in the foregoing, however, is that it is less clear that its rejection will put us on any firmer footing.[27]

25. This, in effect, is Husserl's strategy in his *Logical Investigations*. It is also and more explicitly the strategy proposed by the Neo-Kantian Emil Lask: 'Bolzano's position is characterized by the fact that what is theoretical is spun out of what is objective, even if it still remains in pre-Copernican blindness. Kant has Copernicanism, but not Objectivism. One must unify [vereinigen] Objectivism and Copernicanism' (Lask 1911, Anhang §6, 277). For some discussion, see (Benoist 2006, 13f).

26. Palagyi is led to draw out criticisms of Bolzano along lines quite close to these through his sustained comparison with Kant's views (cf. Palagyi 1902, 36ff).

27. I would like to thank Sandra Lapointe, Charles Larmore, Samantha Matherne, and Timothy Rosenkoetter for discussion of this material and for comments on previous drafts.

References

Benoist, Jocelyn 2006: "Variétés d'objectivisme sémantique". In: Jocelyn Benoist (ed.), *Propositions et états de choses*. Paris: Vrin.

Bolzano, Bernhard 1837: *Wissenschaftslehre*. Sulzbach: Seidel. ('WL')

— 1935: *Der Briefwechsel B. Bolzano's mit F. Exner*. Eduard Winter, ed. Prague: Bohemian Royal Society. ('BW')

— 1972: *Theory of Science*. Rolf George, tr. Berkeley: University of California Press.

— 2004: *On the Mathematical Method and Correspondence with Exner*. Rolf George & Paul Rusnock, eds., trs. Amsterdam: Rodopi. ('CE')

Dummett, Michael 1993: *Origins of Analytic Philosophy*. Cambridge: Harvard University Press.

Frege, Gottlob 1969: *Nachgelassene Schriften*. Hans Hermes et al., eds. Hamburg: Meiner.

George, Rolf 1997: "Psychologism in Logic: Bacon to Bolzano". *Philosophy & Rhetoric* 30(3), 213–242.

Jakob, Ludwig 1902: *Über die Grundbegriffe der Wissenschaftslehre Bernhard Bolzano's*. München: Wolf & Sohn.

Kant, Immanuel 1902-: *Kants Gesammelte Schriften*. Königlich Preussische Akademie der Wissenschaften, ed., Vols. 1–29. Berlin: de Gruyter.

— 1991–: *The Cambridge Edition of the Works of Immanuel Kant*. Guyer & Wood, eds. Cambridge: Cambridge University Press.

Lask, Emil 1911: *Die Logik der Philosophie und die Kategorienlehre*. Reprinted, Tübingen: Mohr, 1993.

Palagyi, Melchior 1902: *Kant und Bolzano: eine kritische Parallele*. Halle: M. Niemeyer.

Prihonsky, Franz 1850: *Neuer Anti-Kant*. Bautzen: Weller ('NAK').

— 2003: *Neuer Anti-Kant*. Edgar Morscher, ed. Sankt Augustin: Academia.

— 2006: *Bolzano contre Kant*. Sandra Lapointe, ed. tr. Paris: Vrin.

Proust, Joelle 1989: *Questions of Form*. Minneapolis: University of Minnesota Press.

Sebestik, Jan 1992: *Logique et Mathématique chez Bolzano*. Paris: Vrin.

Tolley, Clinton forthcoming: "Idealism and the Question of Truth". In: Michael Glanzberg (ed.), *Oxford Handbook of Truth*. Oxford: Oxford University Press.

Grazer Philosophische Studien
85 (2012), 89–129.

KANT AND BOLZANO ON THE SINGULARITY OF INTUITIONS

Timothy ROSENKOETTER
Dartmouth College

Summary

Kant and Bolzano agree that intuitions are non-accidentally singular, but each offers more than one explanation of why this is the case. One model, exemplified by Bolzano's explication of intuitions as "this"-representations, posits a type of representation which is such that it can only have one object. A very different explanation, prominent in Kant's Transcendental Aesthetic, has recourse to the fact that certain classes of objects (spaces and times) can have only one instance, and argues on this basis that some representations with those contents are singular. This paper surveys various versions of these two explanations and uses each philosopher's answers to shed light on the other's.

1. *Introduction*

Famously, one of the signal differences between Kant and his most influential predecessors lies in his insistence that there are two categorically different species of representations, concepts and intuitions. The well-known *Stufenleiter* passage, which contains definitions of each, provides two differentia in order to classify "cognitions" or "objective perceptions" into these species.[1,2] Intuitions are *immediate* and *singular*. Concepts are *mediate* and *general*. It isn't clear how Kant understands these differentia,

1. All boldfacing in quotes has been added; italics are in the original. Outside of quotations italics are used systemically when introducing the name of a position.

2. *KrV* A320. The highest genus is representation, which is divided into those "with consciousness" (perceptions) and those without. "A perception that relates [*bezieht*] solely to the subject, as the modification of its condition, is sensation (*sensatio*), an objective perception is cognition (*cognitio*)" (Kant's merely didactic italicization elided). I take an objective perception to be one which is such as to refer to an object, though it may not actually do so. Accordingly, I understand sensations to be conscious representations without referential purport, *mere* modifications of the subject's conscious state.

just that he believes that intuitions and concepts are the only two species of cognitive representations. There are not, then, any cognitions that are both immediate and general or cognitions that are both mediate and singular, taken in whatever sense these terms are meant to be taken.

Locating intuitions and concepts within Bolzano's taxonomy of representations is more complicated, since *singularity* and *simplicity*, the two defining features of intuitions for Bolzano, do not always travel together, as immediacy and singularity do for Kant. Intuitions are definitionally prior to concepts. The first genus relevant for isolating intuitions is the class of *actually referring*, or "objectual", representations (*gegenständliche Vorstellungen*),[3] which is divided into general and singular representations. In contrast, all representations whatsoever, not merely objectual representations, can be compared with respect to the simplicity or complexity of their content. Simple representations—easily communicated examples are <something>, <not>, <has>, <actuality>, and <ought>[4]—are those that have no proper parts. Complex representations arise from the combination of simple ones, as <nothing> can be formed by combining <something> and <not>. Bolzano selects one of the four 'boxes' yielded by these two orthogonal divisions and calls them "intuitions". Intuitions are singular representations that are also simple. Concepts are then defined as representations that do not contain any intuitions. "Mixed representations" are complex representations that contain at least one intuition.[5] Both concepts and mixed representations can be either general, singular, or non-referring.

Thus, Kant and Bolzano agree that intuitions are singular representations. Moreover, they agree that intuitions are, in a sense that remains to be worked out, essentially singular rather than merely accidentally so. Here is Bolzano noting this additional proviso:

3. Objectual representations do not include accidentally non-referring representations, inconsistent representations, and representations such as <has> whose nature is not to refer. Note that Kant, in contrast, divides conscious representations with referential *purport*, rather than their proper subset of the successfully referring, into concepts and intuitions.

4. I will use offset lines (|I|) to mention intuitions and angle brackets (<F>) to mention concepts and mixed representations.

5. As a matter of taxonomy, complex representations, all of whose parts are intuitions, are mixed representations (*WL* §73), though later in *WL* Bolzano argues that there are no such complexes, since a concept such as <and> is required for the synthesis of two more intuitions into a single representation (§278, III:22).

> ... we must also have a representation such that the fact that it has no object [at all][6] apart from the change that just now takes place in us is not something that holds as it were by chance, or that must be inferred... [*MM* §6.3]

Unfortunately, neither philosopher tells us as much as we might hope about why this singularity is non-accidental or essential.[7] This paper will focus on this question, in part as a way to shed light on larger issues in each philosopher's theory of referring representations. We might reasonably expect, for instance, to better understand their accounts of conceptual cognition if we can reconstruct why some other representations are taken to be non-accidentally singular.

A helpful way to divide the possible strategies is according to whether the explanation proceeds from properties of the representations themselves or from properties of the objects of those representations:

(I). The first group, whose members I will call *Representation-based Interpretations*, takes Kant and Bolzano to hold that a certain type of representation—which type cannot be specified by recourse to the particular contents of its instances—has properties that ensure that these representations can have only one object. This is certainly the style of interpretation that Bolzano's own explanations most readily suggest, since he tells us that the infinitely many distinct intuitions are all of a type that is best conveyed linguistically by "this", and in doing so does not immediately limit this class to certain contents such as <time>.[8] Because this style of explanation also turns out to be promising when applied to Kant, though similarly direct and obvious textual evidence (let alone any development of the idea) is lacking, it will be helpful to approach the question of Kant's commitment to (I) with Bolzano's ideas on the table for comparison.

(II). The second group observes that there are classes of objects, such as spaces, times, and universes, that non-accidentally have only one instance. It then argues that it would be impossible for representations with those objects to be general, for there simply aren't multiple objects to which they could refer. Obviously, this strategy will

6. These brackets contain Rusnock and George's correction of a likely typographical error.

7. These terms will be used interchangeably in this paper.

8. Of course, we might learn subsequently that intuitions are limited to only some contents. For instance, Bolzano will argue against Kant that we cannot have an *intuition* whose content is <the whole of time>.

require that we consider the contents of these representations from the beginning. Let's call these *Uniqueness Interpretations*.

Below we will see that both Kant and Bolzano argue in ways that suggest that they hold some version of (II). The evidence for this is particularly strong in Kant, since one of his central arguments for the existence of non-conceptual (cognitive) representations rests on the uniqueness of space and time. Kant's elusive idea seems to be that if only concepts were available to represent space and time, then our representations would distort the objective fact that each is unique. In one particularly helpful discussion Kant contrasts "metaphysical" space with the spaces that geometers (accurately) cognize *using concepts* as rules for construction: the former, "original" space, "because there are not many [...] cannot be brought under any concept capable of construction".[9] To do so would be to cognize it inaccurately as a one-of-many, i.e., as one of the constructible spaces studied by geometry. Too little has been made of the fact that this argument falls under (II), regardless of whether the constitution of things in themselves or the constitution of our subjective forms of sensibility is ultimately held responsible for facts about space and time. Kant, of course, opts for the latter, but his argument still moves from the uniqueness of a certain object, in a suitably broad sense, to a conclusion about veridical representations of it. Comparison with Bolzano will help us to avoid allowing various unrelated Kantian commitments—some of which are reflected in the fact that Kant does not call space and time "objects"—to obscure the logical structure of that position. This is not to say that there isn't also plenty of less direct evidence that Kant would embrace (I), various versions of which we will canvass below. In doing so we will see that one challenge facing Kant's interpreters is settling whether he is committed to one explanation of essential singularity to the exclusion of the others; he may have been happy to make use of any of several explanations. This paper will not be able to pronounce definitively on that question, but a start will have been made if it can at least provide a map of the main positions available to Kant and how they relate to one another.

A reader looking for resolution will be more satisfied with Bolzano's writings on the topic. Though Bolzano sometimes explains his view in a manner that suggests his dependence on (II), I will argue that there are

9. *Über Kästners Abhandlung*, 20:420 (trans. Friedman 2000). This is taken from the published review, as edited by Schultz. Cf. R. 4425, 17:541.

intractable problems with this strategy. Fortunately, it turns out that a version of (I) will serve him just fine, and there is copious evidence that this is indeed his considered view. This is not to say that no puzzles remain. In a more speculative coda (§6) I will suggest that one of these puzzles could be solved if Bolzano would just borrow a theoretical concept from Kant that is missing from Bolzano's own toolbox: the "matter" of a concept, understood differently than we find anywhere in Bolzano.

Of course, the mere fact that almost six decades after the first *Critique* Bolzano chose precisely the same labels for two broadly similar types of representation should not lead us to simply assume that comparing Kant's and Bolzano's distinctions will be a profitable way to approach the question of why intuitions are essentially singular. As I hope will become clear as we proceed, however, there are distinct advantages to pursuing the present line of questioning in the midst of a comparison of Kant and Bolzano. First, the relevant portions of their theories are sufficiently similar to be sensibly compared, and yet not so similar that Bolzano is simply tidying up Kant's theory. Second, Bolzano's generally clear presentation of available alternatives brings a keen analytical perspective to the topic. Whatever can be said for other parts of Kant's philosophy, the separation of his larger position on intuition into logically independent claims is not something that we find in Kant's theory of intuition. Comparison with Bolzano can help us to see that Kant often treats a number of logically distinct assumptions as if they were one. On this project the current paper makes only a start.

This said, in Bolzano's view there is a virtual chasm separating his entire approach to logic from Kant's, and we might reasonably worry that it will bar us from arranging any illuminating comparison of their positions on the narrower issue in question. Bolzano holds, namely, that a confusion between "subjective" or "thought" representations and "objective" representations (or "representations in themselves") has been endemic to previous philosophy, and that Kant is no exception to this generalization. The former are events, states, or acts that are actual (i.e., have effects) and exist at particular times as the attributes of individual souls. The latter are the "content" (or, synonymously, the "matter [*Stoff*]") of the former (*WL* §271, III:9). They are neither actual, nor temporal, nor the possessions of any soul. Bolzano takes every subjective representation to be the "appearance" or "grasping [*Auffassung*]" of a representation in itself (*WL* §270, III:6). But he does not think that we can safely infer in the other direction. Indeed, he assumes that as a matter of fact there are many representations in themselves that are not tokened by any thinker.

We can dismiss immediately the strongest critique that Bolzano voices, according to which Kant takes the subject matter of logic to be actual acts of thinking or representation (cf. *WL* §12, I:47). Though Kant routinely uses formulations that encourage this misunderstanding,[10] and he certainly fails to introduce terminology that will help readers avoid falling prey to the confusion, the subject matter of *pure* logics (whether pure general or transcendental logic) is not a modification of a particular subject or something that has a particular temporal location, nor is it simply the features that all of these "subjective" or "thought" representations share. Kant's pure logics are instead concerned with representation-*types* and ways in which these types can be analyzed, combined, or compared, as opposed to actual instances of analysis, combination, or comparison.[11] This is sometimes part of what Kant is expressing when he explains that pure logic inquires into the "form" of thought rather than its "matter". Kant's pure logics are not, for instance, concerned with any particular acts of hypothetical judgment, nor even with features that all actual acts of hypothetical judgment share in virtue of being actual (e.g., that they take some time). They are concerned instead with the fact, e.g., that hypothetical judgments are composed of two categorical judgments, where this composition is not a relation between actual entities but rather between "forms".[12] Similar points can be made concerning the subject matter of Transcendental Aesthetic, the pure inquiry that treats intuitions.[13]

For this reason, and despite the fact that the expression is jarring to those accustomed to Kant's terminology, I will feel free to use "representation in

10. One source of difficulties is the term "use", since logic studies the "use" of concepts, faculties, etc. (cf. *KrV* A52), and this language is naturally, but incorrectly, taken to indicate that their subject-matter unfolds *in time*.

11. This is a topic that would benefit from sustained attention, though it won't be my subject here. For the first steps in an attempt (still relatively rare) to develop this sort of point, see M. Wolff 1995, 22.

12. The same cannot be said of applied logics, since they contain, among other things, effective techniques for memorization, etc. and thus depend on the conditions under which the subject matters of pure logics are realized by human beings *in time*. This fits the fact that Kant is willing to classify applied general logic, but not pure logics, as a branch of psychology (*KrV* A55; *JL* 10:18).

13. Why doesn't Kant recognize an in itself correlate for *empirical* intuitions? If I am correct that Kant's version of Bolzano's distinction between subjective representations and representations in themselves is centered on the idea that there are *types* of representations that are tokened by actual subjects, then we should not expect Kant to see a need for something occupying the role of an in itself correlate in cases in which each subjective representation is the *only* instance of its type, as is the case for empirical intuitions (see §4 below).

itself" whenever Kant has a form or a type in view.[14] I don't mean thereby to minimize what, taken at face value, certainly appear to be major differences between the philosophers on the metaphysical status of representations in themselves. Indeed, there is not the slightest hint in Kant's works that he would be inclined to attribute any sort of being to the representation-types that individual thinkers token. Yet I will not take any stand here on how philosophically deep this indisputable surface-level difference cuts. My use of the distinction between the subjective and the in itself is intended merely to minimize the occurrence in Kant of an ambiguity that Bolzano deplored and sought to eradicate. If we wish to learn whether Kant's lack of attention to the distinction led to any consequential errors in his philosophy, then *we* need to keep track of whether the particular claims he lodges concern the possessions of individual subjects or the types that are thereby realized.

I've just argued that we needn't worry that the mere fact that Bolzano distinguishes rigorously between the subjective and the in itself will block fruitful comparison with Kant quite generally. There is, however, a related worry that pertains specifically to the narrower question of this paper. This worry is that Bolzano would be justified in turning any search for an explanation of the non-accidental singularity of intuitions into a virtual dead end by taking it to be correctly understood exclusively as a question about representations in themselves. In this case we can imagine him responding: "There *might* be syntactical constraints that provide limitations on what types of representations in themselves there are. Yet we have no idea whether this is the case. Hence, as far as we are able to tell there simply is no explanation for why there is a class of singular representations in themselves". If this is Bolzano's view, then any comparison of how Kant and Bolzano answer the leading question of this paper is likely to disappoint. Let's call this the *Dead End Worry*.

As a matter of fact Bolzano does not limit himself in this fashion. We will see specific instances of this below, but it is worth mentioning up front a few larger features of Bolzano's methodology that are relevant. First, and quite generally, even though the first Chapter [*Hauptstück*]

14. I choose "representation in itself" in favor of "objective representation", since in the context of Kant commentary it is almost impossible not to hear the latter as "representation of an object". Bolzano's "thought representation" for the contrast class would cause confusion when the representation in question is an intuition. Thus, I use "subjective" representation/concept/intuition below exclusively in Bolzano's sense of the term and avoid using it to express other Kantian theses such as his transcendental idealism.

of the *Wissenschaftlehre*'s Doctrine of Elements (§§47–90) is systematically conceived as a treatment of representations *in themselves*, with the consideration of subjective representations reserved for the third volume (§§270–289), Bolzano frequently treats subjective representations in the earlier chapter, yielding material that certainly seems to amount to more than passing observations. In particular, when Bolzano wishes to argue that there are intuitions in themselves of a certain type, he first endeavors to make plausible that there are actual *subjective* intuitions of that type, only then observing that there must be a representation in itself corresponding to each subjective representation.[15] On reflection, I think that we should not be surprised that Bolzano follows this procedure. The only alternative would be for Bolzano to somehow, arbitrarily, hit upon various differentia and use them to formulate distinctions, without any clue as to whether these in fact correspond to genuine divisions among types of representations in themselves. If we rigorously avoid consulting the realm of subjective representations, then there is nothing to be said against (or in favor of) a division that includes as one of its members simple representations in themselves having, for instance, precisely two objects. Presumably, the reason that Bolzano does not consider this particular class of representations in themselves is that he finds no evidence among our subjective representations that it is non-empty.

The upshot of our consideration of the Dead End Worry for the present investigation is the following. If Bolzano can explain why a certain class of actual subjective representations is always singular, then he will at the same time have found evidence that there are also representations in themselves of that type. We should expect that the former explanation might advert to features of subjective representations, such as their being caused in certain ways, that representations in themselves cannot share. Once these points about Bolzano's actual method are recognized, we can start to recognize certain similarities between Kant and Bolzano that remain obscured if we think, simplistically, of the former as conducting nothing more than an investigation of subjective representations, and of the latter as concerned only with representations in themselves. Consider that a thoroughly uninteresting explanation for the essentially singularity of intuitions is that they are defined as such. As we will see, in the end neither Kant nor Bolzano is satisfied with this answer. Both avoid trivial

15. Cf. *WL* §72, I:326. For the argument that there must be corresponding representations in themselves, see §271, III:8f.

treatments of our leading question, first, by identifying a class of subjective representations in ways other than just by citing intuition's definitional criteria; and then, in a second step, by arguing that the favored definitional criteria apply to that class of representations, including the correlative representations in themselves.

2. *Simplicity*

We have seen that each philosopher offers a definitional criterion for intuitions that the other does not. Kant offers immediacy. Bolzano offers simplicity. Even though this paper will be focusing on puzzles that arise from the criterion that they share, singularity, we first need to look at how the disputed criteria interact with each philosopher's larger views. We will also be asking how much of a dispute is really present.

Simplicity is a property that Kant does not use in classifying representations into types; indeed, it is a property that Kant discusses only rarely. The first thing to be observed is that there is a deep disagreement between the philosophers on the issue of which of our representations are simple. Kant seems to regard only two subjective cognitive representations as simple: the concepts <I> and <something> (or <object>).[16] Now since a strong case can be made that <I> is, beyond this, an essentially singular representation, it can be argued that for Kant it is our representation <I>, and only this representation, that qualifies as an intuition *in Bolzano's sense*. This is in itself a fruitful avenue for comparing these two philosophers. Yet since Kant seems not to have been tempted by the classification of <I> as an intuition, and Bolzano himself does not consider this representation to be simple (*WL* §44, I:142), this path of inquiry would take us into much different territory than we will cover if we stay focused, as I propose to do here, on what each philosopher himself regarded as an "intuition".[17]

16. See *KrV* A345; R. 2870, 16:553. I avoid certain technical questions associated with the latter pair of concepts. More important for our purposes is Kant's view that no concept can have "an infinite set of representations *within itself*" (*KrV* A25). Bolzano argues that Kant thereby commits himself to there being simple concepts in themselves (*WL* §78, I:356); and more than two are clearly required to yield the many concepts that there are. For this reason I suggest that we take Kant's limitation of the list to two to hold only of *subjective* concepts, and perhaps only of subjective concepts that we are *aware of possessing*.

17. I pursue this other line of inquiry in Rosenkoetter (ms), a companion piece to the present article.

Whereas Kant seems to regard only two subjective, cognitive representations as simple, Bolzano thinks that there are infinitely many simple representations, viz., an indeterminate number of simple concepts and infinitely many intuitions. Each of us possesses many simple representations, since "the *next* and *immediate* effect[s]" of changes in the mind are subjective intuitions.[18] It was open to Bolzano to hold that only some of these immediate effects are simple representations, but he commits himself to the bolder, universal claim that I will call:

> *Simplicity Hypothesis:* all subjective representations that are the immediate effects of changes of the soul are simple, though in many cases these intuitions are subsequently combined with one another or with concepts to form complex representations.[19]

Though Kant would agree in calling representations that arise in this way "intuitions", he would do so only because simplicity is not among his own criteria for intuitions. Indeed, Kant would join many other readers in being puzzled by the supreme confidence with which Bolzano defends his Simplicity Hypothesis. Bolzano has two main responses to these critical worries.

First, and what stands out in Bolzano's own explicit defense, is a thesis for which he argues in some detail, viz., that a referring representation need not have as many parts as its object has parts.[20] Let's give a convenient label to a slightly stronger position:

> *Isomorphism Failure Thesis:* there is no regular relation between the number of proper parts in a referring representation in itself and the number of proper parts in its object.

18. *WL* §72, I:326; cf. §74, I:331. Bolzano appears to treat our "noticing [*Aufmerken*]" as a precondition, which, once met, allows a change in the soul (rather than, say, the act of noticing itself) to be the *most immediate* cause of a subjective intuition. Below we will see that we can also notice a representation *R*, in which case *R* is the direct cause of a second representation (a subjective intuition) whose object is *R*.

19. For evidence that Bolzano is committed to a universal claim, cf. *WL* §72, I:326; §277, III:19; and §278, III:22. Though Bolzano does not discuss this, it would seem that the representations formed by synthesizing intuitions using <and> remain essentially singular, though they are no longer simple. These would be close to Kant's empirical intuitions.

20. For Bolzano's clever arguments and counterexamples, all of which concern representations that Kant would classify as *concepts*, see *WL* §§63–64 and *MM* §3.5.

This allows us to consider the hypothesis that Kant's opposition to Bolzano's claim of simplicity for these representations stems at least in part from his assumption that the Isomorphism Failure Thesis, though obviously true of concepts, is *not* true of intuitions.[21] Perhaps the regular relation in this case is still relatively weak. Yet Kant does at least hold that intuitions must be manifolds if they are to represent the manifolds that are their objects.[22] Now, this is a far-reaching thesis, for which I cannot argue here, and it will play no direct, supporting role in the arguments below. It is nonetheless worth flagging as a hypothesis in the present context because it would, if correct, *explain* some differences between the philosophers, including the fact that Kant shows no sign of even considering the possibility that intuitions are simple.[23] So while neither philosopher believes that the *objects* of intuitions are in general simple, only Kant takes this to tell us something about the simplicity or complexity of those *representations*.

The second main point also leads, if not to agreement, at least to a more nuanced perspective on what at first seems to be a brute disagreement between the philosophers. Though Bolzano does not state this clearly, his view seems actually to be that we do not *notice*—I intend this as a commonsense, non-technical term—many, perhaps even the vast majority, of our subjective intuitions. Some care is required in formulating the grounds for this claim. He is quite clear, first of all, that it is not always (and in fact might never be) the case that a simple representation is the *only* subjective representation belonging to a soul at a given moment.[24] This highlights that the simplicity of a subjective representation in no way depends on its being

21. I see no reason to think that Kant assumed that there is a regular relation between the number of proper parts in a referring *concept* and the number of proper parts in its object. His insistence that <something> is simple, together with the fact that it represents everything that is not a "nothing", and thus represents objects with widely varying numbers of proper parts, should make that assumption anathema to him.

22. That the *objects* are manifolds ("every appearance contains a manifold," *KrV* A120, A-only) can be known, according to Kant, already from the fact that they are temporally extended (cf. B155, B-only; B154; cp. *WL* §287, III:95f). Thus, even an awareness of red—a paradigm case of a "simple idea" for Locke (1689, II.iii)—would be a manifold representing a temporal manifold, for Kant.

23. Another difference between the philosophers that might be explained, at least partially, by the above hypothesis is Kant's espousal and Bolzano's rejection of any explication of truth as correspondence of a cognition to its object (cp. *WL* §29 with *KrV* A58; A157; A237; A642; and *Prol* §18, 4:298).

24. Bolzano illustrates this claim with the example of <something>, conceding that we will typically possess it at the same time as we possess a different subjective representation, whose object is the written word or sound that we use to designate <something> (*WL* §277, III:20; cf. §287.5, III:96).

the sole representation that a subject possesses at a particular time, a fact that makes the Simplicity Hypothesis more plausible than it might appear at first glance. Next consider Bolzano's conception of a clear representation, which I take to be his proposed analysis of our commonsense notion of noticing a representation. A subjective representation *R* is clear when we have a second, reflexive intuition whose *sole* object is the representation *R* (*WL* §277, III:28). As Bolzano explains, we have thereby put ourselves in a position where we *could* judge "I have representation *R*", though the further step of actual judging is not itself required for *R* to be clear. Now let's suppose that *R* and *S* are distinct, simple representations that I possess in a single moment. It is a consequence of Bolzano's account that *R* is not a *clear* representation simply by virtue of my acquiring a reflexive intuition whose object is both *R* and *S*, so long as I don't also acquire an intuition whose object is *R* alone. But if I do not notice *R* by itself, if I only notice it as part of noticing a manifold of simple representations, is there are any reason to assume that I would be aware of having a representation that is simple? Though he does not consider this question, I think that it is at least plausible for Bolzano to answer negatively. After all, I am not, on the basis of the clear representation of *R* and *S* together, even in a position to judge anything about *R* by itself. For that I would need a reflexive intuition that has *R* alone as its object. All of this allows us to see that Kant's imagined protest that for all he knows he does not possess, because he does not notice, *simple* (and singular) representations whose objects are changes in the mind need not pose a problem for the Simplicity Hypothesis. Bolzano need merely supplement it with the further hypothesis that Kant, perhaps together with all humans, possesses no reflexive intuitions that manage to capture precisely one intuition as their objects (cf. *WL* §286, III:87f).

3. *Immediacy*

There are several models for understanding Kant's immediacy criterion. Jaakko Hintikka's view that the immediacy criterion is nothing more than a corollary of the singularity criterion is well-known, as is Charles Parsons' rejoinder that "immediacy [...] is direct, phenomenological presence to the mind, as in perception".[25] Something like (i) phenomenological pres-

25. Parsons 1992, 66. I agree with the consensus that Hintikka's view (1968) is not a tenable interpretation.

ence is quite plausibly one thing that Kant is gesturing at when he uses this term to characterize intuitions, as is the further claim that intuitions are (ii) the immediate causal effects of objects. At least on the latter count Bolzano doubtless agrees. Moreover, though Bolzano obviously regards causal immediacy as inappropriate for the *definition* of intuitions, this surface difference between the philosophers is largely a consequence of the fact that it is intuitions in themselves that Bolzano defines, though only subjective representations can be effects. Below I will ask whether we might learn something more about Bolzano's intuitions by using certain properties of subjective representations to specify the relevant type.

Though he might well be gesturing at (i) and (ii) as well, there is strong textual evidence that with "immediacy" Kant intends to define intuition by recourse to what I will call its (iii) *semantic immediacy*,[26] and in so doing he departs from Bolzano. The semantic immediacy of intuitions consists in the fact that *it is true of each subjective intuition that it does not require the existence of any other subjective representations in order to refer to its object.*[27] In this respect intuitions do indeed contrast with subjective concepts, which, as Kant never tires of emphasizing, cannot refer to their objects (in cases in which they do refer) if the relevant subjective intuitions do not belong to the same representing subject.[28] Before we move on it will be necessary to say a word about the semantically immediate objects of intuitions.

---

26. When considering whether (i), (ii), or (iii) belong to the *definition* of intuition—it could be all three or any combination of them—an exclusively semantic reading of the term has the advantage that the non-sensible species of intuition, so-called "intellectual intuition", would also possess immediacy. Attempts to explain the immediacy of intuition by appeal to phenomenology or to the causal dependency of our intuitions on their objects have the *prima facie* drawback that we have no reason to assume that non-sensible intuition will meet these further conditions. If Kant intends an exclusively semantic definition, and does so for this reason, then his thinking is similar to Bolzano's when he concedes that God may be capable of possessing intuitions that are simple and singular even though they are not caused by their objects (*WL* §74, I:331).

27. Cf. *KrV* A19 and A68; cp. along broadly similar lines, Longuenesse 1998, 24; for contrast, Friedman 2000, 191f. Of course, other subjective representations might be causal prerequisites for the very existence of subjective intuitions. The above proposal excludes merely that those other representations make a difference to whether the latter refer.

28. Bolzano criticizes Kant's definitional appeal to intuition's immediacy (cf. *WL* §77, I:345f). The definition of semantic immediacy proposed above allows us, I think, to see that they are talking at cross purposes: Bolzano's critique is formulated with attention to representations in themselves, whereas Kant is not concerned to formulate conditions under which representations *in themselves* refer. His point is rather that in order for *my* tokening of <triangle> to refer I must have correctly related that subjective representation to certain other representations that I also possess (cf. *KrV* A240).

The sense, if any, in which an intuition that has not in some way been combined with a concept—I will call such intuitions *bare intuitions*—is a cognition, and thus has an object, is one of the most controversial flashpoints in contemporary Kant interpretation. In the remainder of this paper I will follow what I take to be the most natural reading of the totality of textual evidence and suppose that bare intuitions do have objects.[29] By this I mean, first of all, that in order that an intuition refer it need not be subsumed under an empirical concept; and, second, that in order for an intuition to refer it need not be brought into relation with other representations in accordance with one of the three original functions of the understanding that give rise to the categories of relation. Kant's technical term for the object of an empirical intuition that has not been modified in one or both of these ways—what he describes as "the *undetermined* object of an empirical intuition"—is "appearance".[30] It is quite important to note that in this use the term does not by itself commit Kant to transcendental idealism.[31] It is in the first place simply the term that Kant uses to describe the objects that a cognizing subject can represent insofar as it makes no use of concepts, either empirical or a priori.

Kant's thesis that bare empirical intuitions refer to appearances forms the basis of the explanation of singularity to be discussed first, (Ia.1), according to which these intuitions do not put a subject that possesses just them in position to grasp the idea of an object that can be represented by other subjects or by the same subject at other times. Representations that refer merely to appearances are singular, we will see, because their very form of representation ensures that each intentional object is cognized

29. Though this is a consequential assumption, I lack the space to defend it here. Note, however, that it is the most natural reading of Kant's definition of intuitions and concepts as two *species* of cognition, since the definitional passage (*KrV* A320) tells us that each is a conscious representation that has objective purport, but does not add that one type lacks this property *unless it is combined with or subsumed under the other*. A reason for trying out the "bare intuition"-hypothesis in the present venue is that without it Kant is much less profitably compared to Bolzano. An interpreter such as Dickerson, who argues that sensible representations must already be subsumed under a concept in order to have referential purport (2004, Ch. 1), thereby takes Kant to hold that we possess no essentially singular cognitions that are not, in Bolzano's language, already at least *mixed* representations.

30. *KrV* A20. I will suggest below that the second case (i.e., the object of an intuition that is unmodified by the relational functions of the understanding) is the dominant sense.

31. Kant obscures this point by using a single term, "appearance", sometimes as "undetermined object of empirical intuition" and other times merely as a synonym for "transcendentally ideal (though determined) object of experience" (e.g., *KrV* A548). I will always use this term in the former sense.

as unique. This first proposal, unlike the others to come, will be roundly rejected by Bolzano. In each of the remaining proposals we will always find, amidst some differences, at least a basic agreement that there is something promising about the general strategy in question.

4. *Default singularity (Ia): lack of categories variation (Ia.1)*

Perhaps subjective representations are, so long as they are cognitions, such that they have just one object unless they are modified in some way. In a slogan, couched in Kantian language: cognitive representations are singular by default, though they can be altered so as to become general. This *Default Singularity* proposal, a genus of Representation-based Interpretation that we can abbreviate as *(Ia)*, admits of different specifications depending on what type of addition serves to convert cognitions from their default singularity to generality. Below in §6 we will consider the possibility that (Ia.2), the introduction of a particular second-order consciousness (Kant), or (Ia.3), causal mediation (Bolzano), might do the trick. This section will develop *(Ia.1)*, the proposal that bringing a semantically immediate representation under the functions of relation, and thereby under the categories <substance>, <cause>, and <reciprocity>, is what is required.

The raw materials for this proposal are most transparently laid out in Kant's development and defense of a thesis, whose boldly universal claim concerning the (perhaps unnoticed) genesis of our subjective cognitions recalls Bolzano's Simplicity Hypothesis: "All of our judgments are at first mere judgments of perception; they hold only for us, i.e., for our subject, and only afterwards do we give them a new relation, namely to an object, and intend that the judgment should also be valid at all times for us and for everyone else; for if a judgment corresponds to an object, then all judgments of the same object must also correspond to one another [...]".[32] Shortly thereafter we learn that these judgments of perception are "merely a connection of perceptions within my mental state, without reference to the object" (*Prol* §20, 4:300). These passages are naturally taken to indicate that these representations do not refer at all, that they are mere modifications of the perceiving subject (mere sensations). I take Kant's point, instead, to be that the constituent representations in a judgment of

32. *Prol* §18, 4:298. More difficult and terminology-laden treatments of the same theses can be found at *KrV* A104ff and A197.

perception do not refer to what we can call, in pointed contrast to Kant's description of appearances as "undetermined" objects, a *determined object.* They do, however, refer to *appearances.*

The text is clear that the representations that compose judgments of perception are, or at least can be, intuitions.[33] Kant could argue (and is perhaps already assuming) the stronger thesis that *all* bare intuitions lack a determined object, since they are either isolated bare intuitions or constituents of judgments of perception. Moreover, he can mollify phenomenologically based doubts by making a move similar to Bolzano's defense of the Simplicity Hypothesis. Kant need merely claim, namely, that by the time we notice most intuitions they have at least been brought under the functions of relation, and perhaps also subsumed under concepts, thereby making them something other than *bare* intuitions. This theory raises many questions, in part because it ventures positions on several issues that are only obliquely addressed by Kant himself. What is most important for present purposes, however, is how the theory just sketched, if it is indeed Kant's, might be taken by him to entail that bare intuitions are essentially singular cognitions.

The evidence indicates that Kant arrived at the distinction between determined objects and mere appearances through his reception of Wolff's ontology, which treats the principle of sufficient reason as a necessary condition for objects being completely determined, and holds that only (i) completely determined objects are (ii) selfsame individuals with determinate identities.[34] I propose that Kant takes appearances to be neither (i) nor (ii). When we represent mere appearances we represent what the

33. Cf. *Prol* §18, 4:298; §20, 4:300; §21[a], 4:304. I can only note here that in my view the best reading of the *Prolegomena*-treatment indicates that there are not only judgments of perception composed wholly of intuitions but also mixed and conceptual versions, in which a concept's combination with an intuition or another concept expresses neither relations of substance to accident and cause to effect, nor strictly logical relations of containment or identity. For other treatments of the topic, see Prauss 1971; Longuenesse 1995.

34. The contrast between mere "dream worlds" and worlds with "transcendental truth" (which they possess because there are sufficient reasons for each of their elements being as it is) is central to Wolff's, Baumgarten's, and Meier's ontologies. In Kant's hands, "dream world" becomes a semantic term, signifying the intentional object of a representation which purports to refer to "something objective" (*KrV* A202), just in case that object is not also represented as belonging to a manifold of substances that are causally and reciprocally ordered. For an argument that Kant's conception of transcendental logic is intricately bound up with Wolff's distinction, see Rosenkoetter 2009, §III. For some relevant texts, cf. "… nothing is distinguished from anything else", *KrV* A198; as well as: A100–103 and A112, both A-only; A146; A202f.; A573; and *Prol* 4:374ff.

Wolffians called "dreams worlds", which do not merely lack actuality (or existence) but, more fundamentally, order.

First, that they should not be completely determined makes sense if Kant thinks that appearances in some way belong to the very act of representing them,[35] since there is no reason to suppose that all possible facts regarding an appearance are determined by the manifold that belongs to that very act of representation. There is not then even in principle an answer to the question "Is it F?" if the intuition does not include within its manifold matter that represents F-ness or its opposite. In contrast, part of what it is to conceive of an intentional object as a substance is to assume that it possesses accidents about which the current act of representation may be silent. Kant in fact assumes with the Wolffians the stronger thesis that "all that exists is *completely* determined", so that "Is it F?" is in principle answerable for any <F> (*KrV* A573, italics elided). All of this might also explain why Kant seems to think that reference-failure is impossible when the object of a representation is an appearance, while he presumably regards such failure as possible when a cognition purports to refer to a determined object. Whereas when we cognize a substance as bearing certain accidents it at least makes sense for us to get those accidents so wrong that we fail to refer, there's simply no room for this to happen when the object is an appearance, since that object *is* only what it is represented as being in that very act of representing.

Second, if Kant holds that appearances are not completely determined, it is natural for him to join the Wolffians in denying that these objects are selfsame individuals with determinate identities. The radical nature of this claim makes it rather difficult to reconstruct. When Kant writes, as quoted above, that judgments of perception "hold only for us, i.e., for our subject", he is not saying merely that the subject-matter of these judgments is inaccessible to other subjects. He is instead denying that the objects of judgments of perception support platitudes about objective facts, such as the fact that "if one speaks of different experiences, they are only so many perceptions insofar as they belong to one and the

35. This recalls Brentano's notion of intentional inexistence, though while Brentano holds that *all* intentional objects have intentional inexistence, Kant maintains his close relative of the thesis only with respect to the intentional objects of bare intuitions. Brentano's thesis apparently entails, problematically, that no two people can intend precisely the same object. The restricted scope of Kant's thesis means that he can use that very problem as an argument for the conclusion that we require the categories of relation in order to cognize even thoroughly ordinary judgments of experience.

same universal experience" (*KrV* A110, A-only). Categories are required for a subject to comprehend objects about which there are truths that are "valid at all times for us and for everyone else". When representing bare intuitions (or associating them with one another in judgments of perception), "different experiences" are always just: different experiences; and their objects are always just: different objects. These objects are unique not by virtue of being in principle comparable to all other objects and different but because the conditions that make comparison possible have not been met. It follows that appearances are *private objects*, where the relevant notion of privacy extends even to the same subject at later times.

Now, the view just sketched is similar to Bolzano's theory of intuition in one respect. Bolzano agrees that the objects of intuitions are private objects in the sense just stipulated. I can never have a subjective intuition whose object is a state of your soul; and you can never have a second intuition whose object is the same as the object of the intuition you had a moment ago (*WL* §75.1, I:334f.). That said, Bolzano does not accept the stronger thesis that Kant uses to infer privacy. He takes it as obvious that the objects of intuitions are determined objects. What is special about them is just that they are objects of which it is impossible to have two distinct subjective representations, if those representations are to remain both simple and singular. But if my goal is instead merely to have a representation that refers to the same state of the soul that was the object of an intuition you just had, then this can very well be done, as Bolzano puts it, "for scientific purposes" (*WL* §75, I:336). Bolzano's suggestion is that person *P*'s objective intuition can be "made known [*kenntlich*]" to others—as well as, I presume, to *P*'s 'later self'—by specifying the time at which it appeared to *P* and the kind of object in question. By the latter Bolzano has in mind what we might do in specifying that the object of the intuition in question was the sound at t_1 rather than the pain at t_1. One wonders why there cannot be two or more pains at t_1 (and similarly for any kind, no matter how specific). But if this is successful, then the resulting representation will be singular, though also complex.

This is enough to illustrate the great distance between Kant and Bolzano regarding (Ia.1). The main idea in Kant's distinction between judgments of perception and judgments of experience, I have suggested, is that the former are not "objectively valid" (*Prol* §§18–19, *passim*). They do not have determined objects, and thus also lack (determined) objects that can be thought by means of a complex description. Relatedly, there is no sign that Kant thinks that a third party's representations could capture "for scientific purposes" all of the facts about *P*'s appearances. In a formulation that will appear again below, we might say: *Kant seems to countenance a species of object that is essentially tied to a certain perspective*. These claims are flatly inconsistent with Bolzano's approach. Confronted with Kant's doctrine, we would expect Bolzano to urge that the mere fact that subjects are sometimes epistemologically not in position to refer to the adherences of a substance is no reason to posit a new realm of shadowy *quasi-objects*, as we can label them for convenience.

It might initially seem that the explanation of singularity just covered deserves to be classified as a Uniqueness Interpretation, since it holds, as we have seen, that bare intuitions are essentially singular because their (quasi-)*objects*, mere appearances, are in each case unique. Looking more closely, however, it should be clear that each appearance's uniqueness is wholly explained by the deficient nature of bare intuitions. Thus, even if (Ia.1) could be interpreted according to its letter as a version of (II), its spirit surely lies within (I). The next section will survey the proposals that, in contrast, belong squarely within type (II). We will see that they are in a first pass comparatively straightforward, since they explain the essential singularity of intuitions by recourse to objects whose uniqueness we can understand without any consideration of types of representation; indeed, these objects might even be unique as a matter of a brute fact.

5. *Uniqueness interpretations (II)*

We can begin with a passage in which Bolzano appears to suggest that intuitions are singular *because* the change in a particular soul at any given time is unique:

> It is impossible to generate for a second time an intuition that we once had, where the two subjective intuitions are to correspond to one and the same objective representation. The reason is that every subjective intuition has its own object, namely the external or internal change which is its immediate cause, and **every such cause exists only once, since every change that takes place at another time**—even in the same subject—**is already a second change**. It follows that any two subjective intuitions have two different objects to which they refer. [...] It makes no difference how similar the just-perceived color, scent, or pain is to a sensation that I had at some other time: they are

> still different, and the objective representation that refers exclusively to one can therefore not have the other for its object.[36] [*WL* §75, I:334f.]

It is obvious that this explanation works only if it is a fact that each temporal position is unique, which in turn requires that there be only one time.[37] Reflecting on this presupposition helps to bring out an underlying similarity between the explanation that Bolzano is advancing and one of the main lines of explanation that we find in Kant. At least on a naïve reading of what should surely count as one of the central texts for the question of this paper, both versions of the Transcendental Aesthetic's Metaphysical Expositions of Space and Time (*KrV* A22ff./B37ff.), Kant is namely advancing a version of (II) in his arguments that some representations of space and time—the "original" ones, whatever precisely that turns out to mean—are intuitions, rather than concepts. The representation of space must be an intuition rather than a concept *because* "one can only represent a single space, and if one speaks of many spaces, one understands by that only parts of one and the same unique space" (*KrV* A25). He makes explicit the assumption that underlies this argument later in his parallel treatment of time: "Different times are merely parts of precisely the same time. That representation, which can only be given through a single object is...intuition" (A32). Kant's idea appears to be that <dog> can be given by experiencing any number of dogs, whereas the original representation of time—matters might be different for other, derivative representations of time—can only be given by the unique object that answers to it. This line of thought deserves to be taken even more seriously because we find similar arguments in the *Inaugural Dissertation* (1770), which supports the hypothesis that whatever other explanations of singularity may subsequently show up in Kant's work, some version of (II) was originally Kant's primary, or perhaps even his sole, reason for holding that some of our representations are essential singular.[38]

Let *(IIa)* be the view that at least some intuitions are essentially singular simply because there is only one space and only one time. The obvious

36. Though this passage allows that the object might be an "external" change, the reader learns later that this is disallowed by Bolzano's strict position (*WL* §286, III:85, 89).

37. Bolzano agrees, cf. *V* §2.1, 206. I don't mean to suggest that all scenarios in which there is more than one time would defeat this (II)-style explanation. What the latter cannot allow is that there are two or more times that are both, so to speak, October 5, 1781.

38. "*The idea of time is singular* and not general. For no time is thought of except as a part of the same one boundless time" (*Diss* §14.2, 2:399; cf. §15.B, 2:402). For a survey of the early reception of Kant's uniqueness claim, see Vaihinger 1892, 211–215.

limitation of this view is that it can only explain the essential singularity of intuitions that represent the wholes of either space or time (though perhaps along with an object, such as the universe, occupying them).[39] One option for constructing an explanation of the full range of intuitions, rather than just that narrow slice, is a *Hybrid View* that supplements (IIa) with some version of (I)—we will encounter more options in addition to (Ia) below—which would then be used to explain the essential singularity of intuitions whose objects do not occupy the whole of space or time. The other option, which is of greater theoretical interest, is to extend the basic idea of the Uniqueness Interpretation so that it also covers particular positions in space and time, as we have seen Bolzano doing with respect to time but not space (a difference that I will ignore). We can call what must be added, a position which Kant would then share with Bolzano, *(IIb)*.

At first the attribution of (IIb) to Kant is likely to strike his readers as downright improbable. We find no direct textual evidence for it, unlike in the case of (IIa). However, the main reason for the reluctance, I suspect, is that Kant is careful to tell us that "time in itself cannot be perceived", meaning that appearances are not given bearing a 'time-stamp' (e.g., October 5), a position that Bolzano shares with Kant.[40] Yet this would be a mistake for two reasons. First, type (II) views in general do not require that the representing subject be aware of the uniqueness of the objects of those representations. Intuitions will possess the property of essential singularity whether or not the representing subject recognizes that the objects are non-accidentally unique. However, it is not inconsistent with type (II) views to *add* that subjects are, or can become, aware of the relevant uniqueness, an addition that we can designate by adding a subscripted-"e", for "evident". Many interpreters who are inclined to attribute (IIa) to Kant will consider $(\text{IIa})_e$ to be even closer to the mark, and this brings us to a second mistaken reason for denying that Kant joins Bolzano in viewing (IIb) favorably. Namely, in order to adopt $(\text{IIb})_e$ it is not necessary to claim that that representing subjects are aware of *which* absolute spatial and temporal positions the objects of intuitions occupy. The latter claim would indeed be a non-starter, but is unnecessary, since it would be sufficient for $(\text{IIb})_e$ for the subject to be aware that the object of an empirical

39. As we will see below, the very existence of subjective intuitions whose objects are exclusively space as a whole or time as a whole—I will call those intuitions "|space|" and "|time|"—is a matter of vigorous dispute among commentators. Yet the distinction between these and |universe| is not important for the above point.

40. *KrV* B233, B-only; and A215. For Bolzano's rejection, see *WL* §303, III:141ff.

intuition occupies *some* unique position within the unique individuals, time and space. And it is surely not implausible that Kant credits us with this awareness.[41]

What, then, can we say about whether Kant and Bolzano actually hold type (II) views? Bolzano certainly presents the simpler case. His commitment to (IIb), unlike Kant's, would not depend on a logically prior commitment to (IIa), since Bolzano does not believe that we have any *intuitions* whose objects occupy, or are identical to, the whole of time.[42] Thus, the way is apparently clear for Bolzano to hold (IIb), a refreshingly straightforward position. Yet this does not withstand closer scrutiny. The problem is that the specification of a particular <moment, soul>-pair is insufficient to determine a unique object. As we saw above in §2, Bolzano routinely indicates that two different intuitions can pick out two different, though simultaneous, alterations of the soul, such as a pain and a sound.[43] The obvious way out of this difficulty is to add a kind-specification to the <moment, soul>-pair. Yet on reflection there seems to be no *a priori* reason to expect that a <moment, kind, soul>-triple picks out only object. Why assume that there cannot be two pains in the same soul at the same moment? This is a signal that we need to reverse course and, adopting a version of (I), begin with representations rather than facts about uniqueness. The problem is that the right sorts of uniqueness facts might not obtain (and if they do, might do so only accidentally). Beginning with representations Bolzano will be able to say that even if there should turn out to be two qualitatively wholly similar pains in the same soul at the same moment, representation *R* has only *its* most immediate cause as its object. Below I will suggest that this is in fact Bolzano's view.

41. Though it is easy to miss, Smit (2000, 256) is one of the few interpreters to have formulated and held a version of (IIb). Dicker might also maintain a version of (IIb), though his brief exposition is ambiguous on the crucial question of whether it is the act of intuit*ing* or the object of the intuition that has the unique temporal position (2004, 36).

42. Cf. *WL* §79, I:361ff. A closely related difference between the philosophers is that Bolzano takes <moment> to be logically prior to the concept that is expressed by phrases such as "the (one) time", in part because he takes that latter concept simply to be <the aggregate of all moments> (cf. Schnieder 2009, 47f.). In contrast, Kant thinks that the representations in themselves of time's parts are "limitations" of, and thus dependent on, the representation in itself of the whole of time (Cf. *Diss* §15.C, 2:402; cf. *KrV* A25). Notice that this is not a claim about *subjective* representations (contrast ft. 28), e.g., about the order in which we must acquire them.

43. E.g. *WL* §69, I:311. Would Kant, interpreted as an adherent of (IIb), face a similar problem? It is not clear that he would, since it is open to him to hold that an intuition of inner sense has as its object *everything* that the subject is aware of during a time-span (cf. *KrV* A99, A-only).

Above I hinted that it might only be a naïve reading of the Transcendental Aesthetic that would attribute a version of (II) to Kant. The reason for that caution emerges when we ask, *Why* is there only one time and one space? Kant obviously does not regard the oneness of time and space as brute facts about the properties of things in themselves or their relations to one another. Instead, he explains the properties of space and time by recourse to our forms of receptivity.[44] Yet though it might at first be tempting to take this as evidence that Kant is committed to a Representation-based explanation, the subjective origin of time and space does not automatically spell trouble for Kant's commitment to (II). This is because the constitution of our forms of receptivity, rather than the constitution of things in themselves, could just as well serve as the brute fact that explains why |space|, |time|, and |universe| are singular representations. What would really compromise Kant's commitment to (II) is if the essential singularity of intuitions *in general* itself somehow explains the uniqueness of space and time. This would enable us to admit in a first step that the essential singularity of |space|,|time|, and |universe| is explained by the uniqueness of space and time, as central arguments of the *Inaugural Dissertation* and the Transcendental Aesthetic strongly suggest, while in a second step explaining that uniqueness as ultimately grounded in the essential singularity that all intuitions share. Let's call this the *End Run* around (II). Either it or a pure commitment to (IIa) and (IIb) has the advantage of parsimony over any number of Hybrid Views. The paper's final section will ask whether any of the new versions of (I) considered below can help Kant effect that End Run. If not, then there is compelling evidence that Kant is committed at least to a Hybrid View, if not to a pure version of (II).

6. *Intuitions as "this"-representations*

Let's begin with the following description of intuitions that strongly suggests a Representation-based Interpretation:

> … it must be a representation of the sort which—if we had to discuss it—could only be expressed by means of words like the following: "*This, which*

44. There are many ways to interpret Kant so that these properties depend not only on receptivity, but also on the unity of apperception, the categories, or even the demands of theory construction. For an example that is of special relevance in the present context because it devotes explicit attention to the uniqueness of space and time, see Friedman 2000, §I.

occurs in me just now, which I just now see or hear or feel" and the like. With such representations it is immediately certain (for it already lies within their nature) that they have only one single object, namely, that change that just then takes place in us, and that nothing else, no matter how similar it might be to this, is represented by the representation. For this reason we employ the word "this", which represents only that which is present just now, and nothing else besides. Were the representation that we possess to represent not just the change that now takes place in us, but something else in addition, similar to it in some way or other, we would have to use the expression "a such and such" (something that is constituted like this) rather than "this". [*MM* §6.3]

Here we see Bolzano identifying a class of subjective representations in some way other than merely as the class of representations having only one object—viz., by appeal to the familiar experience of noticing changes within oneself that one might describe by saying "this-now"—and then arguing that the "nature" of those representations guarantees that each has but a single object.

The previous passage contrasts "this A" with "an A", but it does not mention that Bolzano regards the former as ambiguous between two importantly different types of representations, whether subjective or objective. Taken in what Bolzano regards as a loose sense, "A" is the principal part of "this A". With the addition of "this" to "A", its extension is restricted to just one of all of the many possible As. This is just what Kant has in mind when identifying the "singular function of the understanding in judging", which is one way that we can "use" concepts in judgments.[45] However, in the "more exact" sense of "this A" "this" is instead the principal part. In this case the addition of "A" does not change its object:

> The addition "which is an A" [in "this, which is an A"] expresses an attribute which already belongs to the object associated with "this", and is employed merely for the sake of greater clarity. In this sense we take an expression such as "this fragrance (which I now perceive) is sweet". By the word "this" here, we mean the particular perception we have at this moment. That it is a fragrance is an attribute which already belongs to the object represented by "this".[46] [*WL* §59, I:258]

---

45. *KrV* A70f; *JL* §21, 10:102. We can also "use" them universally (<all A>) and particularly (<some A>).

46. Cf. *WL* §75, I:336f. Notice that Bolzano is here helping himself to the fiction that each soul has only one perception per moment.

Already from this brief sketch the plausibility of attributing a version of (I) to Bolzano should be clear. Intuitions are well-modeled by the expression "this"; and "this"-representations are, like their linguistic analogues, essentially singular. Let's give the label *(Ib)* to the view that subjective intuitions are essentially singular *because* they are "this"-representations; and let's call a specification of it that Bolzano also holds, that such representations are the direct *causal* effects of their objects, *(Ib.1)*.

We noted above that the loose sense of "this A" is very close to Kant's singular function of the understanding in judging. It is worth considering whether intuitions have the same properties (apart for simplicity) for Kant that Bolzano finds in "this A" under the strict reading. This sort of interpretation has been held most influentially by Wilfred Sellars: "To intuit is to represent a *this*, i.e., if I may so put it, to have a representing of the 'this' kind" (1967, 447). This is an eminently plausible reading of Kant, at least if it is restricted to empirical intuitions, as I will assume for the remainder of this section. True, unlike Bolzano Kant does not use "this" to explain his conception of intuition. Yet at least the paradigm cases of empirical intuitions are perceptual demonstratives, and Kant might often be using "immediate" as a label for the immediate perceptual relation between an intuition and its object.[47] That this is a relation of *causal* dependence is supported by this quasi-definitional pronouncement: "Intuition is a representation of the sort that would depend immediately on the presence of an object".[48]

Let's pause and reflect that the theories of cognition and reference that we find in Kant and Bolzano are constructed in the first place as theories of *representation*, with any contributions to the philosophy of language built upon that foundational layer. One of Bolzano's most common refrains, when presenting his theory of intuition, is to tell us that intuitions are representations that, if we had to describe them, we would express using *words* such as "this (that I am just now seeing)". This is tremendously

47. Sellars understands "immediate" in this way (1968, 3). Note that even if Kant's preferred *definition* of intuition does not contain any mention of perception and causation (see ft. 26), most of the passages in which Kant characterizes intuitions as "immediate" occur in contexts in which Kant plausibly has only the sensible variety of intuition in view.

48. *Prol* §8, 4:281; cf. *Met Mrong* 29:800. This reading is supported by the fact that Kant introduces as a foil a broadly Aristotelian proposal, according to which the properties of an object can be transferred, unchanged, to our cognitive faculty. Kant clearly thinks, in contrast, that the matter of a subjective empirical intuition is determined by the causal effect of an object on one's capacity for sensible representation. No more should this matter be assumed to mirror its object than a pain mirrors the motion of the sword that is its cause.

helpful—as one notices especially when comparing Kant's procedure—but it carries with it a particular danger. We know well enough, namely, the linguistic role of "this" and how it differs from non-indexicals, and we can contrast it with close relatives such as "I", "now", "here", "there", and "these". But do we really understand what we are saying when we say that intuitions have the property of being "this"-representations, given that we cannot sensibly be holding that representations are themselves linguistic expressions? Next I want to consider two proposals that have the potential to address this worry by making the still under-determined (Ib) more determinate. Let's begin with Kant.

Having taken the step of interpreting intuitions as "this"-representations, it is natural to think of them as containing something that other types of representation lack. One might assume, perhaps without ever scrutinizing the assumption, that there is a distinguishable "this"-*part* in each subjective intuition, though in a Kantian context one is more likely to think that it is a particular *form* that intuitions have and other representations lack. Perhaps this is what Sellars had in mind in the quote above. However, recalling (Ia) we can reverse our procedure and ask whether representations might instead be "this"-representations by default, which only cease to be such by being modified in some way. In short, perhaps "this" is not something that has been added; it is the starting point. Now the version of (Ia) considered above, (Ia.1), observed that without a change in form provided by the functions and categories of relation we cannot, according to Kant, so much as cognize selfsame objects, which as such are fit to be the objects of distinct acts of representation and to serve as truthmakers for judgments. Yet the deficits inherent in this form of representation are simply too great to make it a plausible reconstruction of (Ib)'s "this"-representations. The core idea of (Ib), which Kant might well share with Bolzano, is that there is a species of representation that is in some sense of the same type as we find in any application of the singular function (i.e., the loose sense of "this A"), though members of this species are not modifications of concepts, as <this A> is a modification of <A>. This core idea is consistent with the claim that the object of such an intuition is a persisting thing, the selfsame object of distinct subjective intuitions, and the truthmaker of multiple judgments. In short, there is no need when developing a Default Singularity interpretation of what it is for intuitions to be "this"-representations to take the radical steps to which Kant commits himself in his theory of judgments of perception.

Kant holds that both concepts and intuitions are jointly constituted by a form and a matter. All concepts share a form, as do all intuitions. Matter is what distinguishes |red| from |purple|, as well as <red> from <purple>.[49] The core idea for a less radical version of (Ia) is that a subjective intuition can be modified so as to become a subjective concept simply by adding a second-order awareness that the intuition might represent something other than its appearance. This proposal, which we can call *(Ia.2)*, contributes to the filling out of (Ib) by suggesting that the "this"-form of a subjective intuition, rather than constituting some kind of addition, consists simply in a lack—specifically, a lack of consciousness that things other than the object of current focus might be subsumed under that very matter. A key feature of the proposal is that the subject wouldn't need to procure a subjective representation with a different matter in order to remedy that lack. She simply needs to adopt a second-order attitude toward the matter that she already possesses. Moreover, the second-order attitude is the same for every case of intuition-to-concept conversion, whether the subject is converting |purple| to <purple> or |red| to <red>. This makes it other than merely arbitrary to call the inclusion of this second-order attitude a "form" of representation.[50]

(Ia.2) tells us that a certain deficit makes a cognition an intuition rather than a concept, but it does not follow that (Ia.2) alone is enough to provide an account of *all* concepts. Indeed, the claim that all concepts can be generated by modifying an intuition is obviously inconsistent with Kant's position on the categories. The same must also be true for pure sensible concepts such as <triangle> and any empirical concepts that themselves incorporate or depend upon schemata, as Kant illustrates with the concept <dog> (cf. *KrV* A320; A141). This tells us that even some empirical concepts incorporate rules that cannot be reduced to the type of rule implicit

49. Cf. *KrV* A20; *Met Mrong* 29:795; *JL* §2, 10:91. Two clarifications: (1) Kant calls attention most often to differences in form between intuitions (e.g., one represents a square, another a triangle), rather than to the form that all intuitions share (singularity). (2) He often tells us that the matter of a concept is its *object* (ibid.), but other passages seem to indicate that he is simply not allowing the existence of co-referring concepts to complicate those discussions (cf. *JL* 10:35). <Morning star> and <evening star> have different matter.

50. "The matter [*Materie*] of all empirical—empirical intuition—is sensation; the form is the figure. Concepts have matter, i.e., content, representations, data, which are given—the form is the reflection of the understanding, by which it brings the sensations together in such a way that it thinks something general thereby [...] The empirical constitutes the matter in perceptions and intuitions and concepts" (*Met Mrong* 29:795). "[...] *conceptus empiricus* [...] The material [*Material*] of all empirical representation is sensation" (Ak. 24:752).

in (Ia.2), viz., "<red> is properly applied to anything like this", where "this" expresses |red|.[51] (Ia.2) tells us what makes some representations non-accidentally singular, even though they are not modified concepts. However, a diversity among concepts might ensure that some concepts cannot be generated merely by eliminating that singularity.

It is also important to be clear that this proposal is not intended as a reductive account of concepts, according to which there is nothing that is distinctive or different in kind about grasping a concept. On such an account, of which Hume provides an example (1739, I.i.7), subjects can possess only one type of representation, instances of which are associated with one another to varying degrees. On the contrary, (Ia.2) holds that to take the second-order attitude described above toward an intuition just is to possess a different *type* of cognition, whose content (in Bolzano's sense) is a different kind of representation in itself. Yet so long as we restrict our view to the in itself, we can do no more than observe that intuitions and concepts are in fact different types of representations in themselves, and this does nothing to settle how it is possible for subjects to realize actual ("subjective") representations that instantiate those types. And on that question (Ia.2) makes the potentially important move of proposing that the same subjective representation that is now serving as an intuition can instead serve, with the addition of a generic second-order consciousness, as a concept. The resulting Default Singularity proposal yields the following larger picture. Some subjective representations are "this"-representations precisely because the subject that possesses them has not reflected upon the standing possibility that their constituent matter might also represent other objects. To have attained that consciousness just is to possess a subjective concept. Then there is a second class of "this"-representations that does not presuppose that lack of consciousness. In these cases of the "singular use" of a concept, a matter that has already been "reflected" as general is referred to just one object.[52]

51. Kant often presents abstraction as requisite for the generation of concepts (e.g., Ak. 24:753), though, again, it isn't clear that this is required for *all* concepts. What do I abstract from in order to generate <red> if my intuition is simply a representation of red? It might be suggested that I must abstract from my consciousness that this red lasts for a particular duration, d_1, which also belongs to the intuition. If so, then the target concept should instead be <red that lasts d_1>. It might be that (Ib.2) can provide a recipe for generating only what strike us as rather unusual concepts.

52. Limitations of space dictate that this view cannot be further developed or compared against key texts for the view (e.g., *KrV* B133f; B136). One of its general advantages is that it can account for signs scattered throughout Kant's works that the very same matter can be

A Default Singularity proposal tailored specifically to Bolzano—one which, moreover, promises to specify Bolzano's central thesis that intuitions are "this"-representations—can be motivated if we first look at a view that cannot be Bolzano's. One might find it natural on an initial reading, namely, to take all of Bolzano's subjective intuitions to be precise analogues of the linguistic indexical "this", all uses of which share precisely the same "character" or "linguistic meaning" (cf. Braun 2010), though they differ in their objects because of different contexts of use. Indeed, one could understand Bolzano's strict reading of "this A" to be making just this point: a representation that is *just like* the pure indexical "this", except that it is non-linguistic, already suffices to fix reference. But this cannot be Bolzano's view, for he would then be guilty of a gross error in holding that there is a different representation in itself for each subjective intuition. Bolzano in effect warns against this mistaken interpretation when he insists that "[...] also things that are simple can be different from one another in infinitely many respects [...] Thus, from the fact that all intuitions are simple representations it does not follow that there cannot be an infinite number of them, such that no two of them are exactly alike".[53] Thus we see that a subjective intuition cannot be *exhausted* by its demonstrative nature, as Bolzano's comparisons to "this" might lead one to assume.

We just rejected a model according to which subjective intuitions are, as it were, all "this" and no matter, to continue to use that Kantian term. Taking our cue from (Ia.2), we might then try out the possibility that they are instead all matter and no "this". More precisely, perhaps the "this"-form that intuitions share is not built into their matter, but a factor that is external to it. (Ia.2) provides one example of how this might look, since the second-order attitude is external to the various matters upon which it can be directed. In Bolzano's case, the obvious candidate is the possibility that the "this"-form of an intuition is *nothing other than* its being the "*next* and *immediate* effect" of its object, an alteration of the soul (*WL* §72, I:326). Bolzano's penchant for explaining intuitions by analogy to "this" ensures that the present interpretation is not the first that a reader

---

cognized either intuitionally or conceptually (e.g. *KrV* A137). Lohmar (1998) and Smit (2000) both give extended defenses of this thesis, though Smit seems to favor (IIb) as an explanation of the singularity of empirical intuitions (see ft. 41).

53. *WL* §72, I:328. Rather than taking intuitions, like the linguistic "this", to be in the above sense always the same (and distinguished only by their contexts of use), we might take them to be in each case a way of representing a pair, $<s_1, t_1>$, composed of a particular soul and a particular absolute time. However, this fails for the reasons that we saw above in our discussion of (IIb).

will hit upon. What's more, one reason to doubt that it matches Bolzano's own understanding is the simple fact that he never explains his view in just this way. A further reason for doubt is that Bolzano does not introduce a theoretical concept that plays the same role as Kant's "matter".[54] Yet this cuts both ways. Perhaps his lack of this concept, which is not surprising in light of his programmatic determination to give explanatory priority to representations in themselves, made it difficult for him to explain his view in the manner just proposed. It is, nonetheless, certainly a view worth considering, not the least because it explains nicely how there could be representations with all of the properties that Bolzano ascribes to intuitions. After all, there should be no particular worries with the claim that alterations of the mind can cause *simple* representations. The difficulties start when we try to understand why such representations are also essentially singular.

If it is a subjective intuition's placement within the causal nexus, and not the particular constitution of its matter, that makes an intuition singular, then there is no obvious barrier to the hypothesis that a precisely similar matter, which differs only from the original intuition in being further displaced from that immediate cause, would be such as to have as its object *any* change of the soul that would have just that effect. So, for example, while *R* represents only the pain that is alteration a_1, since it is the direct causal consequence of a_1, a later subjective representation that is qualitatively precisely similar to *R*, but which is actual also because of some more immediate causal factors, would represent *any* alteration that would have just that representational effect.[55] In short, detach a subjective intuition from its immediate causal dependence on a particular change of the soul and it is freed up, as it were, to represent other qualitatively similar changes as well. In summary form:

> *(Ia.3)* A pure (i.e., unmixed) subjective concept can be formed from the matter of a subjective intuition simply by introducing causal mediation.

54. Instead, he uses that term as a synonym for "content". Two subjective representations have different matter iff they are the appearances of two different representations in themselves (cf. *WL* §48, I:218; §271, III:9).

55. For evidence that an immediate causal connection is needed for a subjective intuition to maintain its status as an intuition, see Bolzano's assumption that the objects of intuitions must *continue* to affect us "in order to maintain [*unterhalten*] an intuition in us" (*WL* §282, III:46; cf. §286, III:84f.).

As with (Ia.2), it is not part of this view that subjective concepts having *all* possible contents can be produced in this manner, as the example of <has> will suffice to make clear.

It should be obvious that with this hypothesis we progress further down the path of attempting to make subjective representations relevant to a full understanding Bolzano's position. In §1 I observed that Bolzano consults his stock of subjective representations in order to identify which possible classes of representations in themselves he has reason to consider non-empty. With (Ia.3) I propose that we might consult the properties of subjective intuitions in order to better understand how one thing can satisfy all of the definitional criteria. This might strike some as contrary to the spirit of Bolzano's project, but it is important to note that Bolzano himself turns his attention squarely to subjective representations at a similar point:

> ... I believe that the object of every humanly attainable (subjective) intuition must be some actual thing. It seems to me that this already follows from the mere concept of an intuition as a simple, singular representation. For if a representation is, notwithstanding its simplicity, to represent only one single object, the representation must have something so peculiar (something that relates exclusively to this object), that its origin in our mind is scarcely to be explained in another way but through the assumption that it relates to this object as an effect to its cause.[56]

Bolzano confesses here that he does not understand how it would be possible for a simple representation to be singular except if it is such as to refer only to its immediate cause. This supports (Ia.3). For if that weren't the case and there were instead always a difference in matter (rather than a mere difference in the context in which the matter is embedded) distinguishing intuitions from concepts, then Bolzano should not be so confident that it is impossible for us to have a subjective intuition of an alteration of the mind that is not also that very representation's immediate cause.

7. *Kant and the Objectivity Problem*

I argued in §5 that Bolzano has good reason to avoid (IIa). If so, then he has good reason to develop (Ib) into the best possible explanation for the

56. *WL* §74, I:331. In a move that is reminiscent of Kant's speculations concerning non-sensible intuition (see ft. 26), Bolzano later concedes that God or the universe might, for all we know, also be objects of *an* intuition, just not any of *our* subjective intuitions (§76, I:341f.).

non-accidental singularity of intuitions. I've just argued that for all of its unfamiliarity, Bolzano would have good reason to consider (Ia.3), complete with its Kantian appeal to the *matter* of subjective representations, as an informative specification of what it is to have a "this"-representation (in the strict sense of "this A"). Perhaps Bolzano would be reluctant to take this course, as it grants subjective representations a significant role in delineating a category of representations in themselves. It suggests that we cannot really understand what it is for a representation in itself to be a "this"-representation if we ignore the causal properties of subjective representations. However, regardless of whether (Ia.3) captures Bolzano's self-understanding, it is clear that he is a committed adherent of (Ib.1), and a consequence of this is that only that which can participate in causal relations, the actual, can be the object of intuitions. One thing that is important about this move is that it allows Bolzano, who clearly has the contrast with Kant in mind, to deny that spaces and times, including the whole of infinite space and the whole of infinite time, can themselves be the objects of intuitions. This follows from the fact, which Bolzano regards as relatively obvious, that space and time are not actual.[57]

Now it is time to remove the previous section's bracketing of a priori intuitions. Once we do so, we immediately begin to see that across this broad spectrum Kant will have a somewhat different way of developing the idea contained in (Ib). The basic source of the difference is that Kant apparently claims more than merely that time is the *form* of all intuitions—a position that would be perfectly consistent with (Ib.1) and (Ia.3). Kant apparently goes beyond this by committing himself to

(ST) Some intuitions have spaces or times themselves as objects.

This is especially plausible for a priori intuitions. Then, while Kant is free to agree with Bolzano in holding (Ib.1) with respect to *empirical* intuitions, he must adopt a different explanation of the essential singularity of *a priori* intuitions. One option is to adopt (Ib), though without any insistence that the demonstrative relation always be cashed out causally (i.e., by declining to adopt (Ib.1) across the board), but otherwise to treat intuitions whose objects are spaces or times just as (Ib) treats intuitions whose objects are dogs or trees. What is common to all of these cases is

57. See *WL* §79, I:362. Bolzano commits himself to (Ib.1) at §74, I:331; cf. §75, I:335; and §77, I:347.

that the object is phenomenologically present to the representing subject. A particularly interesting version of this view holds not just (ST) but also:

> (ST*) Some intuitions have the whole of space or the whole of time as objects.

(ST*) is supported by natural readings of several pivotal passages in the Transcendental Aesthetic, as when Kant tells us that space "is represented as an infinite given magnitude" (*KrV* B39; cf. ft. 9). However, many commentators cannot believe that Kant would actually commit himself to the claim that we can represent in a single intuition anything that is infinite. They interpret these passages differently (cf. Friedman 2000, 189; Bencivenga 2010). This is not the place to assess the defensibility of interpretations incorporating (ST*), or even to work at specifying the various options. Each by itself is a large task. The important point for our purposes is that if it can be made plausible that we have "this"-representations of space as a whole and of time as a whole, then (Ib) will have succeeded in identifying a univocal sense in which intuitions are non-accidentally singular. Perceptual demonstrative representations are essentially singular in virtue of the type of representation that they are, and two special cases of this type of representation are the "this"-representations of the wholes of space and time. Let's call this view *(Ib.2)*.

Notice that according to (Ib.2) the singularity of our cognitions of space as a whole and time as a whole is not in any way owing to the oneness of space and the oneness of time. Instead, our immediate representations of space and time are singular representations *because* those representations are perceptual demonstratives. In respect of the explanation of their singularity they are no different from a "this"-representation of a dog, which is a singular representation despite the fact that there are many dogs. This problem bedevils *all* of the relevant versions of (I), including (Ia.1), (Ia.2), and (Ib.2).[58] On each of those models to have an intuition of *x* is to have a singular representation (though in each case for a somewhat different reason), but this does not tell us anything about whether there are in fact multiple *x*'s. Let's call this:

58. (Ib.1) would be relevant only if it were combined with the claim that time and space as a whole *cause* representations. Neither Kant nor Bolzano accepts this.

The Objectivity Problem: without further argument there is no reason to suppose that the essential singularity of intuitive representations can tell us anything about whether their objects are unique.

Looking closer, it should be evident that *any* version of (I) will be subject to the Objectivity Problem, since the very idea behind Representation-based models is that intuitions 'singularize' their objects, treating the object as one whatever the object itself happens to be. What this shows us is that all versions of (I) are by themselves unable to account for Kant's apparent attraction to type (II) explanations, and this leaves Kant interpreters in the same position diagnosed at the end of §5. Namely, Kant would seem to be committed at least to a Hybrid View, if not to a pure version of (II). In closing, however, I want to propose a route that might work, and that Kant himself might even be endorsing, that would accomplish something like the End Run around (II) that was introduced in §5.

It turns out that in looking to new versions of (I) to effect the End Run, we were looking in the wrong place. Each new version is bound to be unsuccessful precisely because it is a Representation-based explanation of singularity. What we need instead is a different conception of the objectivity of the relevant facts about space and time (in this case, their uniqueness). From the beginning of this paper I have argued that the mere fact that space and time are not "objects", in Kant's most common uses of that term, is irrelevant to the question of whether facts about them are *objective*. In the meantime we considered an application of this point, viz., that (IIa) is a viable explanation of the non-accidental singularity of <space> and <time>, Kant's transcendental idealism notwithstanding. Now as we saw in §4, the objects of bare intuitions *are* only what they are represented as being in that very act of representing. There are no objective facts about these 'objects' in any familiar sense, including any sense that Bolzano would accept. My closing suggestion is this. The best bet for effecting the End Run around (II) is to reinterpret Kant's various assertions that space and time are unique as nothing more than as assertions of quasi-objective fact, which, apart from their apriority, are just like quasi-objective facts about appearances. Here we face no Objectivity Problem, for these quasi-objects just are as they are represented as being.

Now though Kant's underlying commitments regarding appearances and "dream worlds" are, as I conceded in §4, extremely difficult to assimilate because of their radical nature, the hypothesis that I am proposing here is considerably less radical than it might at first seem. This is because our cognitions of space and time are not limited to *bare* intuitions. Much like mathematicians in their proofs, we construct determinate spaces and times in the course of our everyday cognition, and these spaces and times are full-fledged objects. That is, though space and time are "mere forms", there are nonetheless full-fledged objective facts about them. Thus if the fact in question in deciding whether Kant is committed to (II) were the uniqueness of the triangle, then the present hypothesis would be wholly irrelevant, for there are fully objective facts about these constructible objects. Earlier we saw that Kant in effect limits the apparently radical implications of his surprising theory by pointing out that most of our cognitions (especially the cognitions that we notice) are already cognitions of *determined* objects, not cognitions of undetermined appearances. What we have seen in this paragraph is that he can make a broadly similar move with respect to our a priori cognition of space and time.

What, then, is the strictly circumscribed role for which the quasi-objects, space and time, are required? What I will propose, in general terms, is that they justify our allowing a certain assumption to govern our ongoing theory construction—specifically, the assumption that full-fledged objective claims can be about full-fledged objects *that are all related to one another within a single space and a single time*. Kant's own, more specific explanation of this is offered in terms of one specific type of theory, viz., in terms of *geometers'* full-fledged objective claims about full-fledged mathematical objects. The question of justification arises because the geometer assumes that certain constructive acts can be carried out, acts that are demanded of the geometer in "tasks" [*Aufgaben*] or "problems". Kant tells us, namely, that "the geometer grounds the possibility of his problem—to increase a space (of which there are many) to infinity—on the original representation of a single, infinite, *subjectively given* space. This agrees very well with the fact that geometrical and objectively given space is always *finite*. For the latter is only given insofar as it is *made*" (*Über Kästners Abhandlung*, 20:420). All full-fledged objective facts about space have constructed objects as their truthmakers. Kant pointedly avoids claiming that any of those objects is unique. What is unique, and what justifies the geometer in supposing that *any* already constructed line can be further extended, are representations of a single, infinite space. Their role is decidedly modest: if I am justified in judging that two objects stand in a determinate relation to one another, then it will be in virtue of my having constructed a space that extends to and includes both; it will, in other words, be grounded upon constructed objects as truthmakers. The role of the single, infinite space is merely to justify my *antecedent* assumption that I will be able to carry these constructions out.[59]

Kant describes this unique space as "*subjectively given*". Had we not inquired into Kant's theory of dream worlds and appearances in connection with (Ia.1), it would have been natural to assume that with this term Kant points merely to the fact that space and time are not transcendentally real; they are "subjectively given" in virtue of being forms of our receptivity. Yet notice that this provides no grounds whatsoever for Kant's contrast with the "objectively given space" that is produced by the geometer's constructions. Those triangles and other mathematical objects are themselves, after all, every bit as transcendentally ideal. The present suggestion is that with this contrast Kant is pointing instead to the fact that the one, unique space is nothing more than a quasi-object, the object of a bare intuition. Kant lacks a technical term that unambiguously identifies quasi-objects as such. "Subjectively given" is a reasonably accurate description for that status, its disadvantage being that it could mean several other things as well within Kant's philosophy.

This has been only the merest sketch of the context within which it would make sense for Kant to treat |space| and |time| as representations of quasi-objects. A *second* necessary ingredient would be an account of what it is about these representations that makes it a quasi-fact that space and time are each *unique*, rather than, say, *one of several* spaces and times. If it is to qualify, strictly speaking, as a successful End Run around (II), then that account should appeal to a type (I) model. In the case of (Ia.2)—perhaps the most promising model for these purposes—this would involve arguing that the uniqueness of space and time is ultimately explained by the fact that subjects find themselves, again and again, in the position of not having reflected upon the fact that particular bare intuitions, which in each case represent a time (and sometimes also a space), could also refer to *other* spaces and times. As we have seen above, the assumption that each is unique would in a sense be borne out in subsequent practice, as constructed objects are all determinately cognized within a single space

59. The need for such justification can be seen from Friedman's attempt to make do with constructions alone: "Perceptual space is singular or unitary because any outer object **must** be located within the same formal structure of possible perceptual relations: all outer objects **must** be reachable via translation and rotation [...] from a single initial given point of view" (2000, 192). The force of this "must" is never explained. Needless to say, the constructions necessary to reach *all* objects have not already been (and can never actually be) performed.

and time. But the assumption that this *can* be done would be justified by unreflected representations that are singular by default.

There is, however, another way for Kant to go. I'll explain it first and then comment all too briefly on its status. It might, namely, be an evident *quasi-fact* about space and time that each is unique, much as a judgment of perception might make evident the quasi-fact that this dog is moving. One pressing question that I cannot pursue here is how this uniqueness could be evident. Would it suffice, as Parsons considers, for objects always to be "given [...] with a 'horizon' of surrounding space" (1992, 70). This is an issue that "phenomenological" readings of the Metaphysical Exposition, such as Parsons', have already seen the need to address. There is one helpful twist that is added by the present suggestion that the space and time thus evident would be nothing more than quasi-objects. Namely, we should no more assume that the quasi-objects, space and time, will be completely determined than we should assume this of any mere appearance (see §4). In particular, there is no reason to expect that there are other facts, such as the sum of the internal angles of a triangle, about the space and time that are given antecedently to determinate constructions. Indeed, the just-quoted passage positively suggests that it might be *only* their infinity and uniqueness that are "subjectively given" in the relevant sense. All remaining facts about space might well be such that they are made true by full-fledged, constructed objects. True judgments about them are "valid at all times for us and for everyone else" (*Prol* §18, 4:298), whereas each subject's assumption that she can go on extending her ontic theory so that its objects are all related in a single space and time is based on intuitions that each of us has, again and again—whose objects are "subjectively given".

Having sketched this alternative, it is important to note that it does not actually qualify as an End Run around (II), since its explanation of uniqueness bottoms out in evident quasi-facts about space and time, rather than reaching further back to general facts about all intuitions and their singularity. That said, a case can be made that this version of (II) would be importantly different from the versions that we considered above in §5. Bolzano can help make the difference clear. What we considered in §5 was a reading that accepted the constitution of our sensible capacity, including the fact that it determines space and time to be unique, as a brute fact. Though Bolzano himself offers no similar theoretical appeal to a sensible capacity, once it is posited he would be perfectly comfortable with assuming that brute facts concerning its constitution determine the natures of space and time. What we have just considered, in contrast, is a version of (II) that reproduces the key aspects of Kant's theory of appearances that Bolzano would not accept. We saw in §4 that Bolzano sees no problem in principle with a third party compiling all of the facts about *P*'s appearances. (It is merely that this could not be done using intuitions.) Neither should there be a problem with an external investigator making a list of the brute facts about space and time that are explained by the constitution of our sensibilities. The account just considered provides a contrast on precisely this point, for it countenances a species of fact that is in each case essentially tied to a certain perspective, albeit one that all of us are said to share.

Whichever of these two alternatives is selected, a *third* necessary component in pursuing the present interpretative strategy would be some evidence that Kant indeed views space and time as quasi-objects in the sense that has been sketched above. We can close with two points that are promising starting points for this research. First, Kant tells us that space and time are merely *entia imaginaria* (*Diss* §14, 2:401; *KrV* A291). Previous attention to this claim has focused on the role of imagination in particular in bringing about temporally and spatially ordered manifolds[60], but more relevant in the current context is Kant's explication of *entia imaginaria* as "nothings", rather than "somethings", specifically with respect to the *categories of relation* (*KrV* A291). These are, of course, precisely the categories whose involvement or lack thereof determine whether an object is a determined object or a mere appearance. Kant tells us in this crucial paragraph that space and time are "without substance". Each is "in itself not an object, but the merely formal condition of one (as appearance)" (ibid.). As was argued in §4, it is crucial that mere appearances are not substances. Since substances bear accidents, there is a way that they are, which is in principle distinguishable from how they are represented as being, and which any particular subjective representation can get wrong.

A second avenue for future research on the third component is the fact that Kant treats our representations, |time| and |space|, as semantically immediate. As was also sketched in §4, this seems to go hand in hand for Kant with the impossibility of reference failure. That made sense, in turn, because they are nothing apart from the way that they are represented as being, so that there is nothing that an act of representation could get sufficiently wrong for reference to fail. In light of this fact it is notable

60. This is the central thesis of Waxman (1991), who does not appear to consider the interpretation of the *ens imaginarium*-thesis that I propose here.

that Kant does not think that a transcendental deduction of our a priori forms of sensibility is required—at least not for the same reason as we find with the categories. He explains: "Thus a difficulty is revealed here [i.e., with the categories, T.R.] that we did not encounter in the field of sensibility, namely how *subjective conditions of thinking* should have *objective validity* [...]".[61] Here again we see a claim that cannot be explained merely by pointing to transcendental ideality. The empirical objects that we are able to cognize using the categories are also transcendentally ideal, but Kant demands a proof, in the form of a transcendental deduction, that the categories do not simply fail to refer. What is different "in the field of sensibility" is that |space|, |time|, and empirical bare intuitions do not claim for themselves "*objective validity*", which I propose to read as "full-fledged objectivity". The objects of these representations just are immediate expressions of "*subjective conditions*". If the arguments above have been correct, then two important questions for future research will be, first, the exegetical question of whether Kant is indeed gesturing at some way in which facts about determined objects are different from facts (quasi-facts?) about space and time, and, second, whether this move can be philosophically successful. We can already anticipate Bolzano's answer to the latter question.

Bibliography

Abbreviations

Ak. = the *Akademie Ausgabe* of the works of Kant (*Kant's gesammelte Schriften*. Berlin: Walter de Gruyter, 1900ff.). Citations of the form "17:541" will refer to volume and page numbers in Ak. The *Cambridge Edition* has been used as a starting point for all translations, though final results are often quite different from those translations.

GA = Bolzano, *Gesamtausgabe*. Stuttgart: Frommann-Holzboog, 1969ff.

JL = *Jäsche Logik* (Ak. 10).

KrV = *Kritik der reinen Vernunft* (Ak. 3-4). In order to avoid unnecessary duplication, B-edition pagination is provided only if the passage appears only in that edition. When a passage appears only in the A-edition, this is noted explicitly; otherwise, it appears in both.

61. *KrV* A89f. For the reason, unrelated to the current argument, why a transcendental deduction of the *concepts* of space and time, nonetheless, needs to be given, see A88.

Met Mrong = *Metaphysik Mrongovius* (Ak. 29).

MM = *On the Mathematical Method and Correspondence with Exner*. P. Rusnock & R. George (trans.). Amsterdam: Rodopi, 2004. Translations have been emended without noting each instance.

Prol = *Prolegomena* [...] (Ak. 4).

R = *Reflexionen*, from Kant's *Nachlaß* (Ak. 15–19).

V = *Versuch einer objectiven Begründung der Lehre von den drei Dimensionen des Raumes* (1843), in *GA* I.18.

WL = *Wissenschaftslehre*. I use R. George's partial translation (*Theory of Science*. Berkeley: Univ. of California Press, 1972) when possible, though I have emended that translation without noting this in each instance.

Other sources

Bencivenga, Ermanno 2010: "An Infinite *Given* Magnitude". *History of Philosophy Quarterly* 27, 95–100.

Braun, David 2010: "Indexicals". In Edward Zalta (ed.), *The Stanford Encyclopedia of Philosophy*. URL = <http://plato.stanford.edu/archives/sum2010/entries/indexicals/>.

Carson, Emily 1997: "Kant on Intuition in Geometry". *Canadian Journal of Philosophy* 27, 489–512.

Dicker, Georges 2004: *Kant's Theory of Knowledge*. Oxford: Oxford University Press.

Dickerson, Adam 2004: *Kant on Representation and Objectivity*. Cambridge: Cambridge University Press.

Friedman, Michael 2000: "Geometry, Construction, and Intuition in Kant and His Successors". In: Gila Sher et al. (eds.), *Between Logic and Intuition: Essays in Honor of Charles Parsons*. Cambridge: Cambridge University Press, 186–218.

Hintikka, Jaakko. 1968: "On Kant's Notion of Intuition (*Anschauung*)". In: Terence Penelhum (ed.), *The First Critique*. Belmont. Calf.: Wadsworth, 38–53.

Hume, David 1739: *A Treatise of Human Nature*. Reprint, Oxford: Oxford University Press, 2000. (Cited by Book, Part, and Section numbers (e.g., I.ii.5)).

Locke, John 1689: *An Essay Concerning Human Understanding*. Reprint, Oxford: Oxford University Press, 1979. (Cited by Book and Chapter numbers.)

Lohmar, Dieter 1998: *Erfahrung und kategoriales Denken: Hume, Kant, und Hume über vorprädikative Erfahrung und prädikative Erkenntnis*. Berlin: Springer Verlag.

Longuenesse, Béatrice 1995: "Kant et les jugements empiriques—jugements de perception et jugements d'experience". *Kant-Studien* 86, 278–307.

— 1998: *Kant and the Capacity to Judge*. Princeton, NJ: Princeton University Press.

Parsons, Charles 1992: "The Transcendental Aesthetic". In: Paul Guyer (ed.), *The Cambridge Companion to Kant*. Cambridge: Cambridge University Press, 62–100.

Prauss, Gerold 1971: *Erscheinung bei Kant*. Berlin: Walter de Gruyter.

Rosenkoetter, Timothy 2009: "Truth Criteria and the Very Project of a Transcendental Logic". *Archiv für Geschichte der Philosophie* 91, 193–236.

— (MS): "Kant and Bolzano on Self-Consciousness". (Unpublished manuscript).

Schnieder, Benjamin 2009: "Bolzanos Erklärung des Zeitbegriffs". *Archiv für Geschichte der Philosophie* 91, 42–69.

Sellars, Wilfrid 1967: "Some Remarks on Kant's Theory of Experience". *Journal of Philosophy* 64, 633-647. Cited using pagination from reprint in K. Scharp & R. Brandon (eds.) (2007), *In the Space of Reasons*. Cambridge, Mass.: Harvard Univ. Press, 437–453.

—1968: *Science and Metaphysics. Variations on Kantian Themes*. London: Routledge.

Smit, Houston 2000: "Kant on Marks and the Immediacy of Intuition". *Philosophical Review* 109, 235–266.

Vaihinger, Hans 1892: *Commentar zu Kants Kritik der reinen Vernunft*, Bd. 2. Stuttgart: Union Deutsche Verlagsgesellschaft.

Waxman, Wayne 1991: *Kant's Model of the Mind*. Oxford: Oxford University Press.

Wolff, Michael 1995: *Die Vollständigkeit der kantischen Urteilstafel*. Frankfurt a. M.: Vittorio Klostermann.

Grazer Philosophische Studien
85 (2012), 131–149.

FROM ORDINARY LANGUAGE TO DEFINITION IN KANT AND BOLZANO

Waldemar ROHLOFF
University of Missouri, St. Louis

Summary
In this paper I discuss Kant's and Bolzano's differing perspectives on ordinary natural language. I argue that Kant does not see ordinary language as providing semantically organized content and that, as a result, Kant does not believe that ordinary language is sufficiently well-developed to support philosophical analysis and definition. By contrast, for Bolzano, the content given in ordinary language are richly structured entities he calls 'propositions in themselves'. This contrast in views is used to explain Bolzano's criticism of Kant's belief that definition is impossible for philosophical concepts. It is also used to explain Bolzano's criticism of Kant's methods of exposition of philosophical concepts.

0. *Introduction*

Bolzano's reaction to German idealistic philosophy is driven by his insistence on the importance of clarity in philosophical writing. In particular, at many points throughout his works, Bolzano laments Kant's failure to be sufficiently clear and comprehensive in his explanations of philosophical concepts. A major contribution of Bolzano's own work is its rigorous redefinition of Kantian concepts, in particular the a priori, analyticity, intuition and their correlates. In this paper, I aim to show how Bolzano's ideals of clarity and explicitness in definition rest on his views on ordinary language. For Bolzano, ordinary language provides content sufficient to support philosophical definition. This marks an important point of contrast with Kant, for whom the same is not the case. Interestingly, Bolzano's 'linguistic turn' links him to post-Kantian thinkers such as Johann Georg Hamann and Johann Gottfried von Herder. Though the latter writers are far removed from Bolzano in doctrine and intellectual temperament, they share with Bolzano the idea that Kant failed to give sufficient weight to ordinary language in his philosophical inquiries.

In section one, I sketch Kant's views on the relationship between philosophy and ordinary language. As Kant sees it, philosophy and the study of ordinary language are not related in significant ways. This observation helps emphasize the significance of Bolzano's positive contributions. Section two contains a presentation and discussion of Bolzano's views on the relationship between ordinary language and philosophy as conceived in the *Wissenschaftslehre* and in the *Neuer Anti-Kant*.[1] Finally, section three shows how Kant's and Bolzano's differing perspectives on the relationship between ordinary language and philosophy serve to underpin their divergent ideas about the role and proper form of definition and exposition in philosophy.

1. *Kant on language: 'Only a designation and not a concept of the thing'*

Discussing Kant's views on language is all the more challenging given that what he has to say concerning this topic is in short supply. Kant's silence summarizes his views: for him, the study of language has little to contribute to philosophy. At the root of Kant's position is a kind of linguistic pessimism. He does not believe ordinary language displays enough regularity to warrant rule-based syntactic analysis. More importantly, what is expressed by language—what one might call its 'content'—is insufficiently refined to serve as a basis for definition.

Kant's views about the syntax of ordinary language come out in a passage from the *Inaugural Dissertation.* There Kant insists that organizing language in terms of a set of grammatical rules requires that language reaches a certain level of sophistication and development, such as is furnished in rhetoric and oratory. He tells us:

> After science has attained a certain fullness and orderliness, trial and error show what path and what procedure must be pursued if it is to be brought to completion, and made to shine the more purely, once the blemishes both of mistakes and of confused thoughts have been eliminated. It was exactly this way that grammar, after a richer use of speech had been established, and style,

1. (Bolzano & Příhonský 2003), hereafter abbreviated as '*NAK*' was composed by Příhonský but is clearly a compendium of Bolzano's ideas. When discussing ideas and passages specifically from the *Neuer Anti-Kant*, I ascribe them to both figures. All other citations of Bolzano's works are in reference to (Bolzano, 1969–). Citations from the *Wissenshfastlehre* are abbreviated '*WL*'.

after elegant examples of poetry and oratory had been furnished, provided a foothold for rules and method. (AA 2:410)

It is a commonplace of modern linguistic theory to conceive of a speaker's linguistic competence as resting on knowledge of a system of rules. Kant does not share this picture of language. From Kant's point of view, rule-governed syntactic structure is not an inherent aspect of human language. Rather, such structure can only be ascribed to human linguistic behavior once language achieves a certain level of development. While it is unclear what exactly Kant has in mind by a 'richer use of speech', it is clearly something he believes human language ordinarily lacks.[2]

Kant's semantics, i.e. his views on the 'content' of ordinary language, come to the fore in some of the comments he makes in the course of his discussion of definition in the *Critique of Pure Reason*. For Kant, to define properly is 'to exhibit originally the exhaustive concept of the thing within its boundaries' (A727/B755).[3] Kant details his conception of definition in his lectures on logic (AA 24:263–5, 24:912–913). There he explains that definition requires a 'distinct', 'complete' and 'precise' concept of a thing. Distinctness requires that the predicates involved in the definition be a 'clear ground of cognition of the thing' (AA 24:264). A distinct definition provides a clear enumeration of the marks by which we identify things falling under the defined concept. To be complete, a definition must contain enough predicates to allow us to distinguish the things falling under the concept from all other things. For Kant, completeness amounts to giving all essential predicates of a thing (AA 24:919). Finally, a definition is precise if it does not contain any subordinate predicates, that is, predicates that are derivable from other predicates in the definition. Kant illustrates a failure in precision in the definition of a circle as 'a curved line, whose individual points are equally far removed from the centrum' (AA 24:268). Here the definition fails to be precise because the concept of a curved line is already contained in the concept of a line whose points are equidistant from a center.

Interestingly, Kant's account of definition serves him, in the first instance, to argue that empirical concepts cannot be defined. He tells us:

2. In this, Kant is expressing a distinctively anti-Leibnizean bent. Leibniz insists that all instances of natural language, for example the arguments of an orator, have an underlying logical form. See (Losonsky 2006, 52–68).

3. All of my citations of Kant are in reference to the (Kant, 1902) though I do make use of (Kant 1998, 1992a, 2007, 1992b) when quoting from Kant's works. References to the *Critique of Pure Reason* are made using the standard 'A-edition, B-edition' notation.

> An empirical concept cannot be defined *(definiert)* at all but only explicated *(expliziert)*. For since we have in it only some marks of a certain kind of objects of the senses, it is never certain whether by means of the word that designates the same objects one does not sometimes think more of these marks another time fewer of them. Thus in the concept of gold one person might think, besides its weight, color, and ductility, its property of not rusting, while another person might know nothing about this. And in any case what would be the point of defining such a concept, since when, e.g., water and its properties are under discussion, one will not stop at what is intended by the word 'water' but rather advance to experiments, and the word, with the few marks that are attached to it, is to constitute only a designation and not a concept of the thing. (A727–8/B755–6)

Kant believes that a word has the power to bring up certain marks in the mind of an individual. The marks produced may vary on different occasions and from individual to individual. As a result, words have the power to lead us away from the concepts and things about which we wish to communicate. On the view that communication requires the exchange of shared content, this means that language is at times a barrier to successful communication. Kant concludes pessimistically in *Anthropology from a Pragmatic Point of View*, that 'even those who can speak and hear do not always understand either themselves or others. This is to be attributed to the lack of the power of designation … and to the fact that men, especially where reason is concerned, who speak the same language are miles apart with regard to concepts' (AA 7:193).

Kant's views on definition (in philosophy) stand in sharp contrast to his views on the content we get from ordinary language. As we saw above, for Kant, proper definition requires 'an exhaustive concept of a thing within its boundaries'. This requirement ensures that the content given in definitions satisfies well-defined conditions. By contrast, ordinary language gives rise to a motley collection of subjective associations, apparently having no kind of systematic organization. The words and phrases of ordinary language do not have associated with them the kind of content needed to make definition possible. Nor does Kant suggest that the content given through ordinary language satisfies some other general conditions.[4] This is

4. In this respect, we must be careful about ascribing semantic notions to Kant. Though Kant's concerns have been reinterpreted in semantic terms, as in (Hanna 2001, 14–66), we must remember that Kant's semantic concerns, insofar as he has any, have little to do with language. For instance, given what has been said above, the question of the connection between a word and its reference, a question of paradigm importance for twentieth century analytic philosophy,

why words and language are a 'mere designation' and 'not a concept of the thing'. For Kant, ordinary language works successfully when it directs the hearer to the object designated by the speaker. If our individual subjective associations carry us to the intended referent more often than not, then language is successful. But language does not manage to do this in any systematic way.

Despite Kant's tendency to contrast the content given in ordinary language with the content given in definitions, he recognizes important relations between ordinary language and definition.[5] Though Kant does not think that the content given in ordinary language can supply definitions on its own, ordinary language provides primitive data, which when 'completed by reason', gives rise to something with significance for definition. Kant expresses this idea in the following *Reflexion*:

> Human reason has the peculiar property that in addition to what is necessary to make a concept of the thing for a certain aim, it further not only completes this concept by means of everything that constitutes it but also completes the object of the concept in the species of thing to which it belongs. We are not satisfied with what would be sufficient for the ordinary use of words in knowing the concept of the body, a human being, a plant: we try to become aware of it and all of its marks, and from that, if the law of economy is applied, there arises the definition. (AA 18:489)[6]

We again see the emphasis on the inadequacy of the content of words in ordinary language. But here Kant goes further, arguing that human reason 'completes the object of the concept in the species of thing to which it belongs.' What Kant has in mind by 'completion' is linked to the ideas of reason: completeness, unity and the systematic intelligibility of our experience, whose regulative role in inquiry is set out by Kant in the Appendix to the Transcendental Dialectic.

---

probably holds no real philosophical interest for Kant. (Beaney 2002, 456–460) raises related concerns with respect to Hanna's work.

5. One loosely connected place from which a positive view can be extracted is Kant's references to common sense, for example, in the *Groundwork of The Metaphysics of Morals*, where Kant claims to develop philosophical knowledge of morality from a common sense basis. For discussion of the relationship between Kant's philosophy and common knowledge see (Ameriks 2001, 2006).

6. We should not be misled by Kant's use of 'definition' in this context. What he has in mind is something weaker than the proper definitions he describes in the passage where he claims that definition is impossible for empirical concepts. Kant does not believe that proper definitions can be given for concepts such as 'body', 'human being' or 'plant.'

To illustrate this, take, for instance, Kant's 'principle of thoroughgoing determination' (A571/B599). According to this principle, given any possible predicate of an object, either that predicate or its opposite must apply to the object (A572/B600). Clearly this principle is related to the law of excluded middle, but Kant insists that it differs from that 'merely logical' principle.[7] What distinguishes the principle of determination is that it has a transcendental basis, that is, it goes beyond the bounds of any possible experience. Unlike excluded middle, which applies to 'any given pair of opposed predicates', the principle of thoroughgoing determination applies to 'any pair of possible predicates' whatsoever. Kant assumes that the idea that one of any possible pair of predicates must apply to an object is equivalent to comparing that object to the 'sum total of all possible predicates', and since the sum total of all possible predicates is just the kind of thing that cannot be given in experience, the principle is transcendental. In the Appendix to the Transcendental Dialectic Kant explains how principles such as the principle of thoroughgoing determination have a 'regulative', as opposed to a 'constitutive' role in inquiry.[8] Ultimately, what this means is that while the principle cannot be satisfied by any real object (A646/B676), it serves as a guide for rational inquiry by providing the idea of a fully determinate object, that is, an object that satisfies one of every possible pair of opposing predicates. The idea of a fully determinate object is an important assumption for 'definition to arise'. Without the assumption of determinacy, we have no grounds to seek an exhaustive list of predicates, as definition requires. In particular, if an object is indeterminate with respect to some predicate, then any list of predicates will be insufficient to distinguish the relevant object from some other object. For instance, if object x falls under predicates $A_1, \ldots, A_k$, but is indeterminate with respect to B, then for object y falling under $A_1, \ldots, A_k$, which is determinate with respect to B, no list of predicates true of object x will be sufficient to distinguish it from object y.[9] But of course,

7. Kant refers to the logical law as 'the principle of contradiction' rather than 'excluded middle'.

8. Kant does not discuss the principle of thoroughgoing determination specifically in the Appendix, but the principles and ideas of reason he discusses are built around it and related principles.

9. There is, of course, the property of 'being determinate with respect to *B*', which would serve to differentiate object one from object two. However, properties defined in terms of falling under other properties would not be part of a definition for Kant. This is because such defined properties would be derivable from other properties in the definition, thus violating the precision of the definition. For example, 'being determinate with respect to *B*' would be

x and y are distinct because x is indeterminate with respect to B, while y is not. If objects are indeterminate, no list of predicates can be complete, and all attempts at definition are in vain. Thus, the ideas and principles of pure reason are an important prerequisite for attempts at definition.[10]

To summarize Kant's position: Kant believes that expressions of ordinary language do not have as their content an exhaustive list of predicates, nor does their content satisfy any other well-defined semantic conditions. This explains why Kant believes that ordinary language does not provide us with content worthy of analysis, and why he has so little to say about language from a philosophical point of view. Nevertheless, Kant believes that ordinary language provides basic content upon which reason can build. Ordinary language is insufficient, in and of itself, to provide a starting point for definition. But when the content given in ordinary language is enriched in terms of ideas and principles of pure reason, philosophical definition can take hold.[11]

2. *Bolzano on ordinary language*

Bolzano's theory rests on a tripartite distinction between: (i) thought propositions *(gedachte Sätze)* (ii) uttered propositions *(ausgesprochene Sätze)* and (iii) propositions in themselves (*Sätze an sich)* or objective propositions. As Bolzano sees it, thought propositions are types of mental acts. They have real existence, though their existence is dependent on the existence of the agent. Uttered propositions are certain types of instances of ordinary natural language. Like thought propositions, they have a real existence. By contrast, propositions in themselves have a distinctive ontological status for Bolzano. He emphasizes their objectivity and mind independence, and he also stresses that they are not real *(wirklich)*; propositions in themselves do

derivable from B itself. That is, we could infer that an object is 'determinate with respect to B' if we know that it is B.

10. Of course, in many cases, Kant would claim that definition is still, strictly speaking, impossible. The point here is that to even attempt a definition, we must move beyond the content given in ordinary language.

11. It is worth noting an example of Kant's insistence that final authority rests with reason and not ordinary language: 'We were able to remain wholly unconcerned about the agreement of our definition with the ordinary meaning of the word, for the only question which our definition raises concerns a problem which arises in accordance with reason, namely, how it is possible for several substances to coalesce into one thing, and upon what conditions it depends that this one thing is not a part of something else.' (AA 2:389)

not exist in the sense of ordinary, physical objects. According to Bolzano, it is through thought propositions that we 'grasp' propositions in themselves: the content *(Stoff)* of the thought propositions are propositions in themselves. Likewise, uttered propositions have propositions in themselves as their senses (*Sinne*).

The tripartite distinction above is paralleled by another, that between (i) subjective representations *(subjective Vorstellungen)* (ii) signs (*Zeichen)* and (iii) objective representations *(objective Vortellungen)*. Objective representations are sub-propositional parts of propositions in themselves. Just like propositions in themselves, objective representations are not real. Subjective representations are sub-propositional parts of thought propositions. They are certain types of mental states and thus, are real. The contents of subjective representations are objective representations. Finally, in the context of this discussion, signs are those entities that ultimately constitute a language: 'The class of signs which a person is inclined to employ, especially those which he uses to reveal his thoughts to others, is called the language of that person' (*WL*, §285). Like subjective representations, signs are real and have objective representations as their content. The difference between signs and subjective representations is that signs work at one level removed, by first stimulating a subjective representation, such as a mental state, which then is directly associated with an objective representation (*WL*, §285). As a consequence, a single sign, say a word, is capable of eliciting multiple subjective representations, whereas, in general, there is only one objective representation that is truly designated by the word. As Bolzano tells us: 'any word, unless it is ambiguous, designates only one objective representation, but there are innumerable subjective representations which are stimulated *(erweckt)* by it, and their number grows with every moment the word is in use' (*WL*, §48).

As Bolzano sees it, one important task of philosophy consists in determining what is meant by a given set of words, that is, to make explicit the objective content that is expressed by way of a linguistic expression. Bolzano calls this process 'interpretation' (*Auslegung*). He tells us: 'a sentence in which it is asserted that certain signs have such and such a sense is called an interpretation of the sentence' (*WL*, §285). Interpretation however is a challenge: linguistic expressions can elicit multiple, and often extraneous, subjective representations. When engaging in interpretation, philosophers must excise any unintended and irrelevant subjective associations, and in this way, direct us towards the appropriate objective content. For Bolzano, interpretation involves the construction of a paraphrase, namely a sentence in a semi-formal language that serves to excise context-sensitive and ambiguous expressions. This is done by presenting the reader with an alternative set of signs (the new semi-formal language), which does a better job of signifying the objective content linked to the original ordinary language sentence. What is important here is that both ordinary language and formal language are capable of expressing the same objective content.[12] An interpretation in Bolzano's sense serves to disambiguate ordinary language, and it does so by more successfully articulating what should have been grasped through the original ordinary language sentence.

Interpretation is not the only definitional procedure Bolzano resorts to. In order to define logical properties such as analyticity and logical relations such as deducibility *(Ableitbarkeit)* Bolzano resorts to a substitutional procedure. What is interesting about this procedure is how it informs Bolzano's views on interpretation. Bolzano's definition of analyticity and deducibility requires a specification of variable and fixed parts of propositions in themselves. This presupposes that propositions and their components exhibit significant structure. In his elaboration of propositional structure, Bolzano insists that all propositions have the same underlying form: 'A has B' (*WL*, §126–128). Similarly, Bolzano details how objective representations, the sub-propositional units of propositions, have significant structure. They are individuated in terms of their parts (*WL*, §56) and are put together from recursive rules and a primitive vocabulary (*WL*, §58–64).

Bolzano's views on propositional structure and interpretation come together in his account of definition. For Bolzano, to define a concept is to 'indicate its components' (*WL*, §23), along with the 'order of the parts' (*WL*, §58). For instance, Bolzano defines 'earthling' as 'a creature that lives on Earth' (*WL*, §23).[13] In this example, we get an interpretation that makes explicit the components of 'earthling'. Importantly, Bolzano's definition includes logical and non-logical components. In this case, the non-logical components include 'creature', 'living on' and 'Earth'. The

12. Some have argued that the criterion of adequacy for paraphrase is that the sentences in question must express the same proposition (Berg 1962, 52-53). Others have argued that Bolzano accepts a weaker criterion of adequacy, on which identity of truth-conditions expressed in some canonical form, is sufficient for successful paraphrase (Textor 1997, 189-199). For my purposes, I rest with the claim that Bolzano believed ordinary language sentences, and paraphrases of those sentences, can and sometimes do express the same propositions. This need not be a criterion of adequacy for paraphrase. Most importantly, propositions are the content of ordinary language sentences for Bolzano, whether or not the paraphrase of the sentence expresses the same proposition as the sentence itself.

13. See (Lapointe 2011, 38) for discussion of this example.

logical component is 'which has (the property of)', that is a name-forming operation that induces structure by allowing for the apposition of the two noun-clauses. As a result Bolzano's definitions are logically articulate. This marks a significant contrast with Kant, for whom the composition of a concept does not rest on syntactically articulated structure.[14]

We are finally in a position to appreciate an important difference between Bolzano and Kant. For Bolzano, the content given in ordinary language are propositions in themselves. These have significant formal structure, and as a result they are the objects of fruitful analysis, yielding definitions of given concepts. For Kant, as we have seen, the content given in ordinary language does not satisfy any well-defined conditions. Rather, such conditions are imposed by reason in the course of attempting to formulate definitions. This means that, for Kant, the content given in ordinary language, in of itself, is an unproductive object for analysis. This is a significant contrast when it comes to showing how Kant and Bolzano relate to ordinary language. As we will see in the next section, their divergent attitudes towards ordinary language affect their philosophical methodology in important ways.

3. *On definition and exposition in philosophy*

Kant says that to define properly is to 'exhibit originally the exhaustive concept of the thing within its boundaries' (A727/B755). This requires a distinct, complete and precise list of characteristic marks. He claims that definitions are impossible for empirically given concepts such as 'gold', because we can never be sure we have given an exhaustive account of their marks (A728–9/B756–7). New observations can always force us to remove some marks and add others. With empirical concepts, our attempts at definition are at best an imperfect and open-ended enterprise. Kant calls such attempts 'exposition' *(Exposition)*, 'which is always cautious, and which the critic can accept as valid to a certain degree while yet retaining reservations about its exhaustiveness' (A730/B758). According to Kant, definitions can only be provided for constructed concepts, with mathematics providing the paradigm examples of definitions. By contrast, the concepts of philosophy are not constructed, but 'given a priori'. Thus, Kant

14. See (Lapointe 2011, 18–42; Lapointe 2007) for an extensive discussion of Bolzano's understanding of conceptual content and its contrast with Kant.

believes that proper definitions are impossible in philosophy. In philosophy, 'definitions come about only as expositions of given concepts' (A730/B758). These expositions generally take the form of partial definitions, that is, non-exhaustive lists of characteristic marks, along with examples used to test their adequacy. Kant explains that this procedure falls short of true definition since: 'the exhaustiveness of the analysis of my concept is always doubtful, and by many appropriate examples can only be made probably but never apodictically certain' (A728–9/B756–7).[15]

In *Neuer Anti-Kant,* Bolzano and Příhonský fasten onto this aspect of Kant's views, and make it the focus of an extended criticism. They suggest that Kant's dismissal of definition in philosophy has led to a wide-ranging lack of rigor in recent philosophy (*NAK,* 138–139). A major contribution of *Neuer Anti-Kant,* namely its clarification and regimentation of Kantian terminology, is grounded in the belief that Kant ought to have given rigorous definitions where he believes no such definitions are possible. As an example, let us take Kant's exposition of the concept of the a priori. In the introduction to the *Critique of Pure Reason*, Kant explains this concept by giving a limited list of 'marks': necessity and generality, by which we can recognize whether a judgment is a priori (B3–4). Kant goes on to support this exposition with examples, citing mathematics and the proposition 'Every alteration must have a cause'. Bolzano and Příhonský criticize Kant's reliance on examples and opaque characterizations 'which are no less abstruse than the thing defined' (*NAK,* 40–41).[16] Kant resorts to providing marks of the a priori, examples of a priori judgments and partial definitions because he believes the a priori, as a 'given' concept, cannot be completely defined. Bolzano and Příhonský disagree. According to them it is possible to provide a genuine definition of the a priori, and they contend that had Kant properly defined his notion of the a priori, 'he would have hardly stumbled across the many empty

---

15. For Kant's account of exposition see also (AA 24:268, AA24:920–921, AA 9:142–143, AA 24:758–759).

16. Bolzano and Příhonský accuse Kant of having confused exposition with definition of a concept: 'Here he [Kant] clearly confuses the mere agreement on a word—that is, that transition by which we elicit and make known in the mind of the reader that very concept which we associate and refer to with the word—with the definition'. The nature of Bolzano and Příhonský criticism at this juncture is obscure. They seem to believe that Kant thought he was giving definitions when he was merely giving expositions. But this criticism is puzzling given that Kant himself admits that real definitions for philosophical concepts are impossible. Thus, it is unclear in what sense Kant was actually 'confused'. Rather, there seems to be a disagreement about whether genuine definitions are possible in particular cases.

and incorrect explanations which we have seen him stumble on' (*NAK*, 46).[17] Bolzano and Příhonský's alternative definition of the a priori relies on Bolzano's definition of intuition, which in turn relies on the notions of simplicity and singularity as applied to objective representations.[18] A representation is simple if it has no parts (*WL*, §61), while it is singular if it has only one object (*WL*, §68). Intuitions are explicitly defined as a species of representation that are simple and singular. Finally, a priori propositions are defined as those propositions in which no intuition occurs (*NAK*, 36, 46–7).

A question can be raised concerning Bolzano's criticism of Kant. In order to formulate this question we must look at Bolzano's own version of exposition, which he calls 'agreement on a word' *(Verständigung über ein Wort)*.[19] Bolzano's version of exposition is a crucial component of his methodology in the *Wissenschaftslehre*. He discusses the importance of *Verständigung* (*WL*, §668), and often refers to his aim as that of reaching *Verständigung*. Throughout the work we see him employ a common set of strategies when he introduces philosophical terminology. Among these are: looking at the common usage of a term, examining the context of usage, providing a synopsis of various philosophical definitions of the term from other philosophers, discussing closely related concepts, as well as making use of implicit definition. For Bolzano, *Verständigung* involves an informal set of techniques that includes just about anything we might typically do to get another person to truly understand us in a theoretical context. The techniques Bolzano uses are the ones he believes to be most expedient given the aims of science.

Bolzano sometimes resorts to exposition for practical reasons. As he tells us: 'in my view, the definition of a concept *(Erklärung eines Begriffes)*, that is, indicating its components, is not always the easiest or even the most reliable method to reach an agreement (*Verständigung*) about it' (*WL*, §23). At other times, exposition is used because Bolzano is unable to provide an explicit definition. Bolzano admits being unable to define propositions in themselves (*WL*, §23) and representations in themselves (*WL*, §52), both

17. All translations from *Neuer Anti-Kant* are my own.

18. Kant also fails to give a rigorous definition of intuition, eliciting some of the same criticisms from Bolzano and Příhonský. Bolzano and Příhonský co-opt that word but give it an explicit definition which they claim captures the core of what Kant had in mind (*NAK*, 34).

19. As noted by the editors in (Bolzano 2004, 37) '*Verständigung*' presents some difficulties for the translator. There the word is translated as 'exposition' or 'to come to an understanding'. Here I have translated it as 'agreement' as this emphasizes the way in which *Verständigung* must involve at least two parties.

of which are central components of his philosophy. Bolzano's definitions of intuition and a priori propositions, which as we have seen he uses in his criticism of Kant, ultimately appeal to these undefined notions. This latter point should give us pause. Bolzano criticizes Kant for using expositions where Kant believes definitions are impossible. But Bolzano himself seems to do more or less the same in circumstances where he finds explicit definition to be impossible. So our question is: has Bolzano simply traded in Kant's set of indefinable concepts for his own, or is there some other basis for Bolzano's criticism of Kantian exposition? Starting from a different set of indefinable concepts does not necessarily undermine Bolzano's criticism of Kant, for it may be that Bolzano is right, and Kant is wrong, about which concepts are in fact indefinable. But what we want is an explanation of what makes Bolzano right and Kant wrong. Without such an explanation, the worry arises that Bolzano's alternative definitions are a mere redescriptions in alternative terminology, rather than a genuine alternative to Kant's definitions.

One way to provide such an explanation is to turn to the special features underlying Bolzano's understanding of what makes a concept indefinable. In Bolzano's theory, what makes a concept indefinable is its simplicity. We see this, for example, in Bolzano's letter to Exner of December 18th, 1834:

> What can this 'grasp' mean here? I believe that I have already reminded you that the word 'grasp', as well as all the similar words I have used in this attempt to reach *Verständigungen*, are only figurative expressions used in the hope that anyone who understands the language can gather from the whole context which completely simple concepts, or else concepts composed of a very small number of parts, it really was that I designated with these words. Do not therefore demand a definition of the concept I connect with the word 'grasp'; I truly believe it to be simple. (Bolzano 2004, 162)

This passage throws into light the difference between Bolzano's and Kant's conceptions of indefinability: by contrast with Kant, for whom a concept is indefinable if it is 'given', not constructed, for Bolzano, a concept is indefinable if it is simple. With this difference in mind, Bolzano's criticism of Kant is not merely that Kant fails to define this or that concept, but that Kant misunderstands the grounds for indefinability, which then misleads him to forego definitions in favor of expositions. More generally, Bolzano's conception of simplicity and his views on indefinability are linked to his axiomatic understanding of science. Bolzano sees science as an intrinsically ordered system of true propositions. These propositions are ordered in terms of logical relations, the ultimate being the relation of ground-consequence. Relations between propositions are established by showing how some propositions are provable from others. This procedure eventually leads to propositions that cannot be proven. These propositions will contain concepts that are simple, and are thus indefinable.

In light of our discussion, there is another way to explain what motivates Bolzano's criticism of Kant. As we have seen, for Kant, analysis cannot proceed merely with what is given in ordinary language. It is only when reason imposes conditions on the content expressed in ordinary language that the content becomes sufficiently refined for attempts at definition to take hold. However, reason imposes conditions on our concepts in such a way that those concepts are never completed. For example, in our discussion of the principle of thoroughgoing determination, it was noted that we never confront the 'sum total of possible predicates' in experience. When reason refines a concept in terms of this principle, it can never do so by actually completing the concept in accord with the principle. Rather, the principle serves as a guide for how to go about attempting to complete the concept. Since these concepts cannot actually be completed, the exhaustiveness Kant requires of definition can never be achieved. Therein lies one source of Kant's belief that only exposition and not definition is possible in philosophy. By contrast, Bolzano believes that the content expressed in ordinary language is determinate and richly structured. This structured content can be made explicit through interpretation (*Auslegung*). Once the content given in ordinary language has been clarified by interpretation, the philosopher can determine the component parts of a representation, and in this way we arrive at definition (*NAK*, 34). Bolzano's criticism of Kant for failing to give definitions for particular philosophical concept contains an implicit criticism of Kant's views on ordinary language. Had Kant been able to appreciate that the content given in ordinary language is sufficiently rich to support definition, he might have seen his way to definitions for concepts he otherwise only exposited.

Bolzano's criticism of Kant's expositions involves one further element. Bolzano's criticism is not exhausted by the idea that Kant gives expositions where definitions are possible, or even that Kant misunderstands the basis for indefinability. He is also dissatisfied with the particular quality of Kant's own expositions. Bolzano thinks that, as expositions, they are not effective, as they lack clarity and do not make fully evident to the reader what Kant has in mind. This comes out in complaints about Kant's language and style of exposition, which Bolzano finds overly difficult and tedious.[20] This seems like a different kind of criticism, one that does not appear to engage directly with the content of Kant's thought, but rather focuses on his methods of composition. Here I will show that Bolzano's criticism is not a mere matter of stylistic preference. Rather, it has a theoretical basis in the contrast between Kant's and Bolzano's attitude towards ordinary language.

Bolzano's own expositions (*Verständigungen)* show him willing to rely on a rich store of common knowledge embedded in ordinary language. We see, for instance, an emphasis on common usage: Bolzano freely uses examples taken from non-philosophical contexts as an aid to understanding.[21] By contrast, Kant does not generally trust examples to help with understanding (though, as we have seen, they can be used to test partial definitions). Kant even remarks in the preface to the *Critique of Pure Reason* that he omits examples from his book for fear that they will interfere with the kind of systematic understanding he hopes to elicit in his readers (Axviii-ix)! More generally, Kant's tendency towards difficult formulations and his use of uncommon terminology are the result of a real struggle with the limitations of ordinary language in philosophical discourse. Kant's discomfort with natural language is amusingly (and topically) illustrated by his complaint that for the expressions 'Exposition, Explikation, Deklaration and Definition' (Latinate words), the German language has 'nothing more than the one word 'explanation' *(Erklärung)*' (A730/B758). In the Transcendental Dialectic, Kant voices his discomfort with ordinary language in an aside, writing: 'In the great wealth of our languages, the thinking mind nevertheless often finds itself at a loss for an expression that exactly suits its concept, and lacking this is able to make itself rightly intelligible neither to others nor even to itself' (A312/B368-9). Kant's way around this difficulty is not to attempt to clarify ordinary language, but rather to turn to the 'dead and learned languages', Latin and Greek.[22] He believes this helps because the words of a dead language are less liable to cause confusion and more likely to have a fixed and stable meaning than those of spoken natural language.

20. This aspect of Bolzano's frustration with Kant is most clearly evident in (*NAK*, 39–41). In other works, written by Bolzano himself, he shows more patience with Kant, and is more concerned to demonstrate the falsity of Kant's views than to attack his method of exposition. However, see (*WL*, §79, §305) for passages where Bolzano's patience seems to wear a bit thin.

21. For Bolzano's precise instructions on how examples are to be used see (*WL*, §544–550).

22. An exploration of Kant's choice of classical terminology is contained in (Hinske 1974).

The differing use of ordinary language in Kant's and Bolzano's expositions can be explained by their differing beliefs about ordinary language. For Kant, exposition cannot rely on what is given in ordinary language. What is given in ordinary language is too impoverished and unsystematic. So Kant works uneasily with ordinary language, moving away from it wherever possible. Bolzano's willingness to rely on ordinary examples and common usage in his *Verständigungen* reflects his own different attitudes towards ordinary language. For Bolzano, what is given in ordinary language is richly structured content. This content serves as a common focal point for speakers of a language and as a useful tool for philosophical exposition. Thus, Bolzano's criticism of Kant's methods of exposition is an expression of Bolzano's underlying beliefs about the content of ordinary language.

We should not be surprised to see that Bolzano's criticism of Kant's expository methods has a theoretical underpinning. In Bolzano's work, the methodological and theoretical are not separated. Bolzano's *Wissenschaftslehre*, his most extended compendium of theoretical philosophy, is conceived by him as a manual on how to write textbooks *(Lehrbücher)*.[23] He defines *Wissenschaftslehre* as 'the aggregate of all rules which we must follow when we divide the total domain of truths into individual sciences, and represent them in their respective textbooks,' or again 'the science which teaches us to compose other sciences (actually only their textbooks)' (*WL*, §1).[24] Bolzano's recommendations for the writing of textbooks are given in volume four of the *Wissenschaftslehre*, after the more substantive theoretical work of volumes one through three. Bolzano intended this ordering to reflect how his methodological recommendations are grounded in the tenets of his theoretical philosophy. Though it is Bolzano's theoretical philosophy that ultimately carries the argumentative weight, his comments and thoughts on method, Kant's in particular, are important indicators of how his theoretical views are to be applied, both in the criticism of others, and in the construction of future works.

23. I nevertheless am in agreement with Rolf George's assessment that though Bolzano defines the theory of science in terms of the goal of providing guidelines for composition, we should not see those guidelines as the most important contribution of the *Wissenschaftslehre*. Rather, Bolzano's views concerning compositional style are the primary consequence of his theoretical views, with the theoretical views carrying the real weight. See George's introduction to (Bolzano 1972, xliii–xliv).

24. See also Bolzano's discussion in (*WL*, §392).

4. *Concluding remarks*

Bolzano aims to clarify Kantian philosophical concepts and terminology through a process of rigorous definition. What I hope to have shown is that this ambition is supported by certain beliefs about ordinary language. In contrast with Kant, Bolzano believes ordinary language can express content that is sufficiently rich to support definition. In the previous section, I attempted an explanation of Bolzano's criticism of Kant's views on definition and exposition in philosophy. In the first place, Bolzano criticizes Kant for providing mere exposition in places where rigorous definition is possible. This criticism is supported by Bolzano's axiomatic understanding of science and by the related notion of simplicity. But it was also argued that this criticism is supported by Bolzano's belief that ordinary language expresses structured contents, which can lead to definitions. It is precisely Kant's failure to rely on what is given in ordinary language that is being implicitly criticized by Bolzano. Secondly, Bolzano criticizes Kant's methods of exposition. That is, he thinks that as expositions, they are not effective. This criticism is also grounded in Bolzano's conception of ordinary language. Bolzano readily avails himself of the ordinary meaning of a term and examples taken from non-philosophical contexts, because he sees ordinary language as a rich source of common knowledge. By contrast, Kant is pessimistic about the content given in ordinary language, believing that it leaves speakers 'miles apart with regard to concepts.' As a consequence he prefers to avoid common usage and ordinary examples.

The emphasis on ordinary language in this paper is intended to throw new light on the relationship between Kant and Bolzano. A more traditional contrast between them is in terms of a distinction between realism and idealism, in particular, with respect to Bolzano's realism about the proposition in itself *(Satz an sich)*.[25] This way of understanding the difference between them is tempting given Bolzano's frequent asseverations that the 'source of most errors in logic' (*WL*, §12), and especially the errors of Kant, rests on a failure to distinguish the 'subjective' proposition from the 'proposition in itself'. This paper is not an argument against the realism/idealism analysis, but is only intended to show that there is much more to be explored in the contrast between these two philosophers.

The emphasis on ordinary language is also intended to recast Bolzano's place in the history of analytic philosophy. Traditionally, Bolzano's remark-

25. (Simons 1999, 110–115) emphasizes the significance of Bolzano's semantic Platonism.

able contributions to logic and formal semantics take the forefront in discussions of his place in the history of analytic philosophy (Coffa 1994, 22–40; Dummett 1996, 22–26). His definitions of logical analyticity and logical consequence are clear precursors of modern conceptions and are rightly celebrated. My emphasis on Bolzano's views concerning ordinary language is not intended to minimize these contributions. Rather, I wish to bring attention to an underappreciated aspect of Bolzano's role as a founding figure in the analytic tradition by emphasizing his view that ordinary language expresses content worth analyzing. This enables us to see a route by which the study of language became relevant to analytic philosophy, a route that goes by way of an increasing interest in ordinary, rather than formal language.

Finally, Bolzano's views on ordinary language provide an alternative perspective on the source of his ideals of clarity and explicitness of definition. The traditional account of these ideals rests on Bolzano's foundational work in mathematics. Bolzano can be seen as transferring methods and standards from his work in the foundations of mathematics, over to philosophical work in general. There is, of course, much that is correct about this account. But in this paper, we see that Bolzano's expectations about clarity and explicitness also have a source in his confidence about the content expressed in natural language.

References

Ameriks, Karl 2001: "Kant on Science and Common Knowledge". In his: *Kant and the Sciences.* New York: Oxford University Press, 31–52.

— 2006: "A Common-Sense Kant?". In his: *Kant and the Historical Turn.* Oxford: Oxford University Press, 109–133.

Beaney, Michael 2002: "Kant and Analytic Methodology". *British Journal for the History of Philosophy* 10, 455–456.

Berg, Jan 1962: *Bolzano's Logic.* Stockholm: Almqvist & Wiskell.

Bolzano, Bernard 1969– : *Bolzano-Gesamtausgabe.* Stuttgart-Bad Cannstatt: Frommann-Holzboog.

— 1972: *Theory of Science.* Berkeley and Los Angeles: University of California Press.

— 2004: *On the Mathematical Method and Correspondence with Exner.* New York: Rodopi.

Bolzano, Bernard & Příhonský, Franz 2003: *Neuer Anti-Kant und Atomlehre des seligen Bolzano.* Sankt Augustin: Academia Verlag.

Coffa, Alberto 1994: *The Semantic Tradition from Kant to Carnap.* Cambridge: Cambridge University Press.

Dummett, Michael 1996: *The Origins of Analytical Philosophy.* Cambridge, MA: Harvard University Press.

Hanna, Robert 2001: *Kant and the Foundations of Analytic Philosophy.* Oxford: Oxford University Press.

Hinske, Norbert 1974: "Kants neue Terminologie und ihre Alten Quellen". *Kant Studien* 65, 68–85.

Kant, Immanuel 1902: *Gesammelte Schriften.* Berlin: G. Reimer.

— 1992a: Inaugural Dissertation. In: David Walford (ed.), *Theoretical Philosophy 1755-1770.* Cambridge: Cambridge University Press.

— 1992b: *Lectures on Logic.* Cambridge: Cambridge University Press.

— 1998: *Critique of Pure Reason.* Cambridge: Cambridge University Press.

— 2007: "Anthropology from a Pragmatic Point of View." In: Robert Louden & Günther Zöller (eds.), *Cambridge Edition of the Works of Immanuel Kant: Anthropology, History and Education.* Cambridge: Cambridge University Press, 227–429.

Lapointe, Sandra 2007: "Bolzano's Semantics and his Critique of the Decompositional Conception of Analysis. In: Michael Beaney (ed.), *The Analytic Turn.* London: Routledge, 219–234.

— 2011: *Bolzano's Theoretical Philosophy.* New York: Palgrave Macmillan.

Leibniz, Gottfried Wilhelm 1962: *Gottfried Wilhelm Leibiz: Sämtliche Schriften und Briefe.* Berlin: Akademie Verlag.

Losonsky, Michael 2006: *Linguistic Turns in Modern Philosophy.* Cambridge: Cambridge University Press.

Simons, Peter 1999: "Bolzano, Brentano and Meinong: Three Austrian Realists." In: Anthony O'Hear (ed.), *German Philosophy Since Kant.* Cambridge: Cambridge University Press, 109–136.

Textor, Mark 1997: "Bolzano's Sententialism". In: Wolfgang Künne, Mark Siebel & Mark Textor (eds.), *Bolzano and Analytic Philosophy.* Amsterdam: Rodopi, 181–202.

Grazer Philosophische Studien
85 (2012), 151–185.

WILHELM DILTHEYS EMPIRISCHE PHILOSOPHIE UND DER REZENTE METHODENSTREIT IN DER ANALYTISCHEN PHILOSOPHIE

Christian DAMBÖCK
Universität Wien

Summary

Wilhelm Dilthey's epistemology is an attempt at a more cogent and non-reductionist version of empiricism. According to Dilthey the concepts and ideas of the human mind are empirical objects in their own right. Thus, philosophy as the study of these "facts of consciousness" turns out to be an empirical science; its methods are the methods of psychology, sociology, and history. These views of Dilthey are firstly described in the context of their competitive philosophical programs—"metaphysics", empiricism and (neo)-Kantianism—, then they are compared with recent developments of empirical philosophy, namely van Fraassens neo-empiricism, experimental philosophy, and the sociology of philosophical knowledge.

In diesem Aufsatz soll an die wissenschaftsphilosophischen Positionen Wilhelm Diltheys erinnert werden, einem der wichtigsten Vertreter der deutschen akademischen Philosophie, in der Zeit zwischen dem „Zusammenbruch" des deutschen Idealismus und der „Wiedergeburt" der Philosophie in Deutschland, im Kontext von Neukantianismus und Phänomenologie.[1] Drei Gründe sprechen für eine erneute Bezugnahme auf Dilthey: ein pragmatischer, ein philosophiehistorischer und ein systematischer. Der *pragmatische* Grund liegt darin, dass die für seine Wissenschaftsphilosophie relevanten Arbeiten Diltheys zu einem guten Teil erst in den letzten Jahrzehnten erschienen sind und weitgehend noch einer

1. Der Terminus „Wiedergeburt" wurde in diesem Zusammenhang von Carl Stumpf verwendet in (Stumpf 1910). Zum „Zusammenbruch" des deutschen Idealismus und der Darstellung der Philosophie in Deutschland in der Zeit danach vgl. (Köhnke 1986) und (Schnädelbach 1983).

Aufarbeitung bedürfen.[2] Der *philosophiehistorische* Grund liegt in der erst heute erkannten Bedeutung Diltheys bzw. der Dilthey-Schule für die Entwicklung von Rudolf Carnaps Frühwerk und damit, indirekt, auch für die Entwicklung der analytischen Philosophie und Wissenschaftstheorie insgesamt. (Gabriel 2004; Damböck 2012) Der *systematische* Grund schließlich liegt in der Aktualität Diltheys für eine Reihe von rezenten Debatten im Bereich der Erkenntnistheorie und der Philosophie des Geistes. Dieser Aufsatz konzentriert sich auf den letztgenannten Punkt und versucht einen ganz bestimmten Aspekt von Diltheys Philosophie herauszuarbeiten: seine Auffassung von Philosophie als einer *empirischen Wissenschaft*. Diese Auffassung konvergiert mit rezenten Entwicklungen wie van Fraassens Neo-Empirismus, der experimentellen Philosophie und der Soziologie philosophischen Wissens und ist damit in der Lage neues Licht auf den aktuellen *Methodenstreit* in der analytischen Philosophie zu werfen.

1. *Biographisch-historische Vorbemerkung*

Wilhelm Dilthey[3], geboren 1833 in Wiesbaden, studierte erst Theologie, dann Philosophie und promovierte 1864 in Philosophie, bei Friedrich Trendelenburg in Berlin. Nach Professuren in Basel, Kiel und Breslau übernahm Dilthey 1882 die Philosophie-Professur in Berlin und wurde damit Nachfolger von Hegel, Trendelenburg und (dem kurz davor verstorbenen) Hermann Lotze. Dilthey war in Berlin bis 1905 lehrend tätig und starb 1911 in Seis am Schlern (Südtirol). Das nicht nur auf den ersten Blick an äußeren Ereignissen arme Leben dieses Philosophen war geprägt von der Arbeit an einer Reihe von großangelegten Studien, die alle letztlich unvollendet geblieben sind. Das erste dieser Hauptwerke war eine monumentale Schleiermacher-Biographie, deren erster Band 1870 erschienen ist. Es folgte die „Einleitung in die Geisteswissenschaften", Diltheys philosophisches Hauptwerk, dessen erster Band 1883 erschienen ist und dessen Vollendung ihn bis zu seinem Tod beschäftigt hat, wobei

2. Vgl. vor allem die Bände 18 bis 22 von Wilhelm Diltheys „Gesammelten Schriften" (Dilthey 1914–2006), die zwischen 1977 und 2005 erschienen sind. Die Gesammelten Schriften werden im Folgenden nur durch Angabe des Bandes (römische Ziffer) und der Seitenzahl zitiert.

3. Biographische Details zu Dilthey und weitere Literaturhinweise findet man in (van Kerckhoven, Lessing & Ossenkop 2008). Als Einstieg in die Dilthey-Literatur ist (Lessing 2001) geeignet sowie die Bände des *Dilthey Jahrbuchs*, das im Verlag Vandenhoeck & Ruprecht zwischen 1983 und 2000 erschienen ist.

er nach 1895 die Arbeit auf ein drittes Hauptwerk verlagert hat, das 1910 (als Fragment) unter dem Titel „Der Aufbau der geschichtlichen Welt in den Geisteswissenschaften" erschienen ist. Der hauptsächliche Grund für den Abbruch der Arbeit an der „Einleitung in die Geisteswissenschaften" scheint darin gelegen zu haben, dass Dilthey mit seinem philosophischen Programm in den neunziger Jahren des neunzehnten Jahrhunderts zwischen die Fronten geraten ist. Auf der einen Seite wurde seine Arbeit von Psychologen wie Hermann Ebbinghaus gewissermaßen als Einmischung in die inneren Angelegenheiten der Psychologie kritisiert; auf der anderen Seite zog sich Dilthey die Kritik von Neukantianern und Phänomenologen zu, die ihn als „Relativisten" und „Psychologisten" identifizierten.[4] Diltheys Spätwerk war dann von dem Bestreben bestimmt, die Bruchlinien zum Neukantianismus und zur Phänomenologie zu kitten. Für die Diskussion der Frage des Verhältnisses Diltheys zur Phänomenologie und zur Philosophie Heideggers ist eine Analyse seines Spätwerkes somit von großem Interesse. Hier soll jedoch eine andere Seite von Diltheys philosophischem Werk thematisiert werden, die sich nicht durch die „Aufwärtskompatibilität" mit den späteren Entwicklungen in der deutschen Philosophie auszeichnet, sondern wo Dilthey als frühes Beispiel einer *wissenschaftlichen Auffassung* von Philosophie gesehen werden kann, der zufolge Philosophie nur im unmittelbaren Kontakt mit den Einzelwissenschaften funktioniert. Diese Philosophieauffassung, die ihre volle Entfaltung erst in der analytischen Philosophie des zwanzigsten Jahrhunderts gefunden hat, war durchaus charakteristisch für Teile der deutschen akademischen Philosophie zwischen 1830 und 1900. Vielfach wurde in dieser Zeit ein Naheverhältnis zur empiristischen und positivistischen Tradition in Frankreich und England aufgebaut und dennoch versucht, *im Rahmen dieser Tradition,* die im Kern als gescheitert betrachteten Projekte des Kantianismus und des deutschen Idealismus in irgendeiner Form zu retten. Genauer können die Motive für diesen „deutschen Empirismus" etwa wie folgt charakterisiert werden:

Eine Anknüpfung an die philosophischen Traditionen *vor* Kant und dem deutschen Idealismus, wie sie von Bolzano, Brentano oder Lotze vorangetrieben wurde, ist grundsätzlich ausgeschlossen, da Kants „koper-

4. Zur Ebbinghaus-Kontroverse vgl. (Ebbinghaus 1896; Lessing 1985) sowie (Rodi & Lessing 1984, 10–13). Zu dem den Hintergrund dieser Kontroverse bildenden „Psychologismusstreit" vgl. (Kusch 1995). Vgl. außerdem die Dilthey-Kritik Rickerts (Rodi & Lessing 1984, 95–100) sowie die Materialien zur Diskussion zwischen Dilthey und Husserl (Rodi & Lessing 1984, 103–175).

nikanische Wendung“ des Umgangs mit der Vernunft zu der Einsicht geführt hat, dass eine Erschließung abstrakter philosophischer Kategorien nur aus einer Analyse des Denkens selber heraus erfolgen kann. Diese Analyse kann jedoch weder im Kontext einer „transzendentalen“ noch einer „reinen Logik“ im Sinne Kants bzw. Hegels erfolgen, also nicht auf einer apriorischen Ebene. Somit müssen es die empirischen Wissenschaften sein, die uns die Erschließung dieser „geistigen Gegenstände“ ermöglichen, allerdings nicht im Rahmen eines Naturalismus, der geistige Gegenstände auf physikalische *reduziert*, sondern nur in einer um die geistigen Gegenstände *erweiterten* Welt der empirischen Wissenschaften. Wir gelangen so *durch ein konsequentes Weiterdenken*, nicht durch eine einfache Abkehr von der idealistischen Tradition, zu einer neuen Form von Empirismus, die sich dadurch auszeichnet, dass die abstrakten Objekte der Wissenschaften (die „geistigen Gegenstände“) nicht auf physikalische Beschreibungen reduziert werden, sondern einen *eigenständigen* Bereich für empirische Forschung, in Psychologie, Soziologie und den Geschichtswissenschaften liefern.

Die so skizzierte Denkweise findet man bei Friedrich Trendelenburg, dem Nachfolger Hegels in Berlin, und bei einer Reihe von Schülern dieses äußerst einflussreichen Philosophen, von denen vor allem Friedrich Ueberweg, Moritz Lazarus und Wilhelm Dilthey hervorzuheben sind.[5] Im Folgenden wird diese Denkweise exemplarisch, anhand der pointierten Formulierungen in Diltheys frühen und mittleren Schriften zur Erkenntnistheorie illustriert. Insofern fungiert Dilthey hier also nur als *ein Beispiel* für einen bestimmten *Denkstil*, der in der deutschen Philosophie des neunzehnten Jahrhunderts weit verbreitet gewesen ist.

2. *Eine konsequentere Form von Empirismus*

Ein wichtiger Ausgangspunkt von Diltheys frühen und mittleren Arbeiten zur Epistemologie[6] sind die in den Jahrzehnten nach 1830 auch in

5. Weniger in dieses Bild passen würde ein weiterer Schüler Trendelenburgs: Franz Brentano, in dessen „empirischer Psychologie“ die anti-metaphysischen Motive fehlen, die für den „deutschen Empirismus“ bei Trendelenburg, Dilthey, Ueberweg und Lazarus kennzeichnend sind.

6. Relevant sind hier vor allem die Texte aus (I; V, 31–316; XVIII; XIX; XX, A–C; XXI und XXII), wobei die Texte aus den frühen neunziger Jahren, also (V, 90–316; XIX, C–D; XXI, C und XXII, C–D) bereits teilweise den Übergang zur Spätphilosophie erkennen lassen, etwa anhand der stärkeren Abgrenzung von den Naturwissenschaften, die man im Psychologie-Aufsatz (V, 139–240) findet.

Deutschland sehr einflussreichen Philosophien von Auguste Comte und John Stuart Mill, wobei vor allem Mills „Logik" für Dilthey eine wichtige Rolle gespielt zu haben scheint. Dilthey übernimmt von Mill die in der Unterscheidung zwischen Sinneswahrnehmungen und anderen mentalen Zuständen versteckte *hierarchische* Konzeption (Mill 1976 [1843], 849ff.). Es gibt demnach eine Menge von psychologischen Grundzuständen, die sich als Sinneswahrnehmungen identifizieren lassen und daher auf einer rein physiologischen Ebene zu rekonstruieren sind. Diese „einfachen Bewußtseinsphänomene" des „niederen Seelenlebens" (XXII, 76ff.) können, ganz im Sinne Mills, von der Physiologie beschrieben werden. Aber das was eine Psychologie beschreiben muss, die als Grundlage einer Rekonstruktion des menschlichen Geistes dienen soll, sind jene „höheren Bewußtseinsphänomene", die losgelöst von unmittelbarer sinnlicher Wahrnehmung existieren. Wie Mill so verteidigt auch Dilthey die Auffassung, dass wir *für diese* höheren Erscheinungen des Seelenlebens eine eigene psychologische Methode benötigen. Die sogenannte *deskriptive Psychologie*, die für Dilthey diese Aufgabe übernehmen soll, ist also keine Psychologie der Sinneswahrnehmungen (die er als eine im Kern physiologische Disziplin identifiziert), sondern eine Psychologie des „entwickelten Seelenlebens" (V, 169). Und hier liegt bereits der entscheidende Punkt wo Diltheys Psychologie von den Konzeptionen der empiristischen Tradition im Allgemeinen und von der Auffassung Mills im Besonderen abweicht:

> [Die] höheren Bewußtseinsphänomene [...] gehen ohne Zweifel aus den niederen hervor. Die niederen bilden ihre Grundlage. Aber sie sind durchaus nicht bloß zusammengesetzt von Verbindungen, die aus den elementaren gänzlich abgeleitet werden können. Die Ausdrücke Entwicklung, Evolution, Entfaltung sprechen zutreffend die Art von Kausalität aus, welche hier waltet: das Urteil entsteht ja zweifellos, indem die elementaren Operationen eines distinkten Willensaktes getrennte Bestandteile, Allgemeinvorstellungen in sich aufnehmen. (XXII, 12)

Dilthey behauptet hier also eine völlig eigenständige Struktur der Erscheinungen des „entwickelten Bewusstseins", die man weder in einer rein physiologisch-neuronalen Sprache erfassen kann noch in einer solchen Sprache, die diese Ideen des „entwickelten Bewusstseins" auf vergangene „einfache" Wahrnehmungsereignisse reduziert wie dies in der Assoziationspsychologie von James Mill und John Stuart Mill geschieht. Diese Ideen haben für Dilthey einen (zumindest *epistemisch*) völlig eigenständigen

Charakter und sind den physischen Erscheinungen schlicht *unvergleichbar* (XXII, 150f.). Nur im Fall von elementaren Wahrnehmungen und Vorstellungen liegt die *Ähnlichkeit* zwischen Idee und raumzeitlicher Erscheinung, auf der empiristische Psychologien beruhen, vor. Aber bei komplexeren Begriffen, die nicht bloß *Variationen* sinnlich verursachter Ideen darstellen, versagt, so Dilthey, der klassisch-empiristische Standpunkt.

Trotz der in diesem Punkt grundlegend anderen Positionierung Diltheys ist die Nähe seines Programms zum Empirismus ausgesprochen groß. Zwar folgt Dilthey insofern Kant als er den naiven Empirismus Humes verwirft, der die Seele als eine Art Spiegel der Natur hinstellt. Wie für Kant ist auch für Dilthey klar, dass wir es keinesfalls mit einer fertigen Welt zu tun haben, zu der sich die von uns gebildeten Vorstellungen und Ideen etwa wie Zeichnungen oder Fotografien verhalten. „Es muß also etwas sein, was selbst diese Reproduktion der Erscheinungen möglich macht" (Kant 1998, A102), würde Dilthey mit Kant sagen. Diese Grundlage der Erscheinungen bzw. ihrer Reproduktion ist insofern „der Grund a priori einer notwendigen synthetischen Einheit derselben" (ebd.) als wir irgendwelche Begriffe oder Ideen benötigen, um überhaupt in der Lage zu sein, die Wirklichkeit *festzuhalten*. Damit wir überhaupt sehen können und Vorstellungen herausbilden können brauchen wir zunächst für dieses Sehen und diese Vorstellungen eine *konzeptuelle Grundlage*. Aber das Apriori dieser Annahme geht für Dilthey nicht über diese triviale Notwendigkeit hinaus. Es ist für Dilthey, so könnte man sagen, ein analytisches aber *kein* synthetisches Apriori (es ist nur ein Apriori des Prinzips). Das jedoch an diesen *Voraussetzungen der Erfahrung* was tatsächlich synthetisch ist – also das Festmachen dieser Voraussetzungen in ihrem So-und-so-Sein – ist, im Gegensatz zu Kants Annahme, für Dilthey keineswegs a priori. Vielmehr sind diese intellektuellen Grundlagen unserer Erfahrung, genauso wie jeder andere Teil unserer Psyche, etwas, das empirisch entsteht, und das daher prinzipiell stets die Möglichkeit des Anders-sein-Könnens in sich trägt. Dass wir unserer Erfahrung diese und jene Konzepte und Ideen zugrunde legen muss insgesamt von der Erfahrung abhängig sein, diese Konzepte und Ideen müssen *ein Produkt* dieser unserer Erfahrungen selbst sein.

Ein angeblich charakteristisches Merkmal der deutschen Philosophie, das diese in der breiten Wahrnehmung von der angloamerikanischen Tradition abhebt, ist die Ansicht, dass Philosophie bzw. Wissenschaft eine Technik bereitstellen muss, mit der wir aus den psychischen Phänomenen deren wahren Kern – ihr „transzendentales Fundament", ihr „Wesen", die in ihnen versteckte „Idee“ – generieren können. Diese Ansicht findet man bei Kant, natürlich auch bei den deutschen Idealisten, und man findet sie erneut bei den Neukantianern, bei Brentano, Husserl, Heidegger usw. *Der Apriorismus der Philosophie* besteht, dieser Ansicht zufolge, genau darin, dass es nicht auf die „kontingenten“ psychischen Inhalte ankommt, sondern auf einen hinter diesen versteckten „Gehalt“, den es frei zu legen gilt. Der Positivismus und Empirismus behauptet, im Gegensatz dazu, dass ein derartiger hinter den psychischen Inhalten versteckter harter notwendiger Kern ein Mythos ist.

Hier zeigt sich aber, dass das erwähnte Bild von der deutschen Philosophie, zumindest für bestimmte Entwicklungen im neunzehnten Jahrhundert, unzutreffend ist: Dilthey folgt dem Empirismus in dem erwähnten Punkt nämlich zu hundert Prozent und ist insofern (wie viele andere akademische Philosophen in Deutschland, in der Zeit zwischen 1830 und 1900) bemerkenswert „undeutsch“. Die von Dilthey entworfene Programmatik eines *konsequenteren Empirismus* sei im Folgenden anhand von einigen Zitaten aus drei wichtigen Arbeiten aus Diltheys frühem und mittlerem Werk eingehender illustriert: (1) dem ersten (und einzigen zu Lebzeiten Diltheys publizierten) Band der „Einleitung in die Geisteswissenschaften“ von 1883, (2) den Vorarbeiten zum geplanten zweiten Band dieses Hauptwerkes, die in der Zeit vor und nach 1883 entstanden sind, (3) den „erkenntnistheoretischen Fragmenten“ von 1874/79.[7]

(1) Was Dilthey solchen Philosophen wie Comte und Mill vorwirft, ist *nicht*, dass sie in ihrer Empirisierung von allem und jedem die Welt ihrer apriorischen Grundlage berauben, nicht, dass sie dem Reich der Ideen einen empirischen Status zuerkennen, sondern vielmehr, dass diese ihre Empirisierung des Geistes *nicht konsequent genug* ist:

> [E]rfüllt von der unter den Engländern und Franzosen heute herrschenden verwegenen wissenschaftlichen Baulust, ohne das intime Gefühl der geschichtlichen Wirklichkeit [...] haben diese Positivisten gerade denjenigen Ausgangspunkt für ihre Arbeiten nicht gefunden, welcher ihrem Prinzip der Verknüpfung der Einzelwissenschaften entsprochen hätte. [...] Sie haben einen Notbau errichtet, der nicht haltbarer ist, als die verwegenen Spekulationen eines Schelling und Oken über die Natur. [...]

7. Zweifellos haben sich Diltheys Positionen in den gut zwei Jahrzehnten, aus denen die im Folgenden zitierten Schriften stammen, weiterentwickelt. Die Zitate scheinen jedoch zu belegen, dass die Eckpunkte seines „konsequenteren Empirismus“ in dieser Zeit ziemlich stabil erhalten geblieben sind.

> Die Geisteswissenschaften bilden nicht ein Ganzes von einer logischen Konstitution, welche der Gliederung des Naturerkennens analog wäre; ihr Zusammenhang hat sich anders entwickelt und muß wie er geschichtlich gewachsen ist nunmehr betrachtet werden.“ (I, 23f.)

Das Problem ist, für Dilthey, dass „die Positivisten“ versuchen, den „Geisteswissenschaften“ ein naturwissenschaftliches Fundament zu geben, damit aber dem empirischen Charakter der „geistigen Inhalte“, von denen dort die Rede ist, nicht gerecht werden. Dadurch, dass sie *die Historizität* aller geistigen Inhalte ignorieren und stattdessen versuchen, diese anhand von Naturgesetzen zu rekonstruieren, liefern sie für Dilthey letztlich nichts weiter als einen Ableger der „metaphysischen“ Philosophietradition (auch wenn sie ihren Konzepten einen aposteriorischen Anstrich verleihen).

Sowohl die deutschen Idealisten (inklusive Kant) *als auch* die Empiristen (Hume, Comte, Mill, Buckle etc.) behaupten, so ein zentrales Argument Diltheys, dass geistige Inhalte in irgendeinem Sinn *feststehende* Bilder sind, sei es in der Gestalt von ewigen, notwendigen bzw. transzendentalen Ideen (die deutsche Variante) oder im Sinne von aus der empirischen Wirklichkeit, als quasi-fotografische Strukturen, hervorgegangene ideale Abbilder (die englisch-französische Version). *Beide* Varianten sind für Dilthey zu statisch und „blutleer“:

> [D]ie bisherige Erkenntnistheorie, die empiristische wie die Kants [hat] die Erfahrung und die Erkenntnis aus einem dem bloßen Vorstellen angehörigen Tatbestand erklärt. In den Adern des erkennenden Subjekts, das Locke, Hume und Kant konstruieren, rinnt nicht wirkliches Blut, sondern der verdünnte Saft von Vernunft als bloßer Denktätigkeit. (I, XVIII)

Geistige Dinge sind nichts Feststehendes, sondern sie *haben eine Geschichte* und können letztlich immer nur im und aus dem historischen Kontext heraus verstanden werden. „Die Methode des folgenden Versuchs ist daher diese“, schreibt Dilthey im Anschluss an obige Passage aus der Vorrede zur „Einleitung in die Geisteswissenschaften“,

> jeden Bestandteil des gegenwärtigen abstrakten, wissenschaftlichen Denkens halte ich an die ganze Menschennatur, wie Erfahrung, Studium der Sprache und der Geschichte sie erweisen und suche ihren Zusammenhang. Und so ergibt sich: die wichtigsten Bestandteile unseres Bildes und unserer Erkenntnis der Wirklichkeit, wie eben persönliche Lebenseinheit, Außenwelt, Individuen außer uns, ihr Leben in der Zeit und ihre Wechselwirkung, sie alle können

> aus dieser ganzen Menschennatur erklärt werden, deren realer Lebensprozeß am Wollen, Fühlen und Vorstellen nur seine verschiedenen Seiten hat. Nicht die Annahme eines starren a priori unseres Erkenntnisvermögens, sondern allein Entwicklungsgeschichte, welche von der Totalität unseres Wesens ausgeht, kann die Fragen beantworten, die wir alle an die Philosophie zu richten haben. (ebd.)

An die Stelle von Kants Kritik der *reinen* und also *ahistorischen* Vernunft tritt bei Dilthey die Idee einer „Kritik der historischen Vernunft" (I, 116). Das bedeutet weder, wie von Rudolf Makkreel behauptet, eine direkte *Fortsetzung* von Kants kritischem Projekt[8] noch bedeutet es eine vollständige *Abkehr* davon. Dilthey ist, so könnte man sagen, *Kantianer im weitesten Sinn*, weil sein theoretischer Ausgangspunkt zweifellos in Kants kopernikanischer Wende zu verorten ist, also in der Hinwendung der philosophischen Analyse auf das erkennende Subjekt. Dilthey ist insofern Kantianer als er die abstrakten geistigen Inhalte des menschlichen Denkens *nicht*, wie im Empirismus, als etwas zu identifizieren versucht, das durch bloßen Verweis auf die Außenwelt (bzw. als direkte kausale Folge von Ereignissen der Außenwelt) zustande kommt. Geistige Inhalte sind für Dilthey also keineswegs *in einem klassischen Sinn* (wie im Empirismus und Positivismus) *empirisch*. Das Empirische der geistigen Inhalte liegt für Dilthey bloß darin, dass wir selbst, in unserem Denken, diese Inhalte *erfahren* (und dabei gleichzeitig *produzieren*). Diese Inhalte *sind* letztlich nichts anderes als eben diese Denkerfahrungen und sie sind *deshalb* zwangsläufig historisch relativ. Und weil weder die deutsche idealistische noch die französische und britische empiristische Philosophietradition diese Historizität akzeptieren, sind sie beide, im Sinne Diltheys, nicht empirisch genug eingestellt.

(2) Diltheys diesbezügliches Schlagwort lautet: „Empirie, nicht Empirismus" (XIX, 17ff.). Es geht in der Philosophie (und in weiterer Folge in den Geisteswissenschaften) nicht darum, einen Kern von transzendentalen oder naturgegebenen Ideen herauszuschälen; vielmehr geht es darum, geistige Inhalte *aus ihrem Kontext heraus* zu erschließen, und anhand ihres Kontextes zu diskutieren, und schließlich, dort wo es die Diskussion angezeigt sein lässt, zu modifizieren. Das bedeutet eine „kritische Umänderung"

8. Vgl. (Makkreel 1991 [1975], 382f.): „Die *Kritik der reinen Vernunft* wird von Dilthey nur so weit in Frage gestellt, dass Raum für seine Theorie der Geisteswissenschaften geschaffen werden kann. [...] Was seine allgemeineren Erwägungen zu Kants Erkenntnistheorie anbetrifft, führen sie im Grunde den von Kant selbst in seiner *Kritik der Urteilskraft* begonnenen Prozeß des Überdenkens fort."

von Kants „Theorie von den Bedingungen des Bewußtseins“, im Sinne einer Wendung ins Historische.

> Durch diese [kritische Umänderung Kants] wird die Geschichte der Wissenschaft erst verständlich, und an die Stelle des mechanischen Geklappers von Induktion und Deduktion bei Comte, Mill, Spencer tritt die Einsicht in die souveräne Natur der menschlichen Intelligenz, kraft derer sie vermöge der in ihr gelegenen Bedingungen des Bewußtseins sich der Gegenstände bemächtigt, sie konstruiert; zugleich aber auch, und hierdurch unterscheide ich mich hauptsächlich von Kant, an der Vertiefung der Dinge ihre eigenen Voraussetzungen umgestaltet. Das a priori Kants ist starr und tot; aber die wirklichen Bedingungen des Bewußtseins und seine Voraussetzungen, wie ich sie begreife, sind lebendiger geschichtlicher Prozeß, sind Entwicklung, sie haben ihre Geschichte, und der Verlauf dieser Geschichte ist ihre Anpassung an die immer genauer induktiv erkannte Mannigfaltigkeit der Empfindungsinhalte. (XIX, 44)

Philosophie verhält sich so zu den Einzelwissenschaften in geradezu entgegengesetzter Weise wie ihr dies in der „metaphysischen“ Tradition zugeschrieben worden war. Sie ist nicht *apriorischer,* sondern *empirischer* als die Einzelwissenschaften, weil der gesamte Komplex einzelwissenschaftlicher Aussagen und Analysen zu den empirischen Befunden über die raumzeitliche Wirklichkeit hinzutritt und dadurch das Maß an Unwägbarkeiten und Freiheitsgraden in der philosophischen Analyse nicht sinkt, sondern weiter steigt. – Die Naturwissenschaften basieren, für Dilthey, auf apriorischen Annahmen (Konventionen, Axiome), plus einer nicht weiter zu hinterfragenden empirischen Basis von physischen Tatsachen. Diese Wissenschaften sind also weitgehend *objektiv* konzipiert und sie haben eine *breite* apriorische Basis. Die Philosophie hingegen, als „Wissenschaft von den Tatsachen des Bewusstseins“, in der diese apriorischen Grundannahmen und Konventionen *ins Empirische gewendet werden,* weist kaum eine Spur dieser in den Einzelwissenschaften verbreiteten Apriorizität und Objektivität auf:

> Was also vom Standpunkt der einzelnen Wissenschaften als eine Wahrheit letzter Instanz, als ein Axiom erscheint, das ist, als mit Evidenz ausgestattet, für diese umfassende Erfahrungswissenschaft [nämlich die Philosophie, C. D.] eine Tatsche des Bewußtseins, welche in den Zusammenhang der Zergliederung des Bewußtseins tritt und vielleicht in diesem Zusammenhang Aufklärung empfängt, ja vielleicht in ihm noch eine weitere psychologische Zerlegung erfährt. Betrachte ich ein Axiom, sofern es anderen Sätzen, die von

ihm abgeleitet sind, seine Evidenz mitteilt [...], dann bildet dieses Axiom mir eine Wahrheit letzter Instanz; dies ist der Standpunkt auf welchem die Einzelwissenschaften ihre Axiome zugrunde legen und aus ihnen ihre Systeme entwickeln. Betrachte ich aber diese Evidenz in demjenigen Zusammenhang, in welchem sie ursprünglich gegeben sind, in dem Zusammenhang der Tatsachen des Bewußtseins, dann trete ich auf den Standpunkt der allgemeinen Erfahrungswissenschaft, deren Objekt der Zusammenhang der Tatsachen des Bewußtseins ist. (XIX, 82)

Philosophie wird so zu einer „positiven Wissenschaft" und Kants „kopernikanische Wende" vollendet sich auf empirischem Weg:

> Hatte Kant ein Verfahren bloßer Analysis des Subjektes und seines wissenschaftlichen Denkens eingeschlagen, so war damit der Gedanke, das Problem definitiv lösen zu wollen, verbunden, und die philosophische Operation war von den positiven Wissenschaften des Geistes abgesondert.
>
> Erkennt man den Zusammenhang dieser Probleme mit denen der vergleichenden Grammatik, der Mythologie und der Kulturgeschichte; alsdann unterscheidet sich das Geschäft des Philosophen von dem der positiven historischen Wissenschaft weder in bezug auf die Methode noch auf die Hilfsmittel, und in bezug auf den Gegenstand auch keineswegs durchweg. Die Scheidewand zwischen Philosophie und positiver Wissenschaft sinkt hin so gut als sie in bezug auf die Prinzipien der Naturwissenschaft sich nicht hat halten lassen. Sie liegt in der notwendigen Beschränktheit der menschlichen Natur, welche eine Klasse von Hilfsmitteln und Problemen mehr bevorzugt, sie liegt aber keineswegs mehr in einer Verschiedenheit des ganzen Gegenstandes oder der Methode und der Hilfsmittel.
>
> Damit ist die Lösung des Problems dem grenzenlosen Fortschritt des positiven Wissens anheimgegeben – die Grenzen sind nur die unserer Epoche, in denen wir uns halten; es gibt keine absolute Philosophie. (XIX, 89f.)

(3) *Konsequenter Empirismus* bedeutet für Dilthey zunächst, dass die (bei Kant und Locke zu findende) Unterscheidung von äußerer und innerer Erfahrung aufgehoben wird: „Es gibt nur eine Erfahrung." (XVIII, 194) Diese umfasst neben Wahrnehmungen und Vorstellungen auch abstrakte Begriffe, Werte und ästhetische Gefühle.

> Indem [Kant] alle Erfahrung auf die äußere einschränkte, bildete sich in seinem Denken jener Gegensatz zwischen den Wahrheiten der Erfahrung und denen a priori. Indem sonach alle Erfahrung für ihn als äußere auf Induktion gegründet erschien, entstand der Satz, daß Erfahrungswahrheiten niemals Notwendigkeit und Allgemeinheit haben. (XVIII, 193)

Daraus ergab sich, bei Kant, nach Diltheys Auffassung, der (falsche) Schluss, dass alle Feststellungen über der äußeren Erfahrung zugrundeliegende abstrakte Konzepte *synthetische Urteile a priori* sein müssen. Da es aber nur eine Erfahrung gibt und auch diese Feststellungen in dieser enthalten sind, bloß am oberen Ende der Hierarchie, an deren Basis die sogenannten äußeren Erfahrungen angesiedelt sind, fallen die synthetischen Urteile a priori flach. Die *eine* Erfahrung wird vielmehr bloß „in einer doppelten Richtung verwertet [...], und so entsteht die Unterscheidung äußerer und innerer Erfahrung." (XVIII, 194)

Das heißt, in *jeder* Erfahrung stecken naturgemäß die sogenannte äußere und die sogenannte innere Seite, weil jede Erfahrung abstrakte Anteile hat und solche, die sich aus dem Strom des Bewusstseins heraus als deren augenblicklicher Gehalt ergeben. Dieses „Theorem von der völligen Positivität der Welt" (XVIII, 199) grenzt sich insofern sowohl von Empirismus und Positivismus als auch von Kant und vom deutschen Idealismus ab, als in *keiner* dieser Spielarten der Philosophie die Positivität der sogenannten inneren Erfahrung erkannt, sondern entweder wegreduziert oder apriorisiert wird:

> *Letzte Generalisation*: aus der positiv gearteten Welt und dem sie bedingenden Gedankenlauf kann nirgend reine Idealität entspringen. Hier liegt der eigentümliche salto mortale der bisherigen Philosophie. Auch unsere höchsten Ideen sind Generalisationen eines Positiven. [...] Wir müssen die willkürlichen Abstraktionen ganz loswerden. (XVIII, 199)

Der „eigentümliche salto mortale der bisherigen Philosophie" ist die Konstruktion einer hinter den Dingen verborgenen Idealität, „eine ebenso billige als sonderbare Umkehrung, die Wirkung rückwärts als Zwecke, d. h. als Ursachen zu fassen." (XVIII, 200) In Wahrheit sind die logischen Formen keine naturgegebenen ewigen Vorgaben, die Logik ist nicht analytisch oder transzendental, sondern die logischen Formen sind das Produkt „bloßer Anpassung" (XVIII, 199). Die logischen Formen, die Begriffe, die Werte, kurz: die „geistigen Gegenstände" sind empirisch, aber weder in dem naiv-platonischen Sinn der empirischen Wahrnehmung von Ideen noch im Sinne einer materialistischen Theorie, die diese Dinge letztlich physiologisch zu analysieren versucht. Geistige Gegenstände kommen „durch Anpassung" zustande, eine Anpassung, die bei Dilthey jedoch nicht *evolutionsbiologisch* gedacht ist, sondern *soziologisch-psychologisch-historisch*, als das Produkt unserer (über Jahrtausende gehenden) *Erfahrungen des Denkens* bzw. der Geschichte unserer Sprache und Kultur. Insofern könnte man Diltheys Position also als einen *nicht-materialistischen Naturalismus* charakterisieren: Diltheys Epistemologie ist eine „naturalisierte Epistemologie" der zweiten Art, nicht als Naturalisierung im Rahmen von Biologie und Physik, sondern als Naturalisierung *im Rahmen der Geschichte*.[9]

3. *Zwischenbemerkung: Dilthey und die Neukantianer*

Der nicht-materialistische Naturalismus Diltheys kann, wie oben mehrfach angedeutet, als (allerdings radikale) *Weiterentwicklung* Kants gesehen werden. Weil Dilthey der Auffassung ist, dass die abstrakten Kategorien unseres Denkens genuine *Produkte* desselben sind, ist seine Philosophie eindeutig im Rahmen der „kopernikanischen Wende" Kants angesiedelt, also der Auffassung, dass „sich der Gegenstand (als Objekt der Sinne) nach der Beschaffenheit unseres Anschauungsvermögens [richtet]" (Kant 1998, BXVII). Diese erkenntnistheoretische Motivationsumkehr wird von Dilthey, ganz mit Kant, vollzogen. Umgekehrt aber zieht Dilthey aus dem Ansatz der kopernikanischen Ausgangsposition Kants völlig andere Schlüsse als dieser. Während für Kant die kopernikanische Wendung der Vernunftkritik direkt zu der Auffassung führt, dass wir in der Lage sein müssen, alle konzeptuellen Voraussetzungen der Anschauung *a priori* zu erkennen (und zwar in der Gestalt von synthetischen Urteilen a priori), gelangt Dilthey, in der oben skizzierten Weise, zu dem Schluss, dass gerade dieser Bereich von konzeptuellen Voraussetzungen der Anschauung etwas ist, dem wir uns nur auf empirischem Weg nähern können. Damit ist das Verhältnis Diltheys zu den großen Schulen des Neukantianismus – die Marburger Schule um Hermann Cohen, Paul Natorp und Ernst Cassirer und die Südwestdeutsche Schule um Wilhelm Windelband und Heinrich Rickert – bereits klar definiert: Konsens auf der Ebene des kopernikanischen Standpunktes, aber Dissens in der Einschätzung der Art und Weise wie die konzeptuellen Voraussetzungen der Anschauung der Erkenntnis zugänglich gemacht werden können.

Beide großen Schulen des Neukantianismus vertreten die Auffassung, dass die Erschließung des transzendentalen Hintergrundes der Erfahrung grundsätzlich nur a priori erfolgen kann. Zwar räumen beide Schulen ein,

---

9. Naturalisierte Epistemologie *der ersten Art* wäre dann natürlich die materialistische Auffassung, dass Epistemologie im Rahmen der Naturwissenschaften betrieben werden könnte: „Epistemology in its new setting [...] is contained in natural science, as a chapter of psychology." (Quine 1969, 83)

dass die Begriffe, die den transzendentalen Hintergrund der Erfahrung bilden (und damit insbesondere auch die abstrakten Grundbegriffe der empirischen Wissenschaften), *historisch wandelbar* sind, aber das Transzendentale wird dadurch nicht, wie bei Dilthey, insgesamt zu einer empirischen Angelegenheit, sondern es bleibt, wie bei Kant, a priori zugänglich. Für die (auf dem werttheoretischen Platonismus Lotzes aufbauende) Südwestdeutsche Schule liegt zwar eine absolut feststehende objektive Transzendenz der Dinge an sich vor, in der Gestalt transzendenter Normen und Wahrheiten. Diese objektiv feststehende Welt der Dinge an sich ist aber in dem der empirischen Wissenschaft zum Zeitpunkt t vorliegenden Erfahrungsbereich stets nur fragmentarisch zugänglich. Die Begriffe der Wissenschaft zum Zeitpunkt t können nur die objektiven Gesetzmäßigkeiten, die sich im Rahmen dieses begrenzten Erfahrungsbereichs erschließen lassen, ausdrücken. Der transzendententale Hintergrund der Erfahrung ist insofern historisch variabel. Zu einem bestimmten Zeitpunkt t aber sind die aus dem jeweiligen Erfahrungsbereich hervorgehenden Begriffe und logischen Urteile der Wissenschaft absolut und a priori gültig und können auch nur so und nicht anders gebildet werden, vorausgesetzt, die Wissenschaftler machen keine logischen Fehler. Die Begrenztheit der Erfahrung bewirkt also, dass wir uns, in der Bildung naturwissenschaftlicher Begriffe, *von der objektiven Wirklichkeit* der Dinge an sich *wegbewegen. Die Kluft* zwischen den realen Begriffen und den Dingen an sich kann dann nur durch die historische Betrachtung ausgeglichen werden, indem diese das Apriori in seiner nur relativen Gültigkeit identifizieren und lokalisieren hilft. (Rickert 1904, 186–228)

In der Marburger Schule existiert, im Gegensatz dazu, die für die südwestdeutsche Schule grundlegende objektive Transzendenz absoluter Normen und Wahrheiten (und damit der Dinge an sich) nicht. Die naturwissenschaftliche Begriffsbildung, die, für Rickert, in der mathematischen Idealisierung von den objektiven Werten *wegführt* und nur in der historischen Betrachtung wieder mit diesen in Verbindung gebracht werden kann, stellt für Cassirer vielmehr eine *Annäherung* an die Wirklichkeit dar, die wir, ohne diese Begriffe, nie realisieren könnten. Die (mit den Mitteln der Infinitesimalrechnung gebildeten) „Grenzbegriffe" der Naturwissenschaften führen uns also nicht, wie bei Rickert, von der empirischen Wirklichkeit weg, sondern, im Gegenteil, sie *erhöhen* die Objektivität der Erklärung, „[streben] in dieser scheinbaren Abkehr von der Wirklichkeit der Dinge [...] ihr vielmehr auf einem neuen Wege zu" (Cassirer 1994 [1910], 304). Dadurch, dass die Grenzbegriffe das Regelhafte in den Ereignissen, mittels

Induktion, in idealisierter Weise herausheben, umfassen sie einen wesentlich größeren Bereich der Wirklichkeit als ihre singulären historischen Ausgangspunkte. Die Grenzbegriffe *sind* also stets bereits objektiv, in dem jeweils höchstmöglichen Sinn. Bloß ist es so, dass die Erweiterung der empirischen Erkenntnis stets dazu führt, dass sich *bestimmte Teilaspekte* der zuvor gebildeten Grenzbegriffe als nicht-objektiv (also subjektiv) erweisen. Somit gelten auch für die Marburger Schule die naturwissenschaftlichen Begriffe nur relativ zu dem jeweiligen empirischen Erkenntnisstand der Wissenschaften, zu den jeweils vorliegenden empirischen Daten. *Innerhalb dieses Rahmens aber* erfolgt die (natur-)wissenschaftliche Begriffsbildung, für beide Schulen des Neukantianismus, nach streng logischen, a priori erfassbaren Gesetzmäßigkeiten.

Für Dilthey jedoch ist der transzendentale Hintergrund der Erfahrung *nicht nur deshalb* historisch wandelbar, weil sich die empirischen Daten mit der Zeit verändern. Er lehnt grundsätzlich die Möglichkeit ab, dass unsere abstrakten Begriffe aus einer absoluten Gesetzmäßigkeit a priori hervorgehen könnten. Vielmehr sind diese Begriffe, für Dilthey, immer nur *als empirische Resultate* ihres konkreten historischen Hintergrundes zu verstehen, also als Produkte der jeweiligen sozialen und psychologischen Prozesse. Die „Logik der Begriffe", wenn man so will, die für die Schulen des Neukantianismus eine strikt apriorische Angelegenheit ist, kann für Dilthey niemals unter gänzlicher Weglassung empirischer Faktoren verstanden werden, sondern nur unter Einbeziehung der Psychologie und der Soziologie der Begriffsentstehung. Während für die Neukantianer also, gegeben einen Vorrat E von empirischen Daten, nur eine (objektiv gültige) Struktur von Begriffen über E möglich ist, resultieren, für Dilthey, Begriffe über E stets nur als logische Produkte aus E plus dem historischen, sozialen und psychologischen Hintergrund der an der Begriffsbildung beteiligten Personen.

Diese grundlegenden Divergenzen wurden auch von den Neukantianern erkannt. So warf Rickert Dilthey vor, einen „prinzipienlosen Historismus" zu vertreten (Rodi & Lessing 1984, 95–100). Natorp präzisierte den Unterschied zwischen der Marburger Schule und Dilthey anhand der Psychologie: „Wir würden", schreibt Natorp, in einem Brief an Dilthey, über den Unterschied zwischen Diltheys deskriptiver Psychologie und der erkenntnistheoretischen Psychologie des Marburger Neukantianismus, „allerdings Psycho[logie] auf Erkenntnistheorie gründen, nicht umgekehrt, indem wir unter Erk[enntnis]th[eorie] verstehen: Reduktion der großen Kulturgestalten (Wissenschaft – Sittlichkeit – Kunst) auf ihre konstituierenden Funktionen". (Lessing 1985, 203) Es liegt hier also ein völlig entgegengesetztes Begründungsverhältnis vor: die Marburger Schule sucht eine apriorische Begründung für die historisch wandelbaren „Kulturgestalten" und somit insbesondere eine Grundlegung geisteswissenschaftlicher Methoden in einer *a priori* ansetzenden philosophischen Logik, wogegen Dilthey die deskriptive Psychologie zu etablieren versucht, als empirische wissenschaftliche (wenn auch nicht *natur*wissenschaftliche!) Methode, anhand derer Philosophie insgesamt *empirisch*, als ein Teil der Geisteswissenschaften betrieben werden kann.

4. *Nicht-Reduktionismus*

Ein weiterer wichtiger Baustein von Diltheys empirischer Philosophie ist sein *Nicht-Reduktionismus,* im Rahmen des Leib-Seele-Problems und, darauf aufbauend, seine Konzeption einer *deskriptiven Psychologie.* Wie Comte und Mill lehnt Dilthey die dualistische Vorstellung von geistigen Inhalten ab, die losgelöst und unabhängig von ihren physiologischen Repräsentationen existieren. Andererseits lehnt Dilthey aber auch den Reduktionismus ab, der in allen klassischen Spielarten des Empirismus und Positivismus eine zentrale Rolle einnimmt. Zwar *sind* geistige Phänomene für Dilthey, genauso wie für Mill und Comte, zunächst physiologische Vorgänge, die keine *eigenständige* Existenz jenseits dieser physikalischen Instanziierung besitzen. Aber dennoch können wir geistige Phänomene nicht nur aufgrund unseres unzureichenden Wissensstandes, sondern ganz prinzipiell nicht durch die Rekonstruktion ihrer physiologischen Seite *erschöpfend* erkennen.

Die in diesen Grundzügen formulierte Position Diltheys kann als Spielart des *psychophysischen Parallelismus* gesehen werden, obwohl auch zu der klassischen Ausformung jener für das neunzehnte Jahrhundert charakteristischen Konzeption signifikante Unterschiede bestehen (siehe XXII, 122–134). Der Parallelismus, in der von Dilthey kritisierten Form, besagt erstens, dass jedem psychischen Zustand ein neuronaler Zustand eindeutig zugeordnet werden kann und zweitens, dass Kausalität immer nur auf einer der beiden Seiten rekonstruiert werden kann, nie aber zwischen psychischem und physischem Zustand. Dilthey teilt zwar den ersten Teil dieser Auffassung, mit klassischen Parallelisten, wie Gustav Theodor Fechner, er verwirft aber die Idee, dass Kausalität zwischen physischen und geistigen Ereignissen unmöglich sei. Konkret ist für Dilthey, im Fall geistiger

Ereignisse, eine *lückenlose* Kausalbetrachtung auf der Ebene des Physischen im Allgemeinen unmöglich. Diesen vollständigen Parallelismus akzeptiert Dilthey nur auf der Ebene der einfachen Wahrnehmungsereignisse. Dort – im Bereich elementarer Wahrnehmungen – ist offensichtlich eine solche lückenlose physikalische Kausalität gegeben. Im Bereich der komplexeren geistigen Ereignisse dagegen ist, nach der Überzeugung Diltheys, immer davon auszugehen, dass psychische (logische) Konstruktionen physikalische Wirkungen haben und umgekehrt: psychische Ereignisse sind durchwegs von physischen *bedingt*.

> Tatsachen des Geistes sind die oberste Grenze der Tatsachen der Natur, die Tatsachen der Natur bilden die unteren Bedingungen des geistigen Lebens. Eben weil das Reich der Personen oder die menschliche Gesellschaft und Geschichte die höchste unter den Erscheinungen der irdischen Erfahrungswelt ist, bedarf seine Erkenntnis an unzähligen Punkten die des Systems von Voraussetzungen, welche für seine Entwicklung in dem Naturganzen gelegen sind. (I, 17)

Im Gegensatz zur klassischen parallelistischen Anschauung ist der Mensch von der Natur „in einer *zwiefachen Beziehung* bedingt": „Die psychophysische Einheit […] empfängt, vermittelt durch das Nervensystem, beständig Einwirkungen aus dem allgemeinen Naturlauf und sie wirkt wieder auf ihn zurück." (ebd.)

Diltheys Ablehnung des Parallelismus bezieht sich also nur auf die von letzterem behauptete Unmöglichkeit psychophysischer Wechselwirkungen. Jenseits davon, in einem allgemeineren Sinn, ist Dilthey jedoch eindeutig „Parallelist", eben weil er den Dualismus ablehnt und somit psychische Ereignisse für ihn nie *unabhängig* von ihrer physischen Repräsentation existieren können. Diese von Dilthey favorisierte Variante eines „Parallelismus", der psycho-physische Wechselwirkungen erlaubt, kann deshalb auch kaum als „Interaktionismus" charakterisiert werden. Wenn man unter Interaktionismus eine Spielart des Dualismus versteht, in der Leib und Seele als Substanzen aufgefasst werden, die kausal interagieren, dann ist dies offenbar eine von Diltheys Auffassung sehr verschiedene Position. Zwar behauptet Dilthey die Möglichkeit kausaler Interaktionen zwischen Körper und Geist, aber *die Träger* der Interaktionen sind für Dilthey selbstverständlich immer auf der physikalischen Ebene angesiedelt. „[Das] geistige Leben eines Menschen [ist] ein nur durch Abstraktion ablösbarer Teil der psycho-physischen Lebenseinheit." (I, 15) Geist und Körper repräsentieren nur insofern unterschiedliche Bereiche als es *im engeren Sinn* physiologische Aspekte des Seelenlebens gibt (etwa die Sinnesorgane) sowie *im engeren Sinn* geistige (etwa das abstrakte Denken). Die Eigenständigkeit dieser Bereiche des Körperlichen und Geistigen kann, für Dilthey, nur auf der Grundlage ihrer je eigenständigen *physikalischen* Repräsentationen gedacht werden, mit ihren je eigenständigen strukturellen Implikationen, *nicht* aber in dem klassischen Sinn, dass Körper und Geist voneinander unabhängige „Substanzen" repräsentieren.

Seine Position zum psychophysischen Problem formuliert Dilthey auch anhand der folgenden „Drei Grundgesetze über die universelle Beziehung des Physischen zum Psychischen":

> [...] 1. Das erste erkennbare allgemeine gesetzliche Verhältnis zwischen den materiellen und den psychischen Vorgängen ist das der Abhängigkeit psychischer Vorgänge direkt von dem Gehirn- und Nervensystem, mittelbar von den im Körper verlaufenden physischen Prozessen, demnach auch weiter vermittelt von dem ganzen physischen Milieu in welchem der Mensch lebt. [...] Es gibt keine psychische Leistung, welche nicht von der Verfassung des Gehirns und Nervensystems bedingt wäre. [...]
>
> 2. Der Überblick über die organische Welt liefert uns alsdann ein zweites umfassendes gesetzliches Verhältnis. Wir wollen dieses als Korrespondenz oder Parallelismus des Physischen und Psychischen bezeichnen. Der Ausdruck psychophysischer Parallelismus ist vieldeutig geworden. Ich verstehe unter Korrespondenz die Tatsache, daß in der gesamten organischen Welt eine bestimmte Beschaffenheit, Struktur und Differenzierung des Nervensystems mit einer bestimmten Beschaffenheit, Struktur und Differenzierung der seelischen Leistungen verbunden ist. [...]
>
> 3. [...] Es gibt innerhalb der psychischen Welt einen Vorgang der Differenzierung des psychischen Lebens; analog der physischen Differenzierung des Organismus hat sich in der Tierwelt stets auf der Grundlage einer niederen Stufe des psychischen Lebens eine höhere entwickelt. (XXII, 148f.)

Das dritte dieser Gesetze – die Differenzierung des psychischen Lebens – führt dann dazu, dass das Physische und das (hoch differenziert) Psychische am Ende unvergleichbar sind:

> Die seelischen Vorgänge zeigten sich uns den körperlichen parallel. Ebenso wahr und wichtig aber ist die Erkenntnis von der Unvergleichbarkeit beider Gebiete. [...] [B]ei genauerem Zusehen [...] zeigt sich [...] die gänzliche Verschiedenheit der Art und Weise, wie Psychisches untereinander verbunden ist, und der Art, wie wir physische Erscheinungen durch Denkmittel bestimmen. (XXII, 150)

Diese Grundgesetze verweisen erneut auf Diltheys (an Mill anknüpfende) hierarchische Auffassung der psychophysischen Zusammenhänge. Der Geist ist zutiefst abhängig von dem physischen Milieu, in dem er zustande kommt. Im Fall von niederen Tieren (also beispielsweise bei Einzellern oder Fröschen) besteht sogar de facto *Identität* zwischen dem Körperlichen und dem Geistigen. Aber auch im Fall von einfachen Wahrnehmungsvorgängen bei höheren Tieren und beim Menschen liegt diese Identität vor. Komplexere Denkvorgänge sind dann zwar nicht mehr identisch mit physikalischen Vorgängen, wohl aber in unterschiedlichster Weise kausal von solchen determiniert. Das Seelenleben hochentwickelter Tiere und des Menschen ist somit zutiefst abhängig von dem physischen Milieu, in dem es sich abspielt (Grundgesetz 1). Die *Parallelität* zwischen Körper und Geist (Grundgesetz 2) besagt überdies, dass jeder psychische Vorgang ein entsprechendes physisches Korrelat haben muss bzw. eine entsprechende Erklärung im Bereich der Physiologie. Die Evolution führt nicht dazu, dass sich der Geist irgendwann vom Körper „abkapselt". – Die für Diltheys hierarchische Leib-Seele-Theorie typischste Annahme ist aber das dritte Grundgesetz – die prinzipielle *Inkommensurabilität* psychischer und physischer Zustände –, das dadurch zu einer nichttrivialen Aussage wird, dass wir, nach Dilthey, im Fall von einfachen geistigen Vorgängen (Wahrnehmungen, bloße Reiz-Reaktions-Vorgänge) nicht nur Kommensurabilität, sondern nachgerade Identität zwischen physischem und psychischem Zustand haben. Dilthey bringt hier das Beispiel von enthirnten Fröschen, die auf Reize genauso reagieren wie mit Hirn (XXII, 77f.). Erst bei zusehends komplexer werdenden geistigen Leistungen – etwa bei isolierten Vorstellungen oder abstrakten Begriffen – treten die beiden Ebenen auseinander und es entsteht eine zur physischen letztlich inkommensurable Ebene des Geistigen.

Inkommensurabilität bedeutet hier, dass klassisch-empiristische Konzeptionen (die auf der Ebene der Sinnespsychologie für Dilthey durchaus brauchbarer Resultate liefern) genau im Fall einer Psychologie des „entwikkelten Seelenlebens" versagen. Die Konzepte der naturwissenschaftlichen Psychologie reichen nur dort aus, wo es sich um *einfache* seelische Phänomene handelt, bei denen im Wesentlichen psychophysische Identität herrscht. Im Fall von komplexen seelischen Vorgängen hingegen haben wir das Problem, dass die physiologische „Erklärung" dieser Vorgänge etwas völlig anderes erklärt als die Vorgänge selbst. So wäre es sinnlos, einen mathematischen Beweis durch eine Rekonstruktion der physiologischen (bzw. neuronalen) Vorgänge, die in dem den Beweis Durchführenden stattfinden, „zu erklären". Der spezifische Charakter eines solchen Beweises kann nicht auf diese Weise erklärt werden, wohl aber indem man *die logischen Operationen* angibt, die den Beweis konstituieren und indem man erklärt, wie und inwiefern diese logischen Operationen dazu führen, dass ein Beweis die (analytische) Wahrheit eines Satzes etabliert. Diese logischen Operationen aber sind etwas, das nicht nur nichts mit physiologischen Erklärungen zu tun hat, es hat augenscheinlich auch nichts zu tun, mit solchen quasi-naturwissenschaftlichen Gesetzen wie sie die Assoziationspsychologie behauptet. Vielmehr handelt es sich hier um Gesetze, die nur unmittelbar vom logischen Subjekt geschaffen, verstanden, beschrieben und angewendet werden können. *Die Psychologie* also, die sich mit Erklärungen von *dieser* Art befasst, muss *grundlegend* anders funktionieren als die naturwissenschaftliche Psychologie bzw. als alle denkbaren klassisch-empiristischen Varianten der Psychologie (sie wird der Logik mindestens genauso nahe stehen wie den Naturwissenschaften). Diltheys Idee einer *deskriptiven Psychologie* ist genau diese neue Art von nicht-naturwissenschaftlicher, und zwar empirischer, aber dennoch nicht klassisch-empiristischer Psychologie.

Jedenfalls ist diese „neue Psychologie" (XXII, 154–157) Diltheys nicht völlig abgekoppelt von den Naturwissenschaften zu sehen. Ist Diltheys Modus der Empirie seiner „deskriptiven Psychologie" also nicht *nur physikalisch*, so ist es wichtig, zugleich darauf hinzuweisen, dass dieser Modus, zumindest in der Erkenntnistheorie des frühen und mittleren Dilthey, durchaus *auch physikalisch* ist. Diltheys diesbezügliche Feststellung lautet, dass, wie in allen empirischen Wissenschaften, so auch im Fall der als Grundlage der „Geisteswissenschaften" explizierten deskriptiven Psychologie, die *mathematische Statistik* als wesentliches Hilfsmittel fungieren muss:

> Die Bibliotheken sind unsere Archive: sie bewahren den Inbegriff der Urkunden für die Geschichte der Wissenschaften. [...] Die Anwendung statistischer Methoden auf die Schätze der Bibliotheken muß ermöglichen, den Umfang und die Stärke der Richtungen, der Beschäftigung mit einzelnen Zweigen usw., die örtliche Verteilung derselben auf eine quantitative Weise festzustellen. (V, 40f.)

Es handelt sich bei diesen materiellen Fakten aber auch nicht bloß um so etwas wie *externe* Hintergrundinformationen. Tatsächlich ist es das erklärte Ziel von Diltheys deskriptiver Psychologie (zumindest in einer fernen Ausbaustufe) *ihre Grundlage* (gewissermaßen im Sinne der Schaffung einer empirischen Basis) in den Naturwissenschaften zu finden:

> Das Problem, in immer weiterem Umfang psychische Vorgänge den exakten Bestimmungen von Messung und Rechnung zu unterwerfen, ist in seinen Einzellösungen abhängig von dem Auffinden stetiger Verhältnisse zwischen den unmessbaren Größen der psychischen Kräfte und der Beziehung von solchen einerseits und meßbaren Vorgängen der Außenwelt andererseits. (V, 41)

Diese typisch parallelistische Grundanschauung wird von Dilthey zu einer Art *Milieutheorie des Geistes* ergänzt. Es ist nicht nur so, dass wir die Naturwissenschaften benötigen, um exakte Modelle für geistige Vorgänge, anhand ihrer physischen Korrelate, zu entwickeln. Geistige Vorgänge können prinzipiell nicht isoliert von dem physischen Milieu, in dem sich der Geist entwickelt, verstanden werden. Zwar kann man sich eine Geisteswissenschaft *denken*, die, in puristischer Weise, auf „rein geistigen Wesen" beruht, darauf aufbauend, „reine Geisteswissenschaften". Aber obwohl eine solche Wissenschaft grundsätzlich *funktionieren* könnte, würde sie doch von falschen Voraussetzungen ausgehen:

> In Wirklichkeit entsteht ein Individuum, wird erhalten und entwickelt sich auf Grund der Funktionen des tierischen Organismus und ihrer Beziehungen zu dem umgebenden Naturlauf; sein Lebensgefühl ist wenigstens teilweise in diesen Funktionen gegründet; seine Eindrücke sind von den Sinnesorganen und ihren Affektionen seitens der Außenwelt bedingt; den Reichtum und die Beweglichkeit seiner Vorstellungen und die Stärke sowie die Richtung seiner Willensakte finden wir vielfach von Veränderungen in seinem Nervensystem abhängig. [...] So ist das geistige Leben eines Menschen ein nur durch Abstraktion loslösbarer Teil der psycho-physischen Lebenseinheit, als welche ein Menschendasein und Menschenleben sich darstellt. Das System dieser Lebenseinheiten ist die Wirklichkeit, welche den Gegenstand der geschichtlich-gesellschaftlichen Wirklichkeit ausmacht. (I, 14f.)

Damit schließt sich der Kreis, hin zu der oben besprochenen Auffassung Diltheys von einem *konsequenteren Empirismus*. Diltheys *Lebensphilosophie* ist nichts anderes als das Projekt einer *direkten Empirisierung* der abstrakten Produkte unseres Denkens, die als Produkte unsrer konkreten psychischen Aktivitäten, sozialen Interaktionen und unserer Existenz als menschliche Wesen „aus Fleisch und Blut" identifiziert werden. Diese Ebene der Empirie wurde in der Philosophie *vor* Dilthey (bzw. Trendelenburg, Ueberweg, Lazarus etc.) nicht begangen, weil man entweder (wie bei Kant und Hegel) die empirischen „Tatsachen des Bewusstseins" a priori, mit der Brechstange der transzendentalen oder reinen Logik zu erschließen versuchte; oder aber die Analyse wurde zur Gänze auf die Ebene der Gehirnwissenschaften verlagert und die geistigen Inhalte wurden so in der eindimensionalen Objektivität physikalischer Prozesse ihrer wahren empirischen Dynamik beraubt. Aber auch *nach* Dilthey ging dieser empirische Charakter einer Philosophie des Lebens, was hier nur abschließend angedeutet werden kann, rasch wieder verloren. Sowohl im Neukantianismus als auch in Husserls Phänomenologie wurde (wenn auch in je unterschiedlicher Weise) erneut versucht, die geistige Welt einer strikt a priori konzipierten Logik unterzuordnen. Auch wenn spätere Ansätze, von Heideggers Fundamentalontologie bis zu Derridas Poststrukturalismus, teilweise eine Rückkehr zu Diltheys empirischem Standpunkt markieren – die Diltheys Programm auszeichnende konsequente Unterordnung der philosophischen Analyse unter die Methoden der empirischen Geisteswissenschaften wird in keinem dieser Ansätze vollzogen. Stärkere Parallelen zu Diltheys radikaler Programmatik findet man allerdings, wie wir im folgenden Abschnitt sehen werden, in neueren und neuesten Entwicklungen im Rahmen der analytischen Philosophietradition.

5. *Der heutige Methodenstreit in der Philosophie*

Es können eine Reihe von Bezügen zwischen Positionen Diltheys und Entwicklungen im Rahmen der analytischen Philosophie hergestellt werden. Neben den bereits erwähnten direkten Bezügen zu Dilthey in Carnaps Frühwerk könnte man hier vor allem wiederkehrende Motive in der rezenten „Philosophy of Mind“ entdecken.[10] Für die hier herausgestrichene Bedeutung Diltheys als eines Philosophen, der Philosophie *als eine empirische Wissenschaft* versteht, ist aber eine andere Gruppe von Bezügen relevant.

Es geht dabei weniger um ein *inhaltlich* spezifiziertes philosophisches Untersuchungsfeld als um die *methodologische* Auffassung von Philosophie insgesamt. – Die Methodologie der Philosophie war, in der klassischen analytischen Tradition, von puristischen und teils utopischen Ansagen geprägt. So hatten Russell, der Carnap des Wiener Kreises und der frühe Wittgenstein die Auffassung vertreten, dass philosophische Probleme *ausschließlich* mit der Methode der formalen Logik anzugehen seien:

10. Eine solche Herausstreichung von Spielarten eines nicht-reduktionistischen Materialismus, die man bei Dilthey ebenso findet wie in rezenten Debatten, könnte analog verlaufen zu dem von Michael Heidelberger in (Heidelberger 1993) behandelten und in vieler Hinsicht mit Dilthey vergleichbaren Fall Gustav Theodor Fechners.

„Wissenschaftslogik anstatt Philosophie" lautete Carnaps diesbezügliches Motto (Carnap 1968 [1934], § 72). Als Gegenentwurf dazu hat W. V. O. Quine dann die Konzeption einer „naturalisierten Epistemologie" entworfen, der zufolge alle Fragen der Philosophie letztlich auf Fragen der Neurowissenschaften herunter gebrochen werden könnten (Quine 1969). In beiden Fällen wurde die Möglichkeit einer *eigenständigen* philosophischen Methode negiert. Stattdessen *verlagerte* man Philosophie (mit dem hypothetischen Resultat der Selbstauflösung dieser Disziplin) auf einen anderen Schauplatz: die Mathematik bzw. die Naturwissenschaften. Eine konstruktivere Methodendiskussion, im Sinne einer *Wissenschaftstheorie der Philosophie* oder einer *Philosophie der Philosophie*[11], wurde in der analytischen Tradition dagegen erst in neuester Zeit in Gang gebracht. Im Wesentlichen scheinen sich dabei zwei Grundtendenzen (hier mit A und B bezeichnet) herauszukristallisieren: eine *apriorische* und eine *empirische* Methodenkonzeption.

(A) Die eine Tendenz versucht, Philosophie als eine Disziplin zu retten, deren methodologische Vorgangsweise a priori, also „ohne Rückgriff auf die Erfahrung" (Kant) erfolgt bzw. als etwas, „[that] can be pursued in an armchair" (Williamson 2007, 1). Wichtige Spielarten dieser Strategie sind (1) die sogenannte analytische Metaphysik und (2) der (Neo-)Neukantianismus.

(A1) Analytische Metaphysik – Timothy Williamson (Williamson 2007, 19) nennt als typische Vertreter dieser Richtung Saul Kripke, David Lewis, Kit Fine, Peter van Inwagen und David Armstrong – ist eine methodologische Auffassung von Philosophie, wonach diese nicht über Sprachanalyse oder konzeptuelle Analyse funktioniert (Williamson 2007, 21), sondern auf einer eigenständigen Ebene des abstrakten Theoretisierens, auf der philosophische Theorien bereitgestellt und diskutiert werden können, *im Wesentlichen wie in jeder anderen Wissenschaft* – Williamson nennt dies die „unexceptional nature of philosophy" (Williamson 2007, 3) –, bloß mit dem Unterschied, dass Philosophie, für die Vertreter dieser Methodenkonzeption, im Allgemeinen keinen unmittelbaren empirischen Hintergrund besitzt. Es wird gewissermaßen *nur der Modus des Theoretisierens, nur die abstrakte Methode* von der Physik oder auch von der Mathematik

11. Ein Buch dieses Titels wurde von Timothy Williamson verfasst (Williamson 2007), aber auch Wilhelm Dilthey hat sein Spätwerk unter diesem Stichwort zusammengefasst (VIII, insbesondere 206ff.). Die beiden Ansätze stimmen tatsächlich *thematisch* weitgehend überein, in der Sache vertritt Dilthey aber eine zu der von Williamson geradezu entgegengesetzte Position (mehr dazu weiter unten).

übernommen – als *Methode* der Philosophie –, ohne dabei aber gleichzeitig philosophische *Inhalte* als physikalische, psychologische oder sonstwie empirische Inhalte zu identifizieren. Konsequenter Weise spielt dann die Mathematik – namentlich die formale Logik – für die „philosophische Theoriebildung" eine ähnlich zentrale Rolle wie in der Physik etwa, wogegen es, für die analytische Metaphysik, in der Philosophie kein Gegenstück zu den *empirischen* Methoden der Einzelwissenschaften gibt. Analytische Metaphysik ist somit strikt anti-naturalistisch, aber sie negiert auch die Möglichkeit einer eigenständigen philosophischen Methode, jenseits der (formalen) Methode der Wissenschaften. Insbesondere beruft sich analytische Metaphysik nicht auf Intuition o. dgl.

(A2) Im Gegensatz dazu versucht die (neo-)neukantianische Herangehensweise an „philosophy as pursued in an armchair" als ausgezeichnetes Element eben eine solche exklusive (wenn auch sehr „weiche") Methode des Philosophierens zu identifizieren. Michael Friedman (Friedman 2001) beruft sich dabei im Wesentlichen auf drei Kronzeugen. Erstens den Neukantianismus, vor allem den der Marburger Schule um Cohen, Natorp und Cassirer, demzufolge die transzendentalen Bedingungen der Möglichkeit der Erfahrung etwas sind, das mit der Entwicklung der Wissenschaften, in unterschiedlichen historischen und sozialen Kontexten, wechselt, also ein „relativiertes Apriori". Friedman bezieht zweitens Rudolf Carnaps Theorie „formaler Rahmenwerke" für die Wissenschaften, die, im Sinne eines „logischen Toleranzprinzips" (Carnap 1968 [1934], § 17), mehr oder weniger frei gewählt werden können auf diesen neukantianischen Begriff eines „relativierten Apriori" und er interpretiert drittens Thomas Kuhns Theorie über die „Struktur wissenschaftlicher Revolutionen" (Kuhn 1996 [1962]) auf der Grundlage eben dieses Konzeptes. Die Idee ist am Ende, dass die philosophische „Methode" genau in der „Methode" besteht, der sich die Wissenschaften in Situationen revolutionärer Umwälzungen bedienen (Paradigmenwechsel), wo die bewährten Methoden und formalen Rahmenwerke versagen. Die philosophische „Methode" ist die Methode der revolutionären Wissenschaft. Im direkten Gegensatz zum Programm der analytischen Metaphysik steht diese (neo-)neukantianische Herangehensweise also auf dem Standpunkt „[that] it is a folly [...] for philosophy (and for the other humanities) to regret this lack of scientific status, and, even worse, to seek compensation by attempting to strip away such status from the sciences themselves." (Friedman 2001, 24) Während die analytische Metaphysik also die Methode des Philosophierens direkt in der *normalwissenschaftlichen* Methode der Einzelwissenschaften sucht (und zwar in den apriorischen Prinzipien derselben) identifiziert der (Neo-)Neukantianismus als Methode der Philosophie das Denken der Wissenschaftler in revolutionären Situationen.

(B) Dem gegenüber gibt es eine Gruppe von methodologischen Strategien, die Philosophie *empirisch* auffassen bzw. die die Möglichkeit einer „philosophy as pursued in an armchair" negieren. Interessant ist dabei zunächst, dass in diesen empirischen Strategien durchwegs sowohl die rein apriorische Herangehensweise an Philosophie abgelehnt wird als auch die Strategie eines geradlinigen Naturalismus á la (Quine 1969). Während Quines Naturalismus darauf hinaus laufen würde, dass Philosophie durch die im Wesentlichen naturwissenschaftlichen Analysen der Kognitionswissenschaften *zu ersetzen* wäre, belassen die neueren empirischen Strategien (ganz in demselben Sinn wie dies übrigens auch die apriorischen Strategien der analytischen Metaphysik und des (Neo-)Neukantianismus tun) der Philosophie ihren eigenständigen Status. In diesem Kontext erweist sich die analytische Metaphysik als direktes Gegenprogramm zu den empirischen Strategien, weil sie dezidiert versucht, Philosophie in einem allgemeinen Sinn als Wissenschaft zu identifizieren, in einem durch die amerikanische analytische Tradition geprägten Verständnis *als Naturwissenschaft* („science"), und damit die Möglichkeit blockiert, Philosophie als einen Teil der Geisteswissenschaften (Soziologie, Psychologie, Geschichtswissenschaften) fest zu machen wie dies in der neukantianischen Strategie Michael Friedmans geschieht. Die analytische Metaphysik divergiert mit empirischen Strategien also deshalb, weil sie Philosophie in einem geschichtslosen Vakuum ansiedelt, als nur der Rationalität der Logik als Methode verpflichtete Disziplin, in der der „context of discovery" genau so wenig Bedeutung hat, für die Theorienbildung, wie im Fall der (normalwissenschaftlichen) Physik. Diese Auffassung von Philosophie als einer Art von *abstrakter Laborwissenschaft* (das „Labor" wäre hier der Ohrensessel der Philosophin) wird von der neukantianischen Lesart genauso zurückgewiesen wie von allen Spielarten einer empirischen Philosophie. Philosophie ist in all diesen Lesarten *eine Geisteswissenschaft* und also eine Wissenschaft, bei der der „context of discovery" eine entscheidende methodologische Rolle einnimmt. Der Unterschied zwischen der (tendenziell) apriorischen (neo-)neukantianischen Strategie und den verschiedenen empirischen Strategien besteht dann hauptsächlich darin, dass die (neo-)neukantianische Strategie ein anderes und, so die Kritik ihrer Gegner, wenig zeitgemäßes Bild von den Geisteswissenschaften zeichnet, demzufolge diese eine „weiche" Form von Wissenschaft bilden, deren „Methode", überspitzt formuliert, bloß in dem Motto besteht „to let a thousand flowers bloom“ (Friedman 2001, 24). Geisteswissenschaften sind für die Anhänger einer empirischen Auffassung von Philosophie, im Gegensatz zu dieser Sichtweise, nicht bloß die revolutionären Lückenfüller für die Naturwissenschaften, sondern sie sind eigenständige Wissenschaften mit (potentiell) eigenständigen Methoden: und *diese* (empirischen) Methoden (also im Wesentlichen die Methoden der Psychologie und der Sozialwissenschaften) sind eben die Methoden derer sich die Philosophie bedienen sollte. Diese neuen Formen eines Naturalismus reduzieren also Philosophie nicht auf die Naturwissenschaften, sondern empirisieren sie *direkt*, indem innerhalb der Philosophie auf die empirischen Methoden der Geisteswissenschaften zurückgegriffen wird.

Die empirischen Methoden der Geisteswissenschaften (das sind: die statistischen Methoden der empirischen Sozialforschung, die analytischen Methoden der Psychologie und die aus diesem Komplex abgeleiteten Methoden der Geschichtswissenschaften) analysieren die Realität der menschlichen Existenz, in ihren psychologischen, sozialen und (dem übergeordneten) historischen Bedingungen. Nehmen wir diese Bedingungen ernst, so erweist sich Philosophie als eine Angelegenheit, die wir einzig und allein dadurch verstehen können, dass wir sie als das lebendige Produkt der Dynamik psychologischer und gesellschaftlicher Prozesse verstehen.

(B1) Als Vordenker dieser empirischen Konzeption von Philosophie kann der späte Bas van Fraassen betrachtet werden, der dem klassischen Empirismus ein Programm gegenüberstellt, wonach sich Empirismus nicht durch die Orientierung an „empirischen Daten“ auszeichnet, sondern, allgemeiner, durch eine *empirische Einstellung* („empirical stance“), die insbesondere auch die Philosophie selbst betrifft. Dieser empirischen Einstellung zufolge sollten die Wissenschaften und die Philosophie allen Annahmen misstrauen, die auf allzu großen Umwegen aus solchen Annahmen *abgeleitet* sind, die zunächst unmittelbar aus einem empirischen Umfeld hervorgegangen waren. Die empirische Einstellung in der Philosophie ermöglicht eine nicht-metaphysische Annäherung an klassische (vormals nur in der Metaphysik thematisierbare) philosophische Probleme. „I reject metaphysics“, sagt van Fraassen, „and still I'll say that you exist and that you are a person, a conscious being.“ (van Fraassen 2002, 3) Wie kommt es dazu? Philosophische "Theorien", wie der hier angesprochene "Realismus", sind Theorien, die unmittelbar aus der empirischen Wirklichkeit des Daseins heraus entwickelt werden, als Resultat unserer

„empirischen Einstellung" also. *So* sind diese „Theorien" akzeptabel und sinnvoll. Sobald wir aber diese Theorien ihres Kontextes entkleiden und aus der Unmittelbarkeit ihrer Produktion herausreißen, verlieren sie tendenziell ihre Bedeutung:

> Metaphysical theories purport to interpret what we already understand to be the case. But to interpret is to interpret into something, something granted as already understood. Paradoxically, metaphysicians interpret what we initially understand into something hardly anyone understands, and then insist that we cannot do without that. (ebd.)

Die Übereinstimmungen dieser Position mit Michael Friedmans (neo-) neukantianischem Programm sollten nicht übersehen werden. Sowohl van Fraassen als auch Friedman lehnen die analytische Metaphysik ab, wegen ihres von beiden als kontraproduktiv empfundenen artifiziellen Charakters einer *theoretischen Wissenschaft*. Aber van Fraassen verweist in seiner Alternativposition zu dem „blutleeren Theoretisieren" der Metaphysik auf einen Aspekt, der bei Friedman nicht zu finden ist, zumindest nicht in expliziter Form. Es ist dies der Umstand, dass philosophische „Theorien" dadurch *empirisch* sind, dass sie aus der Realität unseres Daseins heraus entwickelt werden, als unmittelbare und insofern unverfälschte Produkte existenzieller Situationen. Während Friedman eine Rolle für die Philosophie nur dort gegeben sieht, wo die theoretischen Rahmenwerke der Wissenschaften *versagen* (beispielsweise in revolutionären Situationen der Wissenschaften), sieht van Fraassen Philosophie als einen substanziellen und ständig relevanten Teil unserer individuellen und gesellschaftlichen Wirklichkeit, als eine ständig gegenwärtige und wesentliche Ebene des menschlichen Denkens, die sich aber dadurch auszeichnet, dass sie niemals den modellhaften und prozeduralisierten Charakter der theoretischen Wissenschaften (also: der Naturwissenschaften) erlangt.

> Philosophy itself is a value- and attitude-driven enterprise; philosophy is in false consciousness when it sees itself otherwise. To me, philosophy is of overriding importance, to our culture, to our civilization, to us individually. For it is the enterprise in which we, in every century, interpret ourselves anew. But unless it so understands itself, it degenerates into an arid play of mere forms." (van Fraassen 2002, 17)

Friedman würde das wohl im Grunde sehr ähnlich sehen. Aber es scheint doch einen feinen Unterschied zu geben, der für die hier angestellten Überlegungen von großer Bedeutung ist, weil er mit dem Unterschied zu konvergieren scheint, der seinerzeit zwischen Dilthey und den Neukantianern bestand. Die neukantianische Zugangsweise siedelte Philosophie außerhalb der Wissenschaften (bzw. nur im Rahmen wissenschaftlicher Revolutionen) an und sah ihre Aufgabe darin, den konzeptuellen (konstitutiven) Hintergrund der Wissenschaften zu schaffen. Aber dieses Schaffen konzeptueller Hintergründe war nichts, das für sich genommen einer empirischen wissenschaftlichen Behandlung zugänglich gewesen wäre. Dilthey öffnete, im Unterschied dazu, das Tor für eine solche empirische Behandlung, indem er die konstituierenden Bestandteile unserer Kultur gleichzeitig *als Bestandteile* unserer Kultur identifizierte und damit als etwas, das für die Kultur- bzw. die Geisteswissenschaften *zur Disposition steht.*

Ist für die Neukantianer (auch für die „Neo-Neukantianer" um Michael Friedman) also Erkenntnistheorie eine Disziplin, die sich als den empirischen Methoden der Wissenschaften unzugänglich erweist und Philosophie damit *gegen jede Form der Naturalisierung* immun, so öffnen Dilthey und – in durchaus ähnlicher Weise und teils mit Bezug auf Dilthey[12] – van Fraassen das Tor für Naturalisierungen, zumindest in der „weichen" Form einer Fundierung philosophischer Untersuchungen in geisteswissenschaftlichen (soziologischen, psychologischen) Analysen. Es ist für die Vertreter dieser empirischen Strategie nicht so, dass die (erkenntnistheoretischen) Analysen der Philosophie in einer Art von transzendentalem Vakuum angesiedelt sind, sondern dadurch, dass die Produkte der Philosophie *unmittelbare* Produkte unserer geistigen Aktivitäten sind, werden sie zugänglich für einschlägige Analysen im Rahmen der Psychologie, Soziologie und Geschichte. – Auch wenn van Fraassen selbst in der zitierten Schrift nicht den letzten Schritt geht und für eine Empirisierung der Philosophie in diesem Sinn eintritt, so bereitet doch die Verschiebung des Philosophiebildes, die er vornimmt, genau eine solche Empirisierung vor.

Konkrete Beispiele, wie eine solche empirische Zugangsweise zur Philosophie, aus heutiger Sicht, aussehen könnte, liefern die sogenannte experimentelle Philosophie und die Soziologie philosophischen Wissens.[13]

(B2) Im Fall der experimentellen Philosophie geht es darum, die intuitiven Bestandteile philosophischer Analysen zu hinterfragen, also die unterschiedlichen Spielarten von Berufung auf Intuition, auf „Common-

12. Van Fraassen erwähnt Dilthey an einer Stelle (van Fraassen 2002, 237 Fn. 31) gemeinsam mit Collingwood als direkten Vorläufer seiner Theorie.

13. Vgl. (Knobe & Nichols 2008) und (Kusch 2000) sowie die dortigen Literaturhinweise.

sense“ oder „folk psychology“. Wenn Philosophinnen, grob gesprochen, behaupten, dass die Annahme X intuitiv, commonsensual oder eine charakteristische Annahme der „folk psychology“ sei, dann liegt es nahe, derartige Behauptungen auf verschiedenen Ebenen zu hinterfragen. In welchen Gruppen gilt diese Behauptung? Gilt sie für Philosophinnen im Allgemeinen oder nur für Philosophinnen einer bestimmten Tradition? Gilt sie für Akademikerinnen, aber etwa nicht für Arbeiterinnen und Angestellte? Gilt sie in europäischen oder amerikanischen Gesellschaften, aber etwa nicht in Gesellschaften in Asien oder Afrika? Überall dort, wo sich die Philosophie darauf beruft, etwas wiederzugeben, das als *Voraussetzung* betrachtet wird (etwa auch im Stil dessen was Kant „transzendental“ genannt hatte), gibt es eine ganze Reihe von empirischen Faktoren, die es zu beachten gilt, um einen derartigen Rückgriff auf Voraussetzungen als stichhaltig oder nicht zu identifizieren bzw. auch um die Abhängigkeit oder Unabhängigkeit der Voraussetzungen von individuellen und sozialen Kontexten zu überprüfen. Kurz gesagt: *überall dort* wo sich selbst noch der (Neu-)Kantianismus auf ein philosophisches Denken a priori und also auf empirische Voraussetzungslosigkeit beruft, findet die experimentelle Philosophie empirische Voraussetzungen, die es zu klären gilt. Diese Klärung soll dann in bestimmten „philosophischen Experimenten“ vorangetrieben werden:

> [W]e would love to know more about the ways in which philosophers differ from ordinary folks, and it seems to us that the best way to find out would be to run some experiments. One could devise a series of questions and then give those questions both to philosophers and non-philosophers, checking to see how intuitions differed between the two groups. [...] our guess is that the overall pattern will be far more complex – and far more interesting – than anyone could have predicted from the armchair. (Knobe & Nichols 2008, 9)

(B3) Sehr ähnlich ist der Ausgangspunkt der Soziologie philosophischen Wissens, die ihrerseits auf dem sogenannten „strong program“ der Wissenssoziologie (Bloor 1991 [1976]) basiert. Das „Starke“ an diesem Programm liegt darin, dass es Wissenssoziologie nicht als einen Ansatz versteht, der soziale Faktoren nur dort als konstitutiv ansetzt wo keine internen rationalen Faktoren vorliegen (also etwa, wie im Fall des neukantianischen bzw. Kuhnschen Ansatzes, nur im Fall wissenschaftlicher Revolutionen). Das „strong program“ versucht vielmehr, Wissenschaft insgesamt aus ihrem sozialen (und psychologischen) Kontext heraus zu verstehen. Wenn wir diesen Kontext ausblenden, so sehen wir *immer* ein verkürztes Bild von den Wissenschaften und wir landen bei Karikaturen bzw. bei verstümmelten Versionen von Wissenschaft. Die Soziologie philosophischen Wissens geht nun von der Annahme aus, dass diese Voraussetzung, die bereits im Fall der Naturwissenschaften kaum von der Hand zu weisen ist, für die Philosophie in noch stärkerem Ausmaß gilt, eben weil Philosophie immer nur verstanden werden kann, als unmittelbare Reaktion auf die *empirischen* Gegebenheiten (also die individuellen und sozialen Umstände), in denen sie entsteht. Martin Kusch illustriert dies anhand des Begriffs einer „philosophischen Tatsache", den man etwa so charakterisieren könnte: Während die Naturwissenschaften, zumindest dort, wo sie empirische Adäquatheit anstreben, von Tatsachen reden, die durch objektive raumzeitliche Daten repräsentiert sind, geht es in der Philosophie und in den Geisteswissenschaften um solche Tatsachen, die wir von Vornherein nicht als raumzeitliche Daten vorfinden, sondern als Konstruktionen der Wissenschaftlerinnen. Philosophische Tatsachen werden nicht *entdeckt*, sondern sie werden von den Philosophinnen *geschaffen*. Natürlich können diese Tatsachen desungeachtet eine Wirklichkeit widerspiegeln, die sich in erster Linie als sehr abstrakte Wirklichkeit bestimmter wissenschaftlicher Theorien identifizieren lässt, also in einem eher „internen" Bild. Aber wie groß auch immer der Anteil solcher „interner Zutaten" sein mag, es bleibt der Sachverhalt, dass die philosophische Tatsache etwas ist, das von Philosophinnen *geschaffen* wird und damit zwangsläufig etwas, das nicht *nur* „interne" Motive haben kann (weil wir nie nur interne Motive haben). So ist es die Aufgabe der Soziologie des philosophischen Wissens, philosophische Tatsachen so in ihrer Entstehung zu rekonstruieren, dass all diese externen sozialen und psychologischen Motive dabei Berücksichtigung finden, nicht um die „interne" philosophische Analyse *zu ersetzen,* sondern um sie *zu ergänzen*:

> [There] is no need to react with too much hostility: Sociophilosophy and the sociology of philosophical knowledge are not in the business of eliminating philosophy altogether. They do not attempt to condense all philosophical clouds into mere drops of sociology. The aim of their work is rather to show the philosophical fly the way out of the fly-trap of sociological naiveté. (Kusch 2000, 35)

6. *Schluss*

Soweit eine Gegenüberstellung unterschiedlicher methodologischer Ansätze in rezenten Diskussionen über die „Philosophie der Philosophie". Die Übereinstimmungen zwischen den Programmen von van Fraassens „empirischer Einstellung", der experimentellen Philosophie und der Soziologie philosophischen Wissens mit Grundgedanken Diltheys, wie sie in den ersten Abschnitten dieses Aufsatzes skizziert wurden, liegen auf der Hand und müssen nicht mehr im Detail ausgewiesen werden. Die Frage, die sich hier stellt, ist viel eher: was tun mit diesen Übereinstimmungen? Es wird immer möglich sein, für bestimmte neue Entwicklungen in der Philosophie, historische Vorläufer zu finden, die diese Entwicklungen in der einen oder anderen Form bereits vorweg genommen hatten. Damit ist aber noch keineswegs klar, dass eine Auseinandersetzung mit diesen historischen Vorläufern automatisch auch *fruchtbar* sein muss, für die Auseinandersetzung mit den neuen Entwicklungen. Im konkreten Fall scheint es aber tatsächlich eine Reihe von Gründen zu geben, die diese historische Auseinandersetzung als potentiell fruchtbar für die rezenten Debatten erweisen. Zum Abschluss dieses Aufsatzes seien daher drei solche Gründe skizziert:

(1) *Die Debatten sind komplementär.* Während es in den rezenten Diskussionen (bislang) in erster Linie darum geht, die empirischen Methodologien in der Philosophie konkret anzuwenden und entsprechende empirische Studien zu produzieren, liegt bei Dilthey der Schwerpunkt auf sehr grundsätzlichen methodologischen und inhaltlichen Erwägungen, mit dem Ziel einer Abgrenzung seines „Paradigmas" von unterschiedlichen philosophischen Traditionen. Dilthey liefert kritische Diskussionen des Empirismus und Positivismus, des Kantianismus und des deutschen Idealismus und natürlich auch der metaphysischen Tradition, die auch für die heutigen Debatten gewinnbringend ausgewertet werden können. Dabei geht es darum, Diskussionen in Gang zu bringen, neue Themen auf die Tagesordnung zu setzen und neue Gesichtspunkte in die Debatte einzubringen, wie etwa in der Gestalt der hier angedeuteten Abgrenzung der empirischen Philosophie von analytischer Metaphysik und (Neo-) Neukantianismus.

(2) *Dilthey liefert einen historischen Präzedenzfall.* Dadurch, dass Dilthey eine Programmatik verkörpert, die weitgehend mit rezenten empirischen Ansätzen in der Philosophie übereinstimmt, wird es interessant, sich anzusehen, wie die Reaktionen auf Dilthey ausgesehen haben und wie sich die

Philosophie aus diesen Reaktionen heraus weiter entwickelt hat (im Sinne einer Suche nach rezenten Entwicklungsmöglichkeiten). So spielen in den unterschiedlichen Spielarten der Lebensphilosophie und der philosophischen Anthropologie bei Scheler, Heidegger, Plessner usw. Überlegungen aus dem Umfeld von Diltheys „empirischer Philosophie" eine wesentliche Rolle. In all diesen Ansätzen geht es darum, die Wissenschaften in einem größeren Zusammenhang zu sehen, nicht als etwas, das wir *wissen,* sondern als etwas, das wir *sind.*[14] Die „kontinentale" Philosophietradition liefert damit potentielle Anknüpfungspunkte für die (eher der „analytischen" Tradition zuzurechnende) empirische Philosophie, die wir so, auf dem Umweg über Dilthey, in neuem Licht sehen und neu entdecken könnten.

(3) *Diltheys Methode erweist sich gerade in ihrem Scheitern als Desiderat.* Diltheys bevorzugte Methode einer „deskriptiven Psychologie" ist gescheitert, da Diltheys einschlägige Entwürfe letztlich weder von den naturwissenschaftlichen Psychologen noch von den Philosophen akzeptiert wurden (den einen waren sie zu philosophisch, den anderen zu naturwissenschaftlich). Geht man davon aus, dass die in der europäischen Philosophie des zwanzigsten Jahrhunderts entwickelten Alternativen zu Diltheys empirischem Ansatz, von der Phänomenologie, über die Fundamentalontologie, bis zur Hermeneutik und Dekonstruktion, alle daran kranken, dass sie den unmittelbaren Konnex zu den Einzelwissenschaften (und damit einen echten empirischen Charakter) vermissen lassen, dann bleibt Diltheys Methode, die diesen Konnex herstellt, gemeinsam mit ähnlichen Methoden wie der Gestaltpsychologie oder der Psychoanalyse, ein Desiderat auch für die rezente empirische Philosophie. Das Scheitern von Diltheys „deskriptiver Psychologie" war im Wesentlichen das Resultat eines Schwenks der Philosophie hin zu einem neuen Apriorismus. Wenn die Philosophie sich heute entschließt, erneut zu einer empirischeren Herangehensweise zu finden, dann könnte Diltheys unvollendetes Projekt, in diesem Zusammenhang neu, aufgegriffen werden.

Danksagungen

Dieser Aufsatz stellt eine theoretische Skizze dar, zu einem größeren Projekt zur Wissenschaftsphilosophie bei Dilthey und Carnap (FWF Projekt

---

14. Diese Formulierung verwendet Hubert L. Dreyfus mit Bezug auf Heideggers Fundamentalontologie. Vgl. (Dreyfus 1991, 3).

P24615, bewilligt am 6.3.2012). Für Hinweise zu verschiedenen Fassungen dieses Manuskripts bzw. für dieses Manuskript beeinflussende Bemerkungen zum Projekt im Allgemeinen bedanke ich mich bei Hans-Joachim Dahms, Richard Dawid, Gottfried Gabriel, Michael Heidelberger, Manfred Kohlbach, Daniel Kuby, Hans-Ulrich Lessing, Christoph Limbeck-Lilienau, Martin Kusch, Matthias Neuber, Friedrich Stadler sowie bei einem anonymen Gutachter der *Grazer Philosophischen Studien*.

Literatur

Bloor, David 1991 [1976]: *Knowledge and Social Imagery*. Chicago: University of Chicago Press.

Carnap, Rudolf 1968 [1934]: *Logische Syntax der Sprache*. Wien: Springer.

— 1998 [1928]: *Der logische Aufbau der Welt*. Hamburg: Felix Meiner.

Damböck, Christian 2012: „Rudolf Carnap and Wilhelm Dilthey: 'German' Empiricism in the *Aufbau*". In: Richard Creath (Hg.), *Carnap and the Legacy of Logical Empiricism. Vienna Circle Institute Yearbook*. Yearbook, 75–96.

Cassirer, Ernst 1994 [1910]: *Substanzbegriff und Funktionsbegriff. Untersuchungen über die Grundfragen der Erkenntniskritik*. Darmstadt: Wissenschaftliche Buchgesellschaft.

Dilthey, Wilhelm 1914–2006: *Gesammelte Schriften* (26 Bände). Göttingen: Vandenhoeck & Ruprecht.

Dreyfus, Hubert L. 1991: *Being-in-the-World. A Commentary on Heidegger's* Being and Time, *Division I*. Cambridge, MA: The MIT Press.

Ebbinghaus, Hermann 1896: „Über erklärende und beschreibende Psychologie". *Zeitschrift für Psychologie und Physiologie der Sinnesorgane* 9, 161–205.

Friedman, Michael 2001: *Dynamics of Reason. The 1999 Kant Lectures at Stanford University*. Stanford: CSLI Publications.

Gabriel, Gottfried 2004: „Introduction: Carnap Brought Home". In: Steve Awodey & Carsten Klein (Hg.), *Carnap Brought Home. The View from Jena*. Chicago: Open Court, 3–23.

Heidelberger, Michael 1993: *Die innere Seite der Natur. Gustav Theodor Fechners wissenschaftlich-philosophische Weltauffassung*. Frankfurt/Main: Vittorio Klostermann.

Holzhey, Helmut 2004: „Der Neukantianismus". In: Helmut Holzhey & Wolfgang Röd (Hg.), *Die Philosophie des ausgehenden 19. und des 20. Jahrhunderts 2. Neukantianismus, Idealismus, Realismus, Phänomenologie (Geschichte der*

Philosophie. Herausgegeben von Wolfgang Röd. Band XII). München: C.H. Beck, 13–129.

Kant, Immanuel 1998: *Kritik der reinen Vernunft*. Nach der ersten und zweiten Originalausgabe herausgegeben von Jens Timmermann. Hamburg: Felix Meiner.

Köhnke, Klaus Christian 1986: *Entstehung und Aufstieg des Neukantianismus. Die deutsche Universitätsphilosophie zwischen Idealismus und Positivismus*. Frankfurt/Main: Suhrkamp.

Knobe, Joshua & Shaun Nichols 2008: "An Experimental Philosophy Manifesto". In: Joshua Knobe & Shaun Nichols (Hg.), *Experimental Philosophy*. New York: Oxford University Press, 3–14.

Kuhn, Thomas 1996 [1962]: *The Structure of Scientific Revolutions*. Chicago: The University of Chicago Press.

Kusch, Martin 2000: "The Sociology of Philosophical Knowledge: A Case Study and a Defense". In: Martin Kusch (Hg.), *The Sociology of Philosophical Knowledge*. Dordrecht: Kluwer Academic Publishers, 15–38.

— 1995: *Psychologism. A Case Study in the Sociology of Philosophical Knowledge*. London: Routledge.

Lessing, Hans-Ulrich 1985: „Briefe an Dilthey anläßlich der Veröffentlichung seiner ‚Ideen über eine beschreibende und zergliedernde Psychologie'". *Dilthey Jahrbuch* 3, 192–232.

— 2001: *Wilhelm Diltheys ‚Einleitung in die Geisteswissenschaften'*. Darmstadt: Wissenschaftliche Buchgesellschaft.

Makkreel, Rudolf A. 1991 [1975]: *Dilthey. Philosoph der Geisteswissenschaften*. Frankfurt/Main: Suhrkamp.

Mill, John Stuart 1976 [1843]: *A System of Logic, Ratiocinative and Inductive, Being a Connected View of the Principles of Evidence and the Methods of Scientific Investigation*. Toronto: University of Toronto Press.

Quine, Willard Van Orman 1969: „Epistemology Naturalized". In: Willard Van Orman Quine, *Ontological Relativity and Other Essays*. New York: Columbia University Press, 69–90.

Rickert, Heinrich 1904: *Der Gegenstand der Erkenntnis. Einführung in die Transzendentalphilosophie*. Tübingen: Mohr.

Rodi, Frithjof & Hans-Ulrich Lessing 1984: *Materialien zur Philosophie Diltheys*. Frankfurt/Main: Suhrkamp.

Schnädelbach, Herbert 1983: *Philosophie in Deutschland 1831–1933*. Frankfurt/Main: Suhrkamp.

Stumpf, Carl 1910: „Die Wiedergeburt der Philosophie. Rede zum Antritt des Rektorates der königlichen Friedrich-Wilhelms-Universität Berlin". In: Carl Stumpf, *Philosophische Reden und Vorträge*. Leipzig: Johann Ambrosius Barth, 161–196.

van Fraassen, Bas 2002: *The Empirical Stance*. New Haven: Yale University Press.
van Kerckhoven, Guy, Hans-Ulrich Lessing & Axel Ossenkop 2008: *Wilhelm Dilthey. Leben und Werk in Bildern*. Freiburg: Karl Alber.
Williamson, Timothy 2007: *The Philosophy of Philosophy*. Malden: Blackwell.
Windelband, Wilhelm 1883: *Präludien. Aufsätze und Reden zur Philosophie und ihrer Geschichte*. Tübingen: J. C. B. Mohr (Paul Siebeck).

Grazer Philosophische Studien
85 (2012), 187–212.

SOCIALLY CONSTITUTED ACTIONS AND OBJECTS[1]

Bernd PRIEN
University of Münster

Summary

This paper addresses the question of how it is to be explained that an action or object *X* counts as *Y*. I argue that John Searle's notion of a constitutive rule should not be employed because it involves a confusion. Instead, I propose an explanatory framework containing the following three elements: 1) The *X*-action or object is involved in certain social practices. 2) These practices confer properties on the *X*-items, in addition to their physical properties. 3) In virtue of these additional properties the *X*-items fall under the *Y*-concept. Moreover, I stress that the social practices mentioned under 1) look differently from case to case and involve various constellations of prescriptive rules and interpersonal patterns of intentions. To illustrate and corroborate these claims, I discuss several cases of socially constituted actions and objects, in particular, the case of money, where I reject Searle's claim that something counts as money in virtue of giving its owner the right to buy something.

It is a well known phenomenon that performing an action *X* in a context *C* can count as performing a different action *Y*. For example, raising your hand at an auction counts as bidding, moving wooden objects across a board according to certain rules counts as playing chess, etc. Moreover, counts-as relations obtain not only between actions, but also between objects. For example, a piece of paper can count as money or, in a different context, as a bus ticket, a line on the ground as a border, etc.[2] Both in the case of objects and actions it is clear that the physical properties

1. Research for this article was funded by the German Research Council (DFG). I would like to thank Michael Kühler, Münster, and an anonymous reviewer for helpful comments.

2. Analytic philosophers have for a long time focused exclusively on actions that count as something, even though the case that an object counts as something is at least as common. Until John Searle's book *The Construction of Social Reality*, published in 1995, the relevant literature on this phenomenon, starting with Searle's book *Speech Acts* (1969), has focused exclusively on actions counting as something.

of the *X*-item (i.e., the action or object that counts as something) are not sufficient to explain that it counts as a *Y*-item, i.e., as a chess-move, as money, or as a border. Instead, it is due to social practices of some sort prevalent in a community that *X* counts as *Y*. For this reason, there is a degree of arbitrariness concerning which actions and objects count as what and, consequently, this can vary from community to community. I think that it is this dependence on social practices that we are trying to express when we say that something *counts* as something, rather than that it *is* something.

Even though the idea that actions and objects count as something evidently has a lot of intuitive appeal, it is on closer inspection not very clear what it means or how it can be explained that *X* counts as *Y*. This is the question I want to pursue in this paper. To do so, I will formulate a general framework for explaining why a counts-as relation obtains in section 1 and, in order to demonstrate its fecundity, apply it to different cases of counts-as relations in section 2 of this paper. In particular, I want to show that different cases of counts-as relations have to be accounted for separately. This means that the general framework proposed in section 1 will have to be filled in in different ways to account for the individual cases discussed in section 2. In section 3, I will summarize and comment on the results of this paper.

1. *A general framework for explaining counts-as relations*

The aim of this section is to formulate a general framework for explaining why an action or object *X* counts as *Y*. Before doing so, however, I want to describe what I take to be the received view concerning this matter, which is due to John Searle's book *Speech Acts* (SA). I think that this view is in part on the right track, though the waters are muddied, rather than cleared up, by Searle's distinction between regulative rules and constitutive rules. After describing and criticizing this view I will develop my own explanatory framework out of what I take to be correct about Searle's view. To conclude this section, I will point to some similarities between my proposal and Searle's more recent views about counts-as relations, expounded in *The Construction of Social Reality* (CSR), a book he published some 25 years after *Speech Acts* in 1995.

1.1 *Searle's view of counts-as relations in* Speech Acts

Before I consider Searle's view, I will briefly discuss a very simple, and clearly unsatisfactory, account of how an action or object can count as something. According to this account, *X* counts as *Y* simply because an additional concept *Y* has been introduced under which the *X*-actions or objects are subsumed. For example, we have introduced the concept 'to place a bid at an auction' to cover all of the different kinds of *X*-actions by means of which this can be done. The introduction of this concept is useful because there are many diverse actions having nothing in common apart from the fact that they all count as bidding. Thus, the introduction of the *Y*-concept greatly facilitates referring to all and only these actions.

An explanation along these lines is unsatisfactory. That raising your hand counts as voting involves more than that an additional concept 'to vote' is available in the language which applies to hand-raisings in certain contexts, the putting of crosses on pieces of paper in certain other contexts, etc.. Similarly, when a piece of paper counts as money or as a ticket, something more than the mere introduction of a new name for this object is going on. Roughly speaking, when something counts as voting or money, this involves that this something has a certain 'import', i.e., certain consequences. This feature of counts-as relations cannot be explained by assuming that an additional concept is introduced. So, even though it is true that *Y* is an additional concept that applies to *X*-actions or objects, this cannot be the whole story.

I will now turn to Searle's view of the nature of counts-as relations. In *Speech Acts*, Searle claims that relations of this kind are due to a special kind of rules, for which he introduces the term 'constitutive rules', in contradistinction to regulative rules. Searle characterizes constitutive rules by the following two formulae:

1) "The creation of constitutive rules, as it were, creates the possibility of new forms of behavior" (SA, 35), whereas regulative rules regulate independently existing ones.
2) "[C]onstitutive rules often have the form: *X* counts as *Y* in context *C*" (SA, 35), while regulative rules "have the form or can be comfortably paraphrased in the form 'Do *X*' or 'If *Y* do *X*'." (SA, 34).

I will briefly explain these two characterizations before I argue that it is hard to see how both can be characterizations of a single entity.

1) Regulative rules regulate actions that can be performed independently of their existence, while constitutive rules first make it possible to perform the actions they are constitutive of. Thus, the rule 'Invitations should be sent out two weeks in advance' is regulative because one can send out invitations even if this rule does not exist. The rules of soccer, on the other hand, are constitutive because one could not play soccer if these rules did not exist. Even if 22 people went through the exact same movements as the players in a soccer game, they could not be said to play soccer unless their actions conform, at least for the most part, to the rules of soccer (cf. SA, 33ff.).

2) The characterization that constitutive rules can be expressed in the form '*X* counts as *Y*'[3] is subject to the following qualifications: First of all, the *Y*-term has to be a specification, such as 'to play soccer', rather than a term of appraisal, such as '(in-)correct', because any regulative rule can be twisted into the form 'Doing *X* counts as incorrect behavior' (SA, 36). Even if this requirement is met, however, expressibility in the counts-as form is not a necessary, but at most a sufficient criterion for constitutive rules. Rules which can be paraphrased in this form are "likely to be constitutive." (SA, 36) On the other hand, many constitutive rules, such as the rule of soccer 'Do not touch the ball with your hands', cannot happily be phrased in the counts-as form without using a term of appraisal. However, "since constitutive rules come in systems, it may be the whole system which exemplifies this form and not individual rules within the system." (SA 36, cf. CSR 28)

For the discussion of Searle's notion of a constitutive rule it will be helpful to have the notion of a prescriptive rule (a prescription, for short) at hand. By a prescription, I mean a rule according to which agents are committed or entitled to perform an action *a* if they are in a situation *S*. Thus prescriptions are expressed by conditionals of the following kind: 'If you are in situation *S*, you should/may do *a*.' Since the requirement in the antecedent can also be empty, prescriptions are sometimes expressed

3. It can be doubted that 'rules' of the form '*X* counts as *Y*' are rules at all because it is impossible to violate them (cf., e.g., Glüer & Pagin 1999, 217f.). Even though one can react to this objection by simply adopting a wider notion of a rule according to which the possibility of violation is not necessary, this observation still throws doubts on Searle's comparison of the fact that *X* counts as *Y* (instituted by a constitutive rule) to the fact that one should do *A* in situation *S* (instituted by a regulative rule).

simply by: 'One should/may do *a*.' Concerning their institution, it seems plausible to assume that prescriptions are not necessarily issued by an authority, nor arrived at by explicit agreement in the community. Rather, prescriptions often emerge out of communally shared ways of acting. In this case, the agents in the community are often unable to explicitly formulate the prescriptions they follow. However, it would be beyond the scope of this paper to address the question of how prescriptions can emerge from communal practices.

Though this characterization of prescriptive rules resembles the one Searle gives for regulative rules (cf. SA, 34), the two concepts do not coincide because the concept 'prescription' does not contain the idea of a contrast with constitutive rules. Thus most, if not all of the rules of soccer, such as 'Do not touch the ball with your hand', 'If a member of one team kicks the ball across the sideline, the other team is entitled to a throw-in', etc. are prescriptions, while Searle would classify them as constitutive, rather than regulative rules.

Having explained the two formulae Searle uses to characterize constitutive rules, I will now present what I take to be the most natural interpretation of his first characterization of constitutive rules (introduction of new forms of behavior). Following that, I will argue that, assuming this interpretation, the second characterization (expressibility in the counts-as form) introduces a hidden ambiguity into the concept 'constitutive rule'.

Though Searle does not explicate it in this way, the most natural interpretation of his characterization of constitutive rules as introducing new forms of behavior seems to be this:[4] The rules of soccer, prescriptions such as 'Each team has 11 players', 'Field players may not touch the ball with their hands', etc., introduce the new form of behavior 'to play soccer' because the content of the concept 'to play soccer' is such that it only applies to actions conforming to or at least subject to the prescriptions known as the rules of soccer.[5] This interpretation gives a clear sense to the idea that the prescriptions known as the rules of soccer first make it pos-

4. It should be stressed that I only think there is a natural interpretation of Searle's *first characterization* of constitutive rules, not of his account of constitutive rules as a whole. My view is that the two elements that form this account do not sit easily with each other, with the consequence that there is no natural way of interpreting the account as a whole.

5. Arguably, the requirement of conformity to these rules is too strong because one still counts as playing soccer if one touches the ball with one's hands, provided this is seen as a violation of a rule and as justifying a sanction, e.g., in the form of a free kick for the opposing team. Therefore, it seems more appropriate to say that the concept 'to play soccer' is such that it only applies to actions *subject to* the rules of soccer (cf. Glüer & Pagin 1999, 221). I will return

sible to perform the action of playing soccer. This proposal has the further virtue of being able to explain why an *X*-action falls under a *Y*-concept, even though the latter typically involve some kind of import, i.e., "introduce further consequences, by way of, e.g., penalties, points, and winning and losing." (SA, 36), while the former have no such import. According to the present proposal, the *Y*-concept does not apply to the *X*-action as such, but only insofar as the *X*-action is governed by certain prescriptive rules. Thus, the *Y*-term 'scoring a goal', which is connected to winning and losing, applies to the *X*-action of kicking the ball across the goal line because this action is subject to the following rule of soccer: If the ball is kicked across the goal line of one team, the other team is entitled to one point.

The interpretation just sketched leads to a clear-cut understanding of what a constitutive rule is, namely a prescription an action *X* has to conform to if a concept *Y* is to apply to *X*. This is a natural, and in my view essentially correct way of understanding how (prescriptive) rules can introduce new forms of behavior. It should be noted that this interpretation of Searle's first characterization of constitutive rules implies that prescriptive rules are constitutive rules only relative to a *Y*-concept. This follows simply because we can, for any prescriptive rule *R* governing the performance of an action *a*, define the concept '*a* performed according to rule *R*'. Then the prescriptive rule *R* will be constitutive of this latter action: it could not be performed if *R* did not exist.[6] Thus, a rule cannot be called constitutive *simpliciter*, but only relative to a *Y*-action (cf. Ruben 1997, 443ff.[7], Warnock 1971, 37f.[8], Black 1962, 123f.). In my view, none of this speaks against this understanding of 'constitutive rule'.

---

to this point in section 2.1 when I discuss the case of playing chess. At this point, however, the difference between conforming to rules and merely being subject to them is not relevant.

6. To take a concrete example: On the suggested interpretation, the prescription 'Take control of the center of the board' has to be regarded as regulative with regard to playing chess and as constitutive with regard to playing chess well.

7. In his response to Ruben, Searle himself rejects the conclusion that rules are constitutive only relative to a *Y*-concept (cf. 1997, 454ff.). However, I am not convinced by Searle's arguments.

8. Warnock's distinction between two different reasons for having rules might help to distinguish genuinely constitutive rules from others. He argues that, though the rules of criminal law create the possibility of new forms of behavior, their purpose is "to 'regulate' in certain respects the conduct of members of society." (Warnock 1971, 38) The rules of soccer, on the other hand, "'regulate' the way in which balls are kicked about in fields" (Warnock 1971, 38), but their purpose is to introduce new forms of behavior. It is not clear, however, how this distinction between two reasons for having rules can be made sufficiently precise.

What is problematic about Searle's account of constitutive rules, however, is that he also takes them to be characterized by the fact that they "often have the form: *X* counts as *Y* in context *C*" (SA, 35). As I will now argue, this introduces a hidden ambiguity into the concept 'constitutive rule'. According to the 'natural way' of understanding of how (prescriptive) rules introduce new forms of behavior, one would describe the case of chess in the following way:

(*) Moving an object across a board according to a certain system of prescriptive rules counts as making a chess move.

In Searle's way of speaking, (*) is a rule of the form 'Doing *X* counts as doing *Y*', in which the *X*-term refers to actions which are subject to the system of the rules of chess. This latter system consists exclusively of prescriptive rules governing antecedently possible actions of moving objects across a board,[9] e.g., the rule 'The object with the starting position e1 or e8 (what is called the king) may move one step in any direction'. Though Searle concedes that the individual rules of chess cannot be phrased in the counts-as from, he insists, as we have seen above, that the system of these rules can be phrased in this way. Now, this claim comes dangerously close to confusing the whole counts-as rule (*) with the system of the rules of chess mentioned in its *X*-slot. However, it is obvious that these two things should be kept apart: On the one hand, there is the system of rules specifying the ways in which objects may be moved across a board, while on the other, there is a statement of a counts-as relation that makes reference to this system of rules. Thus, Searle's claim that the system of the rules of chess can be phrased in the counts-as form can lead to confusion insofar as it might be thought that he is speaking of the whole counts-as statement (*), rather than just of the system of prescriptive rules referred to in its *X*-slot.

The general problem is that the practices in virtue of which an *X*-item counts as *Y* often involve a system of prescriptive rules. Confusion can arise because the concept 'constitutive rule' can either be taken to apply to these systems of prescriptions or to the statement of the counts-as relation between *X* and *Y*, which obtains because *X* is subject to this system of

9. It may be objected that moving objects across a board is not a practice because there does not seem to be much point in engaging in it. However, I am not sure why having a point should be a condition for qualifying as a practice. Moreover, with sufficient ingenuity one could imagine circumstances in which moving objects across a board serves some purpose.

prescriptions. By proposing his two formulae to characterize constitutive rules, Searle seems to want to have it both ways.[10]

If this argument is accepted, one way to react would be to understand the term 'constitutive rule' in accordance with the interpretation of Searle's first formula proposed above, i.e., as prescriptive rules an *X*-action has to be subject to if a *Y*-term is to apply to it. This, however, would have the consequence that Searle's second formula (expressibility in the counts-as form) no longer applies to constitutive rules. Accepting this consequence would be a considerable concession for Searle to make.

Even tough I think that understanding the term 'constitutive rule' in the way just described has a lot to be said for it, Searle's understanding of 'constitutive rule' in terms of *both* of his formulae is extremely well entrenched in analytic philosophy. Thus, assuming a modified understanding of this term is likely to engender confusions. In my own account, I will therefore eschew using the term 'constitutive rule' and its counterpart 'regulative rule' altogether. Instead, I will speak of prescriptive rules governing *X*-actions and of counts-as relations obtaining between an *X*- and a *Y*-action or object.

1.2 *A framework for explaining counts-as relations*

It is the central aim of this paper to introduce and to defend a general framework for explaining why an action or object *X* counts as *Y*. This framework can be arrived at by explicating and suitably generalizing certain features of the above explanation of why kicking the ball counts as playing soccer. Roughly speaking, this was explained by the fact that the action of kicking the ball is subject to the rules of soccer, that because of this the action of kicking the ball acquires additional, non-physical properties, and

10. Glüer and Pagin seem to see a tension between Searle's two formulae as well: They distinguish between two alternative readings of what constitutive rules are, which seem to correspond to the two formulae Searle uses to characterize them. On the one hand, there is what they call the traditional reading, according to which constitutive rules are characterized by the counts-as form (cf. Glüer & Pagin 1999, 215ff.). On the other hand, there is what they call the non-traditional reading, which Glüer and Pagin summarize as follows: "a practice is constituted by a set of rules if it is possible to engage in that practice only insofar as the rules of that set are in force for the agent." (Glüer & Pagin 1999, 221) If I understand them correctly, the non-traditional reading agrees with the way I have interpreted Searle's claim that constitutive rules create "the possibility of new forms of behavior" (SA, 35). Now, the fact that Glüer and Pagin introduce the traditional and the non-traditional as two "alternative" readings (Glüer & Pagin 1999, 208) does not necessarily mean that they take them to be incompatible, but it strongly suggests such a view.

that the concept 'to play soccer' applies to actions with these properties. The following three elements can be discerned in this explanation:

1) The *X*-action or object is involved in social practices of a certain kind.
2) These practices confer new properties on the *X*-action or object, i.e., properties in addition to their physical properties.
3) The content of the *Y*-concept is such that it only applies to actions or objects with these properties.

These three elements combine into an explanation of the following shape: *X* counts as *Y* because *X* is involved in social practices of a certain kind, which practices confer additional properties on *X*, and *Y* applies to actions or objects in virtue of these properties.

I have arrived at this explanatory framework by *generalizing* certain features of the explanation why kicking the ball counts as playing soccer. Before proceeding to apply this framework to different cases in section 2 of this paper, I want to comment on the nature and the point of generalizing in just the way I have done.

a) The social practices mentioned in elements 1 and 2 are to be conceived of as consisting of prescriptive rules, of social patterns of individual intentions, and of collective or shared intentions (conceived of as irreducible to individual intentions).[11] Thus, the social practice the *X*-action or object is involved in need not always and exclusively consist in a system of prescriptive rules. Indeed, even in the case of soccer, it can be argued that, apart from being subject to the rules of soccer, both teams also have to pursue the goal of winning[12] if their *X*-actions of kicking the ball are to count as playing soccer. In this case, an interpersonal pattern of intentions is part of the practices the *X*-actions and objects are involved in.

b) Moreover, assuming that the practices in virtue of which something counts as playing soccer contain two quite different elements, a system of rules and a pattern of individual intentions, the idea suggests

---

11. It is a crucial feature of the explanatory framework proposed here that the list of elements comprising the relevant practices is open. Even if I can't think of any candidates beyond those listed above, nothing speaks against further elements comprising our practices. One element the practices cannot be taken to consist of is, obviously, that of counts-as relations.

12. Winning can be spelled out as kicking the ball across the opposing team's goal-line more often than the opposing team.

itself that the practices accounting for different counts-as relations might look quite differently. In section 2 of this paper, I will discuss a number of different socially constituted actions and objects (making a chess-move, money, bidding at an auction, tickets, and borders) to argue that this is indeed the case. Thus, different counts-as relations have to be analyzed separately.

c) Element 3 of the explanatory framework invokes the content of the *Y*-concept. This can either be done in the form of giving an explicit analysis of the *Y*-concept (e.g., the explanation of why pieces of paper count as money offered in section 2.2 starts from the assumption that the concept 'money' conceptually involves being a medium of exchange) or it can be done in the form of relying on our implicit grasp of this content, as in the above explanation of soccer. Even though this explanation contains no explicit analysis of the concept 'soccer', it relies on our implicit grasp of the content of this concept in order to judge whether the involvement in practices of a certain kind is necessary and/or sufficient for actions of kicking a ball about on a field to count as playing soccer.

d) Concerning the explanation why the *X*-item falls under the *Y*-concept, the following restriction has to be made: Sometimes, the properties conferred by the social practices explain only in part why the *X*-item falls under the *Y*-concept because *Y* also contains as one of its conceptual marks that objects falling under it have certain physical properties. Thus, in his discussion of baseball, Ransdell argues that "the application of the game-term 'Bat' [a *Y*-concept in my terminology] to a given object *connotes* that the object has a certain size, shape, construction, etc., but the *import* of the term concerns what can and cannot be done with the object within the game" (1971, 388). Ransdell concludes that "the meaning of game-terms [of baseball, such as 'bat', 'field', 'ball'] (with a few exceptions) is dual comprising both connotation and import" (1971, 389) or, in my terminology, both physical properties and socially conferred properties.

I do not want to commit myself to the claim that the meaning of *Y*-concepts in general is dual in this sense, but an adequate explanatory framework has to allow for the possibility of *Y*-concepts of this kind. And insofar as this is the case, explanations conforming to the framework presented here can explain only in part why the *X*-item falls under the *Y*-concept, namely as far as its import, i.e., the socially conferred properties are concerned. That the *X*-item also satisfies the connotation of the *Y*-concept, i.e., has the appropriate physical properties, will be presupposed here.

1.3 *Searle's more recent account of constitutive rules*

Before I turn to the application of the explanatory framework to individual cases, I want to give credit for some of its features to Searle's book *The Construction of Social Reality* (CSR). There, Searle proposes an account of counts-as relations which introduces a couple of considerations not present in *Speech Acts*. Searle does not take these new considerations to contradict his earlier analysis, but rather to constitute a "more thorough account" (CSR, 29). However that may be, I just want to note that my proposal of an explanatory framework is in part inspired by these more recent considerations. The considerations I have in mind are summed up in the following quote:

> Furthermore, because the physical features specified by the X term are insufficient by themselves to guarantee the fulfillment of the assigned function specified by the Y term, the new status and its attendant function have to be the sort of things that can be constituted by collective agreement or acceptance. (CSR, 44)

In this quote, Searle presents the idea that a status-function is assigned to an *X*-object which it cannot serve purely in virtue of its physical properties. Later in the book, it emerges that *X*-objects in general are supposed to serve their status-functions by means of deontic powers, i.e., by means of the fact that certain commitments and entitlements pertain to their use. Thus, a piece of paper counts as a ticket roughly because it entitles its owner to enter a certain room, and a line on the ground counts a border because it may not be crossed. In correspondence to this, the explanatory framework proposed here more generally invokes properties going beyond the physical properties of an *X*-item. I prefer this more general formulation because I do not think that the additional properties conferred by social practices can always be understood in normative terms. In particular, I will argue in section 2.2 that Searle's favorite example 'money' cannot be analyzed in terms of deontic powers.

So, apart from the fact that Searle still uses the two formulae to characterize constitutive rules (cf. CSR, 27f.), there is much that I find inspiring in Searle's more recent account of counts-as relations. However, with regard to both the practices and the properties conferred on actions and objects, I would allow for greater variability than Searle does.

2. *Different cases of socially constituted actions and objects*

In order to let the explanatory framework just outlined prove its mettle, and to illustrate the differences between the kinds of social practices underlying different counts-as relations, I will now discuss the cases of 1) making a chess move, 2) money, 3) bidding at an auction, and 4) tickets and borders. The first two cases are the more important ones, so those who have little time (and who doesn't?) can skip sections 2.3 and 2.4 and go directly to the conclusion in section 3.

2.1 *Making a chess move*

Let us start by considering the social practices in virtue of which moving objects across a board counts as making a chess move. In my view, three different constituents should be distinguished here: First, there are two opposing players each pursuing the goal of taking the opponent's king. Secondly, there are the rules of chess specifying the ways in which the different chess pieces may be moved.[13] Moving objects across a board counts as making a chess move only if this is done in a context in which these moves are subject to these rules and if the movements at least by and large conform to these rules. Thirdly, the moves in a chess game form a systematic whole in which each move constrains which future moves are allowed and advisable for both players, given their goals of taking the opponent's king.

Concerning the second constituent of the practice, the following remarks should be made: In footnote 5, I have explained with regard to the case of playing soccer that the requirement that *X*-actions should *conform to* the rules of soccer to count as playing soccer is too strong, and that the requirement that they should be *subject to* them seems more appropriate. The same can be said about chess: One can violate a rule of chess, i.e., move a chess piece in an incorrect way, and still count as playing chess, provided such a move has consequences that mark it as a violation, e.g., that one should correct the move, that one loses the chess-piece in question, or that one has lost the game. It seems wrong, in my view, to say of

13. Chess pieces cannot be defined in terms of their shape because the king, the bishop, etc. can in principle have any shape. However, chess pieces can be identified by their position on the board at the beginning of the game. Therefore, the rules of chess have to be of the form: The chess piece with the starting position a1 is allowed to move in such and such ways: …

a player in a chess game who violates a rule of chess that he wasn't playing chess after all.[14]

This description of social practices, with its three constituents, serves to fill in the first element of the explanatory framework. With regard to its second element, I want to argue that these social practices confer an additional property on the actions involved in them. In the case of chess it is, admittedly, difficult to give an independent description of this property. All that can be said is that in virtue of the practices just described (with the three constituents of opposing goals, constraint by the rules of chess, and forming a systematic whole) the action of moving an object across a board acquires the qualities characteristic of making a chess move.

The third element of the general framework consists in an analysis of the *Y*-concept 'making a chess move'. In the present case, we have to be content to rely on our intuitive grasp of this concept in order to judge that actions which are to count as chess moves have to be involved in the practices with the constituents just described.[15]

In sum, the fact that certain actions of moving objects count as chess moves is explained by the fact that the content of the concept 'chess move' is such that it only applies to actions of moving objects across a board which are involved in certain social practices in virtue of which they acquire the characteristic qualities of chess moves. According to this explanation, it is also clear that 'making a chess move' is not simply a new name for certain acts of moving objects across a board. The concept 'chess move' does not apply to actions of moving objects across a board *simpliciter*, but only insofar as they are involved in practices of the kind just described.

I want to close my discussion of the chess case with the following remarks concerning the practices relevant in this case: The social prac-

14. Against this view, an anonymous reviewer has proposed to consider the following case: Suppose that someone who after violating a rule of chess is asked to correct her move simply refuses to do so. In such a case, the objection goes, we would say that she simply is not playing chess (any more). In response to this objection I want to argue that my theory, according to which an *X*-action counts as *Y* if it is *subject to* the prescriptive rules in question, can accommodate this intuition. To do so, it has to be taken into account that not acting in accordance with prescriptions can in some cases lead to one's no longer being subject to them either. This can be so especially if one was subject to them by voluntary decision in the first place, e.g., by agreeing to play a game of chess. Simply no longer obeying the rules of chess is then a way of saying: I no longer recognize these rules as binding me.

15. Furthermore, I think that game concepts such as chess, soccer, etc. are vague in that they allow for some degree of modification of the rules the corresponding *X*-actions have to be subject to. Thus, one can play soccer, not with one fixed player occupying the role of the goalkeeper, but with the rule that the player closest to her own team's goal-line occupies this role.

tices conferring an additional property on the action of moving an object across a board contain prescriptive rules governing actions and individual intentional states. Even though most readers will probably concede that it can be assumed as clear in the context of this discussion what intentional states are, it may be objected that my account of phenomena of the kind that *X* counts as *Y* presupposes the notion of a prescriptive rule. It may be objected that the notion of a prescription is itself rather problematic and in need of much more elucidation than has been given above in section 1.1. Even though I have to concede this, I still think that explaining cases of counts-as relations partly in terms of prescriptive rules is a considerable step ahead because the notion of a prescriptive rule is in any case much clearer than that of *X* counting as *Y*.

2.2 *Money*

As a second example, I want to discuss the case of money. In accordance with the explanatory framework presented above we need, as one element of an explanation of why something counts as money, a rough analysis of the *Y*-concept, in this case of 'money'. There are several properties generally taken to be characteristic of money: being a medium of exchange, a unit of account, and a store of value. In what follows, I will take 'money' to be defined by being a medium of exchange which seems to be its central function.

For the discussion of the practices in virtue of which objects can serve as a medium of exchange, i.e., fall under the concept 'money', it is helpful to distinguish, as Searle does, between three types of money: commodity money, contract money, and *fiat* money (cf. CSR, 42f.). In the first case, the objects used as money, coins made of gold or silver, are themselves regarded as valuable. However, because it is inconvenient and dangerous to carry around large amounts of gold, contract money has been introduced. In this system, certificates are issued by banks in exchange for coins made of gold and silver. These certificates are then used to buy and sell goods, as placeholders for coins, as it were. Moreover, if such a certificate is presented to the bank that issued it, the bank is committed to give the presenter a certain amount of gold in exchange for it.

My main interest, however, concerns *fiat* money. As Searle explains, *fiat* money has developed out of contract money once it was recognized that it is not necessary to deposit gold at banks for the certificates to serve as a medium of exchange. Since people most of the time use certificates only to buy and sell goods and only rarely to withdraw the gold they have deposited for them, the bank can start to issue more certificates than it has gold available. In a second step, the deposition of gold for certificates can be abolished completely. Certificates (pieces of paper) are used to buy and sell goods, but one can no longer go to a central bank and demand to be paid a certain amount of gold.[16]

Let me now turn to the social practices relevant in the case of fiat money. I will start by discussing and rejecting Searle's proposal as to what these practices look like. I hope that my critique of Searle helps to make the motivation behind my own proposal clearer. Searle wants to account for the fact that bits of paper can serve as a medium of exchange by pointing to certain deontic powers.[17] In his view, bits of paper of a certain kind count as money because they give their owner the right to buy something (cf. CSR, 70f.). However, Searle does not go into any details about what the right to buy something amounts to. Three possibilities of what he might have in mind suggest themselves. On all three interpretations, however, Searle's claim that money gives its owner the right to buy something turns out to be untenable.

1) The 'right to buy something' one has in virtue of owning money might consist in the authority to make a contract. This is a right children under a certain age do not have. But whether or not one has this right depends on one's age, not on one's possession of bits of paper. If I am old enough, I can enter into a binding contract whether or not I own money. The question is whether I will be able to honor the contract. So, 'the right to buy something' that money gives its owner cannot be understood in this way.
2) The 'right to buy something' could also be understood in the sense that handing over the money to somebody commits that person to give me certain useful goods. But in this sense, owning money does not give me the right to buy anything. Others are only committed to give me something in exchange for my money if they have agreed on a contract with me to do so. However, nobody is

16. Of course, one can still buy gold on the market, but one can no longer demand to be given a *certain amount* of it. The price of gold may rise and one may end up getting very little gold for one's money.

17. Searle uses 'deontic power' as a general term for commitments, entitlements, obligations, permissions, authorizations to do something (cf. CSR, 100).

obliged to enter into an agreement with me according to which she has to give me useful goods for my money.

3) Finally, that one has the 'right to buy something' can be understood in the weak sense that it is not forbidden to buy something. In this sense, people do in fact have this right, but this right is nothing people enjoy in virtue of owning money. It is not forbidden to buy something whether or not one owns money. Moreover, it remains unclear how 'the right to buy something' in this sense can confer the property of being a medium of exchange on something.

So, on reflection, it is not at all clear what the deontic power associated with owning money could amount to.[18] In view of the difficulties to specify a deontic power money gives its owner, I will argue that the power or ability to buy something that money gives its owner should rather be understood in terms of interpersonal patterns of intentional states. In other words, I will argue that the practices in virtue of which bits of paper can serve as a medium of exchange essentially consist in an interpersonal pattern of intentional states, rather than in prescriptive norms and the deontic powers they give rise to.

I think that the social pattern of intentional states in virtue of which pieces of paper serve as a medium of exchange contains two elements: The first element consists in the fact that there are many agents who are prepared or willing to exchange their useful goods for useless bits of paper of a certain kind.[19] People are often prepared or willing to do this, but they are not committed to do so, simply because nobody is committed to sell anything to anybody. The second element of the pattern of intentional

18. The reason why Searle thinks money has to have a deontic power is probably the following: I can buy something (in the sense of making a sales contract *and* honoring it) only if I own enough money. So, owning money gives me a certain power, in the sense that it puts me in a position to perform an action I could not otherwise perform. This action can be characterized as acquiring the property rights to a specific object. Moreover, this ability to do something is not explained by the physical properties of money. While I accept all of this, I do not accept the conclusion Searle seems to draw from it, namely that this non-physical property of money has to be a deontic power, similar to the authority of a president to appoint somebody as vice-president, say. Searle's conclusion would be valid only if having deontic powers was the only way in which objects can have properties going beyond their physical ones. Only in that case would we be forced to understand the ability to acquire property rights that money gives its owner as a deontic power. As we will now see, however, this is not the case.

19. Not everybody is prepared to sell her house, furniture, car, or whatever, even if offered a reasonable price. What seems crucial for our money-practices is that for a large class of goods, one can be sure to find someone who is willing to sell them.

states consists in the shared belief that there are many agents prepared to exchange their useful goods for useless pieces of paper.[20]

Thus, there is a wide-spread preparedness to give away useful goods in exchange for useless pieces of paper and it is common knowledge that this is so. The fact that agents are aware of this wide-spread preparedness explains why they themselves are often prepared to give away useful goods for useless pieces of paper (which considered in isolation seems rather stupid): They are prepared to do so because they can be sure to find another agent who is prepared to take their useless pieces of paper in exchange for (other) useful goods.

Now, the social practice consisting in this interpersonal pattern of intentions confers a property on pieces of paper which can be characterized as 'generally known wide-spread acceptance in exchange for useful goods'. In virtue of this property, the pieces of paper can serve as a medium of exchange and thus fall under the concept 'money'.[21]

It is also in virtue of this interpersonal pattern of intentional states that owning money gives agents the power to buy something in the sense of making a sales contract *and* honoring it. Generally, there will be someone in the economic community who is willing to sell an object of the kind I want to buy, and on this assumption owning money puts me into a position to acquire the property rights to an object of this kind. So, the fact that the power to buy something that money gives its owner cannot be due to its physical properties does not imply, as Searle seems to assume, that this power has to be a deontic power.

Even though the assumption of a social pattern of intentions takes up center stage in this account, one cannot entirely eschew prescriptive rules in the social constitution of money: For the practice of exchanging useful goods for useless pieces of paper to work, it is important that agents

20. Two remarks concerning the uselessness of these pieces of paper should be made. First of all, they are not completely useless because one could, for instance, use them to take notes on. But in comparison to many other goods their use is rather limited. Secondly, an anonymous reviewer has remarked that pieces of paper are not useless if they can be exchanged for useful goods. In response to this I want to clarify that 'useless' should here be understood as 'useless in itself' or useless apart from being exchangeable with other goods.

21. An anonymous reviewer has objected that this is a mere Cambridge property and that Cambridge properties are explanatorily void. I concede that the property in virtue of which bits of paper counts as money is a Cambridge property because in the end, this property just consists in being the object of intentional states. However, when it comes to the rational explanation of actions, properties of this kind often do explanatory work: For example, the fact that I hate a certain painting might be the reason why somebody takes it to the next room.

cannot simply produce or manufacture these bits of paper. A prescriptive rule is needed according to which these bits of paper may only be produced by a special agency that also keeps an eye on the amount of money in circulation. Even so, the assumption of social patterns of intentional states constitutes the most important part of the explanation of why bits of paper count as money, while in the explanations of chess, bidding, tickets, and borders, the assumption of prescriptive rules was or will be central.[22]

2.3 *Bidding at an auction*

In order to account for the case of bidding, the explanatory template proposed here has to be filled in in a slightly different way, in particular with regard to the practices the action of raising one's hand has to be involved in in order to count as bidding. Before I describe these practices, I will propose an analysis of the *Y*-term 'placing a bid', i.e., I will start with the third element of the explanatory framework. In my view, placing a bid conceptually involves bringing about certain normative consequences. More specifically, this action in part consists in acknowledging a commitment to buy the auctioned object if no other bids are placed.

The general idea that an action conceptually involves modifying the normative situation can be illustrated by the example of the action of walking the batter which can be performed in the context of a baseball game. By the normative situation, I mean the collection of all commitments and entitlements that persons in a given context have. In the case of baseball, the normative situation is quite complex because there are different social positions in this game that are occupied by different players, such as the pitcher, the batter, the fielders, and the runners. The rules of baseball specify the different commitments and entitlements of the players occupying these different positions. This means that, according to their position in the game, the players are committed and entitled to do different

22. I think that in the cases of commodity money and contract money the practices in virtue of which objects count as money are essentially the same as those described here: pieces of gold and certificates for gold count as money because many people are prepared to accept them in exchange for useful goods and because it is generally known that many people are prepared to do so. The difference is that while in the case of *fiat* money, this preparedness is grounded exclusively in trust, it is in the other cases supported by other considerations as well: People's preparedness to accept money for useful goods is supported by the consideration that even if they can't find anybody who accepts their money in exchange for useful goods, they still have the gold or the certificates which are 'as good as gold'.

things: One player is the pitcher; he has to stand on the pitcher's mount and throw (pitch) the ball at the batter. The batter has to stand at home base; he is entitled (and expected to) swing his bat so as to hit the ball. Then, there are fielders who have to stand somewhere on the field and are entitled to catch the ball and throw it to the baseman. Finally, there are runners who are entitled to run from one base to the next.

Now, this normative situation can be modified by certain moves in the game: For example, there is a rule in baseball according to which the batter, when he has allowed four balls to pass, has to leave home base and go to first base. When this happens, it is said that the batter has been walked, or that the pitcher has walked the batter. He is then no longer the batter, but a runner at first base. This means that new commitments and entitlements arise for this player: He is no longer allowed to use a bat to hit the ball. Instead, he is allowed, and also expected to try to reach second base, etc. Moreover, this change in the normative situation is what is essential to a batter being walked. So, the action of walking the batter conceptually involves modifying the normative situation in the way just described.[23]

Something similar happens in the case of placing a bid at an auction: When someone performs this action, she acquires a commitment and an entitlement to buy the auctioned object at the price in question, if no higher bids are placed. Bidding *conceptually* involves this normative consequence, as can be seen from the fact that an acquisition of this commitment *cum* entitlement is what is common to all cases of placing a bid, in which ever way this may be done (raising your hand, shouting out a number, pressing a key, etc.). Moreover, it should be noted that to place a bid at an auction is to generate a *conditional* commitment *cum* entitlement, in the sense that this deontic status only takes effect if no higher bids are placed.

So far, I have analyzed the *Y*-concept 'to place a bid' and argued that this action conceptually involves modifying the normative situation. What it is to perform this action is, essentially, to add to one's existing commitments and entitlements to act a conditional commitment and entitlement to buy a certain object at a certain price. Having analyzed the *Y*-concept in accordance with element 3 of the explanatory framework, I will now turn to its first element, the description of the social practices *X* is involved in.

23. After presenting this example, Ransdell concludes: "Thus instead of saying 'he was Walked' we could say something like: 'In virtue of having been Pitched four Balls he is now subject to the following rules: ______.'" (Ransdell 1971, 390) Here, the blank is to be filled by rules of the form 'Runners at First Base have the following commitments and entitlements: …'. That a player is now subject to these rules means that they now apply to him.

In the description of these practices, I will again presuppose the notion of a prescriptive rule. In particular, it will be important here that prescriptive rules often are of the conditional form 'If condition *C* obtains, action *A* should/may be performed', i.e., that they require for their applicability that a certain condition *C* is fulfilled and state that in this case certain commitments or entitlements arise.

Now, two cases can be distinguished with regard to the condition *C*: On the one hand, *C* can be the condition that an event has occurred or that a state of affairs obtains, as, respectively, in the prescriptions 'When a phone call has been cut off, the original caller should try again and the other person should wait' and 'When it is dark, drivers should turn on their lights'. On the other hand, *C* can be the condition that a certain action is performed in a certain context, as in the prescriptions 'When you enter a church, you should take off your hat' or 'When the king enters the room, everybody should stand up'. This second possibility is the one that is relevant for the case of bidding at an auction. If someone performs the action (in the appropriate context) mentioned in the antecedent of such a prescription, the commitments and entitlements mentioned in the consequent of the prescription will arise. This means that the action mentioned in the antecedent will not only have causal consequences, but also normative consequences, i.e., it will result in the modification or creation of commitments and entitlements.

To apply this general idea to the case of bidding, we have to assume that in the context of an auction the following prescription is in force:

> If you raise your hand and no one else raises his or her hand after that, then you are committed and entitled to buy the auctioned item at the price in question.

This is a prescriptive rule of the kind just described, i.e., a rule requiring for its applicability that certain actions are or have been performed, and which states that in this case certain further actions may or should be performed. In virtue of this prescription, the action of raising your hand has the normative consequence that you acquire the conditional commitment and entitlement to buy the auctioned item, contingent on the fact that no one else raises his or her hand.[24]

24. There is a crucial difference between the rules 'When you enter a church, then …' and 'When the king enters the room, then …' on the one hand and the rule 'If you raise your hand at an auction, then …' on the other. In the latter case, the action mentioned in the antecedent

The social practice the *X*-action is involved in consists in the prescription just formulated being in force. This constitutes the first element of the explanatory framework. According to its second element, a property these practices confer on the *X*-action has to be identified. In the case of bidding, the social practices confer on the action of raising your hand the property of having the above-mentioned normative consequence.

After filling in all three elements of the explanatory framework, the analysis of the *Y*-concept, the description of the social practices *X* is involved in, and the properties conferred by these practices, we can now draw these elements together in order to explain why raising your hand counts as placing a bid: In virtue of the norm formulated above, raising your hand at an auction has the normative consequence that you are committed and entitled to buy the object in question at the price in question if no one else raises his or her hand afterwards. And the concept of placing a bid at an auction is characterized precisely by these normative consequences.

2.4 *Tickets and borders*

As my final examples, I want to discuss the cases of tickets and borders. Like money, these are cases in which an *object* counts as something, whereas playing chess and bidding are cases in which an *action* counts as something. According to the explanatory framework proposed here, we have to specify practices conferring non-physical property on a piece of paper or on a line on the ground in virtue of which they count as a ticket or a border. I will now argue that new properties are conferred on these *objects* because certain *actions* involving them are governed by prescriptive rules.

Let me start with the case of a concert-ticket: I think that an analysis of the *Y*-concept 'ticket' shows that something is a ticket only if possessing it has the normative consequence that one is entitled to attend the event (e.g., a concert) to which the ticket is a ticket. Another necessary condition for being a ticket is that it has been issued by a person who is in a social position to determine who is entitled to attend the event and who is not, i.e., the organizer of the event. Thus, a police ID-card may entitle its bearer to enter the room or area where the event takes place, yet it is not a ticket because it has not been issued by the manager of the concert hall.

---

is usually performed in order to bring about the normative consequences described in the consequent clause, while in the first two cases, the action is usually performed for other reasons and its normative consequences are merely a by-product.

The property that has to be conferred on a piece of paper in virtue of which it falls under the concept 'ticket' can thus be characterized as 'giving entitlement to attend an event, issued by the organizer of the event'. The social practices conferring this property on an *X*-object consist in the acceptance of the following prescriptive rule: 'If a person possesses a piece of paper of a specific kind that is issued by a specific person, he or she is entitled to attend a certain event'. Because of this prescriptive rule, the piece of paper can be said to acquire the normative property just described, in addition to its physical properties. And in virtue of this normative property, the piece of paper falls under the concept 'ticket'.

As a second case of an object that counts as something, I want to consider a line on the ground that counts as a border. This case contains an interesting twist, at least if the *Y*-concept 'border' is understood as 'something that serves the function of separating two areas'. This case is interesting because objects can fall under this concept both because of their physical properties and because of their normative properties.[25] As Searle remarks, an object can serve this function because it cannot be climbed over easily, as in the case of a wall. However, objects that present no physical obstacles to crossing, such as lines on the ground, can serve this function as well if it is collectively accepted that they *may* not be crossed (cf. CSR, 39f.).

Now, in my view, the fact that a line on the ground counts as a border can be accounted for in the following way: A line on the ground counts as a border insofar as the action of crossing it is governed by a prescriptive rule according to which people are only entitled to do so under special circumstances. Because of this prescriptive rule, the line on the ground acquires a normative property, in addition to its physical properties, which we might call 'not-to-be-crossed-ness'. In virtue of this property, the line falls under the *Y*-concept 'border', understood as 'something that serves the function of keeping people in and/or out'.[26]

Objects can serve the function of keeping people in and/or out both in virtue of their physical properties and in virtue of their normative properties. So, in the case of the line counting as a border, the property

25. Alternatively, one could propose to understand the concept 'border' as 'something that may not be crossed'. In that case the account of why something counts as a border does not differ much from that of why something counts as a ticket.

26. It is assumed here that the prescriptive rule has at least some influence on the actual behavior of people. The more people actually heed this rule, the better the line will serve its function of keeping people in and out.

conferred by the social practices (i.e., not-to-be-crossed-ness) is one special case of the property that was taken to be contained in the *Y*-concept (serving the function of keeping people in and out), whereas in the other cases considered here, these properties were identical. For example, I have argued that the content of the concept 'to place a bid' involves the idea of modifying the normative situation in a specific way, and that the practices relevant in this case confer just this property on the action of raising one's hand.

These observations give rise to the following clarification concerning the general explanatory framework proposed here. Since this framework is intended to allow for *Y*-concepts such as 'border', i.e., *Y*-concepts objects can also fall under because of their physical properties, it should be understood as requiring only that the property conferred on *X* by the social practices is sufficient for *X* to fall under *Y*, but not that this property is also necessary. In other words: Though in many cases, the property socially conferred on *X* is both necessary and sufficient for *X* to fall under *Y*, the general explanatory framework only requires that this property be sufficient.

3. *Conclusion*

It was the main objective of this paper to introduce and argue for a particular framework for explaining counts-as relations. According to this framework, there are social practices that confer properties on actions and objects,[27] in addition to their physical properties, and it is in virtue of these properties that *X*-items fall under *Y*-concepts. Moreover, the *Y*-concepts cannot simply be regarded as new names for the *X*-items because they do not apply to the *X*-items *simpliciter*, but only insofar as the *X*-items are involved in the social practices.[28] I have tried to argue for this

27. I have argued that our practices can confer additional properties on actions *or* on objects. In many cases, however, practices simultaneously confer properties on both actions and objects. This is the case when the conferring practices involve both actions and objects, as in the case of making a chess move, tickets, money, etc.

28. I have relied on an intuitive understanding of what it is for an *X*-item to be *involved in* a social practice. The difficulty can be illustrated by Glüer and Pagin's example of "queat" as a *Y*-concept which applies to the *X*-action of eating while "the rule 'Don't smoke' exists (somewhere)" (Glüer & Pagin 1999, 220). This is a case in which a *Y*-concept applies to an *X*-action, but only insofar as a prescriptive rule is in force. But one would not say that the action of eating is *involved in* the social practice consisting in the general acceptance of the rule 'Don't smoke'.

explanatory framework by applying it to a number of cases in which an action or object counts as something.

A second objective of this paper was to argue that the practices which confer additional properties on objects and actions differ from case to case. In order to explain why an *X* counts as a *Y*, one thus has to look at the particularities of the case at hand. In particular, the practices do not always and exclusively have to consist of prescriptive rules; they can also contain interpersonal patterns of intentional states, or yet other elements, such as collective intentions (understood as irreducible to patterns of individual intentions). So, as much as the fact that *X* counts as *Y* can be due to prescriptive rules concerning *X*, this can also be due to patterns of individual intentions concerning *X* or due to collective intentions concerning *X*.

A third objective of this paper was to argue that one should eschew using the concept 'constitutive rule', at least if this concept is explained in terms of Searle's two formulae, as creating the possibility of new forms of behavior *and* as being of the counts-as form. I have argued that both of these formulae together make the concept 'constitutive rule' ambiguous between the whole statement 'Doing *X* according to a system of prescriptions counts as doing *Y*' and the prescriptions mentioned in this statement.

A fourth, though by far less important aim of this paper was to defend the particular explanations of counts-as relations proposed here. This aim is of lesser importance because one can reject any or all of these explanations, while taking it that the first three objectives of this paper have been reached. In particular, followers of Searle's *The Construction of Social Reality* will probably be dissatisfied with the proposed explanations because they take collective intentions to be an indispensable ingredient of property (or as they would say: status-function) conferring practices, while I have exclusively invoked the acceptance of prescriptive rules and interpersonal patterns of individual intentions. So, someone could argue that, in any one or all of the cases considered here, the practices conferring the additional property on the *X*-item should be taken to include irreducible collective intentions, while still accepting the explanatory framework proposed here.

I want to close by noting that I have proposed no general criterion for the types of actions and objects that can be socially constituted, i.e.,

---

A first step of spelling out what it is for *X* to be genuinely involved in social practices consisting in a prescription or a pattern of intentional states would be to require that the prescription has to govern *X* and that the intentional states have to be directed at *X*.

for concepts that can figure in the *Y*-slot of the '*X* counts as *Y*' formula. It would be desirable to have such a criterion, but I have been unable to find one. However, I would like to discuss Searle's and Hindriks' proposals concerning this question. Searle claims that the contents of *Y*-concepts can in general be spelled out in terms of deontic powers:

> My aim is to try to state the general form of the content of the Y status-function [...] and I am suggesting that for a large class of cases[[29]] the content involves some conventional power mode in which the subject is related to some type of action or course of actions. (CSR, 104)

As the examples discussed in this paper should make clear, this is only so in some cases. In particular, I have argued that 'money' cannot be understood as something that gives its owner the right to buy something, but rather as something that is often accepted in exchange for useful good and generally known to be so accepted. Moreover, the concepts of chess and many other games contain not only the idea that game-actions are subject to certain prescriptive rules, but also that they should be performed with the goal of winning. So, these concepts cannot be spelled out entirely in terms of deontic properties, either.

This critique also applies to Hindriks who proposes to spell out this aspect of Searle's revised account of counts-as relations in terms of what he calls *YZ*-rules, i.e., rules determining the deontic import of *Y*-terms: "The idea, then, is that *YZ* or status rules explicate the normative attributes that come with having a particular status." (Hindriks 2009, 262) For example, "the status rule for property explicates the rights involved in that status, notably the right of use, the right of exclusion, and the right of transfer." (Hindriks 2009, 264)[30]

29. The remaining cases are "degenerate cases of the deontic, because the rights and obligations that typically go with the status-function have shrunk to the point that the status is valued or disvalued only for its own sake." (CSR, 110) In his latest book, *Making the Social World*, Searle treats these cases as limiting cases of deontic powers so that he can now say that "all status functions create deontic powers." (2010, 24)

30. I am not sure if Hindriks, by introducing the notion of status-rules, really succeeds in "develop[ing] the notion of a constitutive rule *beyond* the characterization Searle has provided" (Hindriks 2009, 262, my emphasis). A status-rule such as 'If you own an object, you have the right of use, exclusion, and transfer' which "explicates the rights involved in that status" (Hindriks 2009, 262) in the end just seems to be a semantic rule, spelling out what the concept 'private property' means. And we have seen that Searle himself takes the analysis of the content of *Y*-concepts in normative terms to be an important part of his account of social reality. Thus, what Hindriks writes here does not seem to go beyond what Searle writes.

References

Black, Max 1962: "The Analysis of Rules". In: Max Black (ed.), *Models and Metaphors*. Ithaca, N.Y.: Cornell University Press, 95–139.

Glüer, Kathrin & Pagin, Peter 1999: "Rules of Meaning and Practical Reasoning". *Synthese* 117, 207–227.

Hindriks, Frank A. 2009: "Constitutive Rules, Language, and Ontology". *Erkenntnis* 71, 253–275.

Ransdell, Joseph 1971: "Constitutive Rules and Speech-Act Analysis". *The Journal of Philosophy* 68, 385–400.

Ruben, David-Hillel 1997: "John Searle's The Construction of Social Reality". *Philosophy and Phenomenological Research* 57(2), 443–447.

Searle, John R. 1969: *Speech Acts*. New York: Cambridge Univ. Press. (SA)

— 1995: *The Construction of Social Reality*. New York: The Penguin Press. (CSR)

— 1997: "Responses to Critics of *The Social Reality*". *Philosophy and Phenomenological Research* 57(2), 449–458.

— 1999: *Mind, Language, and Society*. New York: Basic Books.

— 2010: *Making the Social World*. Oxford: Oxford University Press.

Smith, Barry 2003: "The Ontology of Social Reality". *American Journal of Economics and Sociology* 62, 285–299.

Warnock, Geoffrey J. 1971: *The Object of Morality*. London: Methuen.

Grazer Philosophische Studien
85 (2012), 213–236.

TWO DOGMAS OF COHERENTISM*

Daniel Enrique KALPOKAS
Universidad Nacional de Córdoba &
Consejo Nacional de Investigaciones Científicas y Técnicas

Summary
This paper discusses two dogmas attributed to Davidson's coherentism. The first dogma says that perceptual experience is only a causal link between the world and beliefs. The second one says that only beliefs can justify other beliefs. Against these two statements it is argued that the conception of perceptual experience as a mere causal link between the world and our beliefs makes the world unknowable. Moreover, the article presents some additional reasons against that conception: it misses the phenomenological and perspectival character of perception, and its independence from belief. Finally, Davidsonian externalism is considered. It is shown here that Davidson's conception of experience makes it impossible to individuate the contents of beliefs. The article concludes rehabilitating the empiricist idea according to which perceptual experience can be used to justify beliefs.

1. *Introduction*

Any theory which attempts to explain how we can get to know our environment has to take into account the role that perceptual experience plays in empirical knowledge. Traditional empiricism claims that perceptual experience is essential to explain the *source* as well as the *justification* of empirical knowledge. As a theory about the source of empirical knowledge, empiricism affirms that all our knowledge comes from experience. As a theory of justification, instead, it claims that all knowledge about the world is justified, lastly, by experience. The first strand of empiricism is practically uncontroversial nowadays. It seems obvious that without perceptual experience we could not know the empirical world. However,

* I would like to thank Laura Danón, Mariela Aguilera, Fabián Mié, Darío Scattolini, José Giromini, Federico Castellano, Nahuel Recabarren, Carolina Ames and the anonymous referee of *Grazer Philosophische Studien* for useful suggestions to an earlier version of this paper.

the second strand of empiricism—as a theory of justification—has been severely criticized and rejected by distinguished philosophers in the second half of 20th century. For instance, Rorty has considered the linguistic turn as an opportunity to abandon the notion of experience as an epistemic tribunal between mind and world (Rorty 1998, 150), and, consequently he has developed an explicitly anti-empiricist variety of pragmatism (Rorty 1999). Davidson has also tried to definitively defeat empiricism (in its epistemic strand) by criticizing what he calls "the third and last dogma of empiricism"—the schema-content dualism (Davidson 1984a, 189). Finally, following Rorty and Davidson's anti-empiricist "impetus", Brandom proudly claims that the term "experience" has no place in the inferentialist semantics that he articulates in *Making It Explicit* (Brandom 2000, 205).[1]

The two aforementioned aspects attributed to experience—as a source and justification of empirical knowledge—remained indissolubly united in classic empiricism. For the recently mentioned philosophers, however, those two aspects can and must be separated, for they understand that the idea that experience justifies our beliefs is a dogma of empiricism. For them, the term "experience" has only a causal meaning: experiences are causal intermediaries between beliefs and the world, not reasons for holding beliefs.[2] As a result of the affection of our sense organs, we have a belief; but experience in itself is not an episode with cognitive content. Experience causes beliefs, but it does not justify them; it put us in direct contact with the world, not in a cognitive way but in a causal one. It is argued that experience cannot justify a belief because it lacks propositional content. In Davidson's words:

> The relation between a sensation and a belief cannot be logical, since sensations are not beliefs or other propositional attitudes. What then is the relation? The answer is, I think, obvious: the relation is causal. Sensations cause some beliefs and in *this* sense are the bases or ground of those beliefs. But a causal explanation of a belief does not show how or why the belief is justified. (Davidson 2001a, 143)[3]

In this paper I will focus my attention on Davidson's work, and I will consider the thesis expressed in the quoted passage, in which the

1. Cf. also Brandom 1994, 1998 and 2002. In the case of Rorty, Davidson and Brandom, their anti-empiricism attitude comes from Sellars. See Sellars 1997 and 1975. BonJour, in other times a champion of coherentism, has recently abandoned it. See BonJour & Sosa 2003.

2. Cf. Rorty 1998, 140f.

3. Cf. also Davidson 2005.

epistemic character of experience is denied[4], as the first dogma of Davidson's coherentism. This dogma is closely associated with a second dogma, according to which "nothing can count as a reason for holding a belief except another belief" (Davidson 2001a, 141). Both dogmas constitute the heart of Davidson's coherentism. In what follows, in order to provide new reasons in favor of certain attempts to defend the epistemic character of experience,[5] I will try to make the serious problems that those dogmas generate evident. Since Rorty's and Brandom's conception of perceptual experience is very similar to Davidson's, I think that many of the objections that I will present against the latter can be directed against the former as well.

2. *Coherentism, truth and experience*

In "A Coherence Theory of Truth and Knowledge" Davidson defends a coherentist point of view about knowledge based on certain epistemological and semantic remarks. His purpose is to show that "coherence yields correspondence" (Davidson 2001a, 137).[6] If coherence is a proof of truth —Davidson argues—then we have a reason to think that many of our beliefs are true, for we certainly have a reason to believe that many of those beliefs cohere. What brings truth and knowledge together is meaning. If meanings can be given by means of truth conditions, then the knowledge that sentences satisfy those conditions gives us knowledge about their meanings too. However, the knowledge that those conditions have been satisfied cannot be obtained—Davidson says—by confrontation: "No such confrontation makes sense, for of course we can't get outside our skins to find out what is causing the internal happening of which we are aware" (Davidson 2001a, 144). Instead of confrontation, Davidson appeals to coherence: if coherence is actually a proof of truth, then it is proof of the judgment that truth conditions have been satisfied as well. Davidson's slogan that synthesizes this point of view is "correspondence

4. It is useful and usual to distinguish the terms "perceptual experience", "perception" and "sensation". However, since Davidson does not make an explicit distinction between those terms, in this paper I will use them interchangeably. Mainly, I will use "perceptual experience" and "perception" as equivalent. The usual differences between those terms are not relevant in the following discussion.

5. I think of McDowell (1994) and Brewer (1999).

6. In "Afterthoughts" Davidson withdrew some statements from his original "A Coherence ...". His later point of view does not affect what I will say in this paper.

without confrontation" (Davidson 2001a, 137). Now, even though they are in a coherent system, certainly some of our beliefs are (and can be) false. Thus, mere coherence cannot prove that *each* belief is true. All that coherentism lets us hold is that, in a coherent set of beliefs, *most* of our beliefs are true (Davidson 2001a, 138f.).

Here, it is important to remember that Davidson's coherentism pretends to be an alternative conception to foundationalism. As it is widely known, foundationalism—in its empiricist version—aims to justify the building of knowledge by appealing to entities such as sensations, perceptions, experiences or sense data.[7] Davidson thinks, however, that the idea that those items can justify beliefs is wrong because how items which do not have propositional content can enter into logical relations with items which do have propositional content cannot be explained. As a consequence, Davidson says that knowledge and meaning depend only *causally* on experience (Davidson 2001a, 146), and that only beliefs can justify other beliefs (Davidson 2001a, 141). Here we have the two dogmas that I have attributed to Davidson's coherentism.

Now, Davidsonian coherentism aims to be an epistemological outlook which—in contrast with empiricism—does not make skepticism about the external world "inevitable" (Davidson 2001b, 286).[8] If—as it is usually said—epistemological intermediaries between us and the world are responsible for skepticism, we should get rid of them and only preserve causal intermediaries (Davidson 2001a, 144). Thus, according to Davidson, when we causally interact with the environment, it causes beliefs in us. Beliefs—perceptual and non—are justified, not by perceptual experience, but by other beliefs also caused by the world (Davidson 1999, 105f.).

Even though Davidson thinks that there is no way to "get outside our beliefs and our language so to find some test other than coherence" (Davidson 2001a, 141),[9] he affirms—in contrast with Rorty—that we need to explain how we can know and talk about an objective world which we have not produced. That is why Davidson is interested in showing that coherence yields correspondence. However, in this very aim, which is to explain how we can know the objective world, the problems of Davidsonian coherentism arise evidently. Indeed, if it is true that—having gotten rid of epistemological intermediaries—the relation between the world

7. In the paper I am commenting on, Davidson does not make any distinction among those different items.

8. Cf. also Davidson 1999, 105.

9. Here Davidson quotes approvingly Rorty (1979, 178).

and us is exclusively causal; if it is true that sensations, perceptions or experiences can only cause beliefs, but not justify them; and finally, if it is true that "we can't get outside our skins to find out what is causing the internal happening of which we are aware", then, how can we ***know*** that objects and events which cause our sensations are the ones we think they are?[10]

Experience—according to Davidson—does not put us in cognitive contact with the world either in a direct or in an indirect way. It is only a causal intermediary. Davidson says, for instance:

> We look, hear, smell, and touch, and we are caused to believe there is an elephant before us (…) To have an experience in the cases in which we are interested is just to be caused by our sense organs to have an empirical belief. (Davidson 2001b, 289f.)

It is this causal, non-cognitive conception of experience which generates what I will call "the problem of cognitive accessibility". To know that there is an elephant before us we have to perceive it. However, if looking, hearing, smelling and touching were not ways of knowing the environment (but only causes of perceptual beliefs), we would never be able to know what caused our belief that there is an elephant in front of us[11]. In other words, if those modalities of perceiving were only causal intermediaries between beliefs and reality, as Davidson suggests, we would not have any cognitive access to the entities that affect our sensibility[12].

10. This is not the same problem considered by Brewer and Campbell, that is, the problem of how to identify an object in the world Cf. Brewer 1999 and Campbell 2002. Rather, my objection points out the fact that, according to Davidson's conception of experience, we are blind about how the world is. In Davidson's view, perceptual experience does not open us to the world, but it only affects our senses. In doing so, Davidson makes something similar of the world to the Kantian *Ding an sich*: the incognizable source of our sensory affections.

11. In fact, there is the temptation to say that, if perceptual experiences of an elephant are not more than perceptual beliefs, caused by our sense organs, about how it smells, how it is heard and how it looks, then, strictly speaking we do not perceive the elephant at all. We have only perceptual beliefs about it.

12. It has been objected to me that Davidson could argue that, in a situation in which by looking at an elephant I acquire the belief that there is an elephant in front of me, I may also acquire the belief of second order that I acquired my belief of first order by looking (i.e. by visual experience and not by hearing). This objection does not solve—I think—the problem of cognitive accessibility. If we could not know what the causes of our first order beliefs are—as I argue—why to suppose that the belief of second order can do the job? How could my belief of second order tell me that my belief of first order was caused by an elephant if perceptual experience cannot reveal to me the cause of my original belief? At best, I could know that my belief of first order was caused by visual experience (and not, for instance, by hearing), but nothing at all

In "Empirical Content", in the context of radical interpretation, Davidson suggests an answer to the objection according to which coherentism makes the world unknowable:

> Our basic methodology for interpreting the words of others necessarily makes it the case that most of the time the simplest sentences that speakers hold true *are* true. It is not the *speaker* who must perform the impossible feat of comparing his beliefs with reality; it is the *interpreter* who must take into account the causal interaction between world and the speaker in order to find out what the speaker means, and hence what that speaker believes. Each speaker can do no better than make his system of beliefs coherent, adjusting the system as rationally as he can as new beliefs are thrust on him. But there is no need to fear that these beliefs might be just a fairly tale. For the sentences that express the beliefs, and the beliefs themselves, are correctly understood to be about the public things and events that cause them, and so must be mainly veridical. (Davidson 2001c, 174)

However, this answer does not solve the problem of cognitive accessibility. Here again, if it is true that we cannot make any confrontation between what we believe and the world (for "we can't get outside our skins to find out what is causing the internal happening of which we are aware"), if perceiving is not more than being "caused by our sense organs to have an empirical belief", why would the interpreter have a sort of access to the world that the speaker could not have? What else than "make his system of beliefs coherent, adjusting the system as rationally as he can as new beliefs are thrust on him" can the interpreter do? If, in general, perception does not provide us with a cognitive access to the world, then it seems obvious that this applies to the interpreter in the situation of radical interpretation as well: the best the interpreter can do is to compare her *beliefs* about the linguistic behavior of the speaker—caused by the speaker—with her own *beliefs* about the world—supposedly caused by it. As it is manifested, the world in itself here is not a term of comparison which beliefs are contrasted with.[13]

---

is said about the specific object which caused my belief. The problem of cognitive accessibility derives from the lack—in Davidson's philosophy—of recognition of the intentional character of experience. On that feature of experience, Cf. Anscombe 1965 and Searle 1983.

13. At this point it could be argued that the nature of radical interpretation by itself warrants that beliefs are about their typical causes, and then, that the direct perception of those causes is not necessary. As it is known, Davidson thinks that the principle of charity recommends attributing to the speaker a large number of true beliefs, consistent among themselves and with the interpreter's beliefs. However, it has been objected that the fact that the interpreter should

Thus, there is indeed an undesirable consequence of Davidson's coherentism which follows from the two dogmas that I have attributed to it: the cognitive inaccessibility of the external world. Then, if the remarks presented above are correct, the direct but exclusively causal link posited between reality and us does not prevent the skeptical question; on the contrary, it turns that question into a more demanding one, for the question about what the cause of our beliefs is arises inevitably from the inside of Davidsonian coherentism without any possibility of answering it suitably. The combination of the first dogma that I have attributed to Davidson's coherentism (the relation between experience and belief is causal, not logical), with the second dogma ("nothing can count as a reason for holding a belief except another belief") is what makes—paradoxically according to Davidson's aims—skepticism about the external world inevitable.

Another way to express this point is the following: Davidson thinks that justificatory relations can only have their place inside a body of beliefs. This means that there are not (nor can there be) justificatory relations between beliefs and the world. Notwithstanding, and although he thinks that we cannot get outside of language and of our beliefs, Davidson wants to explain how we can know and speak about a world "which is not our own making" (Davidson 2001a, 141). However, it is worth noting here that there is a striking and unexpected similarity between representationalism and coherentism. While representationalism posits epistemic intermediaries between us and the world, coherentism eliminates those intermediaries in order to posit only a causal link with reality; but in both cases a dichotomy is supposed between what is inner and outer to the network of beliefs (or representations). That dichotomy is what makes the problem of cognitive accessibility arise. The skeptical question arises again naturally: if our contact with the world is only *causal*; if justification cannot take place but *into* a body of beliefs, how can we know that those beliefs describe the world "which is not our own making"? From a historical point of view, this coherentist view about justification can be understood as a radical-

---

attribute to the speaker a large amount of beliefs which the interpreter *holds* to be true does not imply that those beliefs *are* true. It should be demonstrated that the interpreter's beliefs are true. In order to fill in that gap, Davidson introduced the argument of the omniscient interpreter. The argument, however, has been criticized by many interpreters and Davidson himself has abandoned it. That is why Davidson tries to solve the problems generated by his coherentism by means of his externalism. Davidson introduces the omniscient interpreter argument in Davidson, 1984b, 201. Cf. also Davidson 2001a, 150f. For objections to the omniscient interpreter argument, see for instance Génova 1999, Haack, 1993, chap. 2; Stroud 1999, 148; Ludwig 1992. Davidson abandons the argument in Davidson 1999b, 192.

ization of the Modern epistemological model which Davidson aims to leave behind.

Whether this objection is correct or not, there are some additional reasons for rejecting the Davidsonian conception of experience. We should remember that, in Davidson's view, since experiences, sensations or perceptions lack propositional content, they cannot have logical relations with beliefs and, for that reason, they cannot be used for justifying them. Notwithstanding, as we have seen, experience causes beliefs. Consider what Davidson says in response to an objection formulated by McDowell:

> For him [McDowell], first there are the appearances and then we decide whether things are as they appear; for me, there is no distinction between things appearing to me a certain way and my taking them to be that way [...] Where he has appearances I have perceptual beliefs. (Davidson 2001b, 289)

Reasonably, McDowell distinguishes between appearances and beliefs (McDowell 1994). The first are the immediate results of our interaction with the environment and they are the first cognitive occurrence which discloses how the world is. Since, for McDowell, the experience—the appearance—is conceptualized from the beginning, it can be adduced as a reason for justifying a belief. Davidson, in contrast, rejects the appearances—epistemic intermediaries inserted between nature and belief—and then he states that the first epistemic result of our causal interaction with the world is perceptual belief.

Now, in contrast to what Davidson holds, and in agreement with McDowell on this point, I think that the epistemic distinction between belief and perceptual experience is crucial, not only because the idea of experience as a cognitive way to access to the world seems recommendable when facing skepticism, but also because there are some reasons for thinking that both belief and experience constitute different states with their own contents.

First, it is worth noting that, when we ordinarily talk about "seeing", "touching", "hearing", etc., we imply that those acts are about objects with respect to which senses provide us some kind of information. That is, when we talk about those acts, we not only understand that perceptual experience causes beliefs, but also that in experience the world is *presented* as being thus and so.[14] What the ordinary use of verbs of perception sug-

14. In ordinary speaking, when we say, for instance, that we see an apple, we do not say merely that the apple *causes* the belief that there is an apple there, but also that we *perceive* the apple in itself from a certain point of view.

gests is that the world *appears* in front of us, in different conditions and according to some perceptual modality, as being thus and so. We usually say that we touch—for instance—an apple, that we feel its texture, its weight, etc., and not merely that we have certain beliefs about the apple caused by our touching it. Likewise, we normally say that we see an apple, with its particular color, form, shade and brightness, and not merely that we are caused, by looking, to have certain beliefs about the apple. I do not think that this argument is conclusive, but I think that it certainly has certain weight. Since our ordinary linguistic practices do not speak in favor of Davidson's conception of perceptual experience, I believe that it would be necessary to explain—from Davidson's point of view—how and why the use of perception verbs is misleading.

Secondly, in accordance with the grammatical remark made above, there is an essential aspect of perceptual experience usually stressed by philosophers of perception: its phenomenological character. In perception—it is usually said—the world appears to us as being thus as so. Davidson's conception of experience misses the phenomenological character of perceptual experience. This has a very important consequence. The fact that the world reveals itself in perception under some modality and in determined conditions explains the fact that we can be conscious *of* the world. There is a crucial difference between being conscious of an accepted proposition which refers to an object or event in the world, and being conscious (by perception) of that object or event in the world. The Davidsonian conception of perceptual experience not only forgets the phenomenological character of experience, but it also seems to imply—consequently—that we can at most be conscious of beliefs that refer to the world, but not of the world itself.[15]

Let me clarify this point. Let's assume—for the sake of the argument—an intuitive conception of experience according to which, in perception, the world is presented directly to us as being thus and so. Taking McDowell's expression, we can call it the conception of experience as openness to the world.[16] It is not that we perceive the world through certain representations (sense data, Humean ideas, sensations, etc.) which refer to the world. That idea has been rightly criticized by Davidson and others. Instead, according to

15. Rorty, who has a very similar conception of perceptual experience, gets to say that "there is nothing to be known about an object except what sentences are true of it". Rorty 1999, 55.

16. Cf. McDowell 1994 and 1998. On the adverb "directly" used in this context, cf. Snowdon 1992.

the conception of experience as openness, when we perceive, for instance, a tomato, we perceive the tomato with its properties (color, form, etc.) from certain point of view. The idea that the perceived object—the tomato—constitutes, in part, the content of the experience is sometimes associated to the conception of perceptual experience as openness.[17] Now, according to this conception, when we perceive an object, we are conscious of the object by means of being conscious of its appearances, by means of how it looks from our point of view. In this conception of experience, to perceive something is not to have a perceptual belief. Empirical beliefs are representations of the world (even perceptual beliefs);[18] to perceive—from the conception of experience as openness—is not to represent the world; rather, it is for the world to be *presented* to one as being thus and so. There are not epistemic intermediaries between mind and world—in the case we are considering, perceptual beliefs—by means of which we are conscious of the world. As I said, it is because of the world manifests itself in perception, that we can be conscious directly of it.

Consider now a third argument against the Davidsonian conception of perceptual experience. Davidson says that to have an experience is "to be caused by our sense organs to have an empirical belief" (Davidson 2001b, 289f.); however, it can be argued that belief is different from perceptual experience because doxastic content does not pick up every aspect contained in the content of experience.[19] When we pay attention to what we see, it is clear that things are always presented to our sensibility from a certain perspective. We not only perceive how things *are*, but also how things *look* from the position in which we are. We perceive that a plate is round, for instance, and that it looks elliptical from here. Those apparent properties which constitute the base of our perceptual comprehension of the world are not mere subjective sensations, but objective properties—though relational—of things, which depend not only on how things are, but also on the position into the space from which we perceive them.[20] As

17. This thesis has been stressed by disjunctivists. Cf. Snowdon 2009; McDowell 1998a and 1998b; Martin 2009.

18. I am not attributing to Davidson the notion of representation criticized by him. My use of the term "representation" here is harmless; by that term I just mean something which is in the place of other thing (an object or event) for somebody. On a harmless notion of representation in Davidson's work, cf. Neale, "On Representing", with Davidson's "Reply to Stephen Neale", Hahn (ed.) 1999.

19. I pick up here an argument presented by Noë against the causal theory of perception in order to use it against Davidson's view of experience. Cf. Noë 2004, 169 and ff.

20. Noë calls them "P-properties". Noë 2004, 83.

Noë has stressed (Noë 2004), the perceptual content has a dual aspect: on the one hand, in perception the world is presented as it is, independently of any perspective (the "factual dimension" of the perceptual content); on the other hand, however, we always perceive things from a certain particular perspective (what Noë calls the "perspectival dimension" of the perceptual content). It is this last perspectival dimension of the perceptual content which cannot be accommodated in Davidson's framework for experience-belief relationships.[21]

Fourthly, some cases have been adduced in favor of the independence of perceptual experience with respect to belief, which can be interpreted as drawing a distinction between the content of experiences and the content of beliefs. As we have seen, Davidson considers that an experience can never justify a belief. For the same reason, it cannot contradict the content of a belief nor conflict with it. However, there are situations in which we know that things are not as they look. According to my visual experience, it seems to me that the two lines of the Müller-Lyer illusion differ in length even though I know that they are of the same length.[22] In a similar way, my visual experience can show me that a stick submerged in water is broken even though I considered that the stick has not been broken by the mere fact of being introduced into water. In neither of both cases are we obliged to believe that things are as they look according to our experience of them. Now, if it is true that perceptual experience can *contradict* (or to be in conflict with) what we believe, then it surely can also be in *accord* with the content of a belief. And if this is possible, then it cannot be true that experience is only a causal link between beliefs and the world, as Davidson says. In fact, if experience can contradict or accord with the content of beliefs, it can be stated that perceptual experience has an epistemic content; then it can enter into rational relations with beliefs.

21. For the sake of the argument, here I am making the concession that Davidson can explain the factual dimension of the perceptual content; but I do not actually think so. As we saw, for Davidson experiences do not have content at all. The point of the argument is this: even though Davidson wanted to explain certain facts about perceptual experience by appealing to the propositional content of beliefs caused by experiences, this misses the perspectival dimension of the perceptual content.

22. Noë presents this example to show that perceiving is something different from judging. Cf. Noë, 2004, 188f. McDowell uses the same example to show that it must be possible to judge if experience accurately represents how things are. In that sense he assumes that there is an epistemic difference between perceptual experience and belief. Cf. McDowell 1994, 11. The original use of this example comes (at least it seems to me) from Gareth Evans, who uses it to show that experience is independent from belief. Cf. Evans 1982, 123.

In this case, the appeals to experience could serve to justify or criticize what we believe about the world.[23]

Davidson could argue that those examples show not a conflict between what is perceived and what is believed, but a conflict between a perceptual belief and a non perceptual belief. For instance, he could reply that what the Müller-Lyer's example shows is a conflict between the perceptual belief "The two lines differ in length" and the non perceptual belief "The two lines are equal in length". Or he could say that the observer of the Müller-Lyer lines is caused by her sense organs to have the perceptual belief "It looks as if the two lines differed in length", even though she does not believe "The two lines differ in length" (the observer would actually believe "The two lines do not differ in length").[24] In the first case there would be a genuine conflict between beliefs, in the second, there would not.

However, this answer does not work. There is a general reason for rejecting those possible replies. Beliefs are vulnerable to reasons, they can

23. How exactly can experiences be in agreement or disagreement with beliefs? The answer depends on the sort of content attributed to experience. There are several possibilities to be considered. For instance, in *Mind and World* McDowell assumes that perceptual content is propositional. I am not sure whether perceptual content is propositionally articulated, but if this thesis were accepted, it would be easy to explain how experience can contradict or accord with beliefs. It would just be a matter of confronting propositions. However, McDowell has recently changed his view. In 2009 he says that experiences have intuitive conceptual content, which is different from discursive content expressible in a judgment. Intuitive content can be verbally expressed by means of demonstrative expressions such as "This red cube". Expressed in that way—McDowell thinks—the intuitive content can be a part of a judgment (i.e. "This red cube is the one I saw yesterday". 2009, 270). If perceptual content is what McDowell now calls "intuitive content", how can experience contradict or accord with beliefs? Well, in my opinion, the intuitive content "This red cube" not only can provide the subject of a judgment, but it also can be in agreement with the belief "This cube is red", and in disagreement with the belief "This cube is blue". Take our example: the intuitive content "These two different lines in length" can be in disagreement with the belief "These two lines are not different in length". According to my development of McDowell's ideas, the intuitive content can be expressed by a judgment, but this fact does not imply that the very intuitive content is a judgment or a proposition. The intuitive content is a non discursive content which, however, can be unfold into a judgment, as in the examples above. Even though the intuitive content lacks a predicative element, it is obvious that not just any judgment based on it is adequate. The purpose of my use of the Müller-Lyer lines and similar cases in the text is quite general: I use it to make clear that we can make vivid a distinction between experience and belief by using an intuitive notion of experience. I think that both McDowell's first and second thesis about the conceptual character of perception are promising ways to explain the relationships between experiences and beliefs.

24. This latter possibility was suggested to me by an anonymous referee. Besides the general objection I present in the text, it is possible to make a particular and previous remark related to the very sense of the sentence "It looks as if the two lines differed in length". What does "looks" mean in that sentence? As Austin once noticed (1962, 36), the use of the verb "looks" is

be defeated by other beliefs; perceptual experiences, in contrast, are not susceptible to that kind of modification; that is, while we can abandon a certain belief when we recognize that there are good reasons for thinking that the belief is false, we do not stop perceiving that things are thus and so simply by the fact that we have good reasons for thinking that things are not thus and so (in our case, we continue perceiving the difference in length of Müller-Lyer lines even though we are told that the lines are equal in length). Note that this argument is sound even in the case of the belief "It looks as if the two lines differed in length". If the Müller-Lyer illusion effectively involved that belief (and not a contrast between the visual experience and the belief about the lines, as I originally presented the case), it would be possible to defeat it by means of adequate reasons. However, that is not the case. For all I know, every empirical belief can be questioned and abandoned if we have convincing reasons against it.[25] Interpreting the Müller-Lyer illusion as consisting of a contrast between two beliefs does not make justice to the nature of the visual illusion. Thus, as in the case of previous arguments, it can be concluded that perceptual experiences are not beliefs, but occurrences with their own content.[26]

---

restricted to the general sphere of vision. But according to Davidson, experiences—in this case, visual experiences—only cause beliefs; they do not present (or open to) the world to us. Strictly speaking, in his view of visual experience nothing can look anyway because visual sensations lack any content, they are just causal linkages between our beliefs and the world.

25. One final—although not Davidsonian—alternative can be suggested: it could be argued that beliefs about appearances are immune to reasons because it is not possible to doubt their truth. Some logical positivists held this point of view. However, they did not hold that experiences are mere beliefs.

26. It is interesting to consider here the following possibility: perhaps, by adequate training a subject who knows that the Müller-Lyer lines are of equal length might learn to see the lines just as they are. In that case, the mentioned illusion would disappear for the experienced observer. (I give thanks to an anonymous referee for the suggestion of this possibility). However, that plausible case does not undermine my argument. Still it would be true that there is a certain illusion at first sight, for the non-experienced observer. For her it would be true that it is not the case that things always look the way they are. Besides, even in the case of the experienced observer it would be true that her new ability to perceive the Müller-Lyer lines as equal in length would not be the result of reasoning (as if reasons would have refuted her original perception), but of a certain kind of training (if to perceive involves certain kind of practical abilities—as Noë says—to perceive is a matter of *know how*, not of *know that*). Finally, the case of Müller-Lyer illusion is only one sort of case that can be adduced. There are other cases where there is a clear difference between experience and belief and where there is not any illusion. Cf. McDowell 2009, 158 and 2002, 277.

In consequence, I think that it can be said that, since we do perceive the world as being thus and so,[27] it makes perfect sense to judge wether the world is such as experience exhibits it. Usually, when perceptual experiences conflict neither with each other nor with beliefs, no doubt about how the world is arises. In contrast, when a conflict takes place between experiences or between a perceptual experience and a belief, we have to solve that kind of cognitive conflict and decide which of the cognitive states mentioned is correct. No doubt, it can happen that we should revise our trust in a perceptual experience because of a belief, or vice versa, that we should abandon a belief in virtue of what we experience. The general presumption in favor of the veridical character of most of our experiences does not force us to give them an epistemic status of infallibility. Our trust in experiences is always revisable on the light of further experiences and beliefs. As it can be seen, it is very important to recognize the epistemic character of perceptual experience because, otherwise, it could be incomprehensible how experience could be the source of empirical knowledge if it did not have any content to transmit.[28]

Thus, once the epistemic difference between perceptual experience and belief is recognized, and it is also recognized that the former has a content that can be correct or not in virtue of how the world is, a different conception of justification can be introduced: not only a belief can justify another belief (as Davidson says), but also a perceptual experience can justify (or be justified by) a belief. The fall of the first dogma of Davidson's philosophy causes the fall of the second dogma.

To sum up, I have offered different reasons against Davidson's conception of perceptual experience (and, in consequence, against his thesis about what kind of states can justify a belief) which make it clear that his coherentist conception of justification rests on two unsupportable principles. I have argued that Davidson's conception of perceptual experience a) makes the world inaccessible, cognitively speaking; which b) makes skepticism inevitable. Furthermore, c) I have presented four reasons to think that perceptual experience and belief are two different kinds of epistemic states that have their own content. Because of that, those two kinds of states are able to enter into logical relations. Now, even if my objections expressed

27. We perceive that the Müller-Lyer lines are different in length, that the stick in water is broken, etc.

28. In this sense, I think that the two aspects attributed to perceptual experience at the beginning of this paper—as a source and as a justification of empirical knowledge—should not be separated.

in a) and c) were correct, it can be argued that Davidson can avoid the fatal consequence expressed in b) because he has an explanation (at least that is his aspiration) about how beliefs are linked to the world in a constitutive way that does not appeal to the epistemic character of experience. Thus, in the next section I will examine Davidson's account of the empirical content of beliefs in order to show how my objections also affect his externalism.

3. *Externalism and empirical content*

Davidson's argument in favor of externalism starts from the observation that we think, and then from there it aims to explain two closely related questions: 1) what explains the fact that thoughts are *objective*, that is, that their truth is independent of their being believed; and 2) what accounts for the *contents* of these thoughts (Davidson 2001a, 2001d, 2001e, 2004a, 1999a, 1999c, 1999d).

In his externalist account, Davidson pays attention to two kinds of situations: the language learning situation, and the radical interpretation one.[29] Consider the first one. In Davidson's story, the adult ostensively teaches the child very simple sentences such as "mama", "water" and "red". The reward or the absence of it, from the adult is the indication for the child of the correctness or incorrectness of her answers. The adult and the child find objects in the environment similar, and they find their responses to those objects similar as well. Because of that fact, they can classify together the objects which cause their reactions, on the one hand, and their own responses to those objects, on the other. Without that shared classification they could not note that they respond to the same object and we would not have any ground for saying that the adult and the child are doing that. By conditioning, the child learns the extensions of the words and picks up the notion of truth. The child cannot doubt the meaning of the words or the content of beliefs from the beginning because the ostensive learning is what it gives content to the words and beliefs that she learns. For the child, whatever thing that is pointed out constitutes the content which she should learn. There is no room for skeptical doubts. Only after thoughts and words have been anchored to reality, that is, only after beliefs have

29. Davidson's externalism involves what he calls "triangulation". In order to be brief, I have omitted any reference to it. The omission does not affect my argument.

gained content, is it possible to doubt their truth. In this way, according to Davidson, the causal history of our interactions with the environment and other people determines the content of our beliefs. Before that process, skepticism is not possible because there is no propositional content to doubt; but afterwards, global skepticism about the external world is not possible either, because doubting already presupposes that beliefs and words, which have content, have been anchored to the world.

Something similar happens in the case of radical interpretation. In her intent to understand the sentences uttered by the speaker, and on making the speaker's behavior intelligible, the interpreter should correlate the speaker's utterances with the world and with her own utterances. The objects that cause the speaker to assent to given perceptual sentences constitute the contents, for the interpreter, of the speaker's beliefs which are expressed by the uttered sentences (Davidson 1999a, 107). In the simplest cases, thus, the object and event which cause a belief also determine the content of that belief. In this way, Davidson's externalism aims to account for the objectivity and the content of thoughts without appealing to the epistemic role of perceptual experience.

Now, since "we can't in general first identify beliefs and meanings and then ask what caused them" (Davidson 2001a, 150), the appeal for the causal link between beliefs and reality not only accounts for—in Davidson's view—the objectivity and the empirical content of thought, but also offers an argument against skepticism:

> What stands in the way of global skepticism of the senses is (...) the fact that we must, in the plainest and methodologically most basic cases, take the objects of a belief to be the causes of that belief. And what we, as interpreters, must take them to be is what they are. Communication begins where causes converge: your utterance means what mine does if belief in its truth is systematically caused by the same events and objects. (Davidson 2001a, 151)

In spite of the merits of Davidson's externalism,[30] I think that—as it has been remarked by McDowell—it becomes a mystery how thought acquires its empirical content. From McDowell's point of view, Davidson does not succeed in explaining how beliefs can have empirical content because he disallows the idea of a rational constraint from the experience. For Davidson, the constraints which come from outside of thought are only

30. I consider Davidson's thesis according to which the nature and individuation of mental events depend on the causal relations between those events and the objects which they are about, *in part*, correct.

causal. But that is not enough—McDowell argues—because, if thought is to be a world-view adopted correctly or incorrectly in virtue of how the world is, beliefs and perceptual experiences are to be related in a *rational* way (McDowell 1994, 14, 17 and 35).[31] Now, I think that this objection is correct, but it rests on a premise unshared by Davidson, which is that the justificatory relation between beliefs and the world (or experience) is constitutive of the content of the former. Thus, I would like to add a more direct objection to Davidson's account of the contents of beliefs which does not presuppose McDowell's controversial premise.

My objection picks up what I have said in the previous section and it is very simple. It has been established that, according to Davidson, the contents of beliefs are constituted by their causes. Davidson affirms also that "we can't get outside our skins to find out what is causing the internal happening of which we are aware" (Davidson 2001a, 144). We do not directly perceive the causes of our beliefs; perceptual experience is, in Davidson's view, a mere causal intermediary between our beliefs and the world. But then, if those premises are true, we can conclude that we cannot know what objects cause our beliefs.[32] This is a very important problem for Davidson's externalism because it means that we have beliefs whose causes are unknown and unknowable for us. However, as it has been presented, Davidsonian externalism aims to individuate beliefs in virtue of their causes (to which beliefs supposedly refer). In consequence, if the causes of our beliefs cannot be known by perceptual experience, it becomes impossible to identify the contents of beliefs and to say to what objects our beliefs refer. The Davidsonian slogan "correspondence without confrontation" (Davidson 2001a, 137) falls down. Once again, an inadequate conception of perceptual experience leads to the problem of cognitive accessibility.

This problem can be observed in the language learning situation as well as in the case of radical interpretation. Without cognitive access to the conditions of application of the first linguistic expressions that only perceptual experience can provide, there is no way of giving content to those

31. In (1994, 134) McDowell expresses his main objection to the idea that exercises of 'conceptual sovereignty' are only causally affected by the course of experience, and not rationally answerable to it. Davidson's answer is in 1999a and 2001b.

32. This conclusion assumes that perceptual experience is the only source of empirical knowledge. If it were proved that human beings have a faculty different from the senses by means of which they can know the empirical world, that conclusion of course would not be followed. Davidson has not argued, however, in that direction and it is not easy to do so.

expressions because it is by connecting rightly words and sentences to the appropriate objects and events that the linguistic expressions can have a reference for the child. To do that, the referents of the linguistic expressions have to be accessible to the child herself; and perception as openness to the world makes possible the required access. In a more detailed way: Words and sentences are certain kind of representations.[33] We understand those representations when we understand—at least in part—what those representations represent. Then, to understand and learn simple sentences such as "mama", "table" and "red" the child has to be able to have perceptual access to the items represented by those sentences. That kind of access is intelligible only if we think of the notion of experience as openness to the world.[34] What is represented by the simplest words and sentences should be given in perception if the child is to be able to connect rightly those words and sentences to the empirical world. I do not want to say that to know the referents of the linguistic expressions is to learn a language. Rather, what I am pointing out is that to learn what linguistic expressions represent is an essential element in the learning process of a language, and that perceptual experience as openness to the world has an unavoidable role in it. Thus, the crucial point here is that without a perceptual apprehension of the world the child cannot learn to correlate linguistic expressions to things, and to give the usual meaning to those expressions.[35]

Likewise, from the radical interpreter's point of view, the objects which cause the speaker's beliefs are to be perceived by the interpreter. If the interpreter has to correlate the speaker's conduct of assent or dissent to a sentence uttered with the objects and the events of the environment, it seems obvious that the interpreter needs to be able to perceive and perceptually identify those objects and events which cause the speaker's conduct. Only in that way can the interpreter also contrast her interpretative hypothesis with the speaker's behavior. This applies for the speaker's situation too. The speaker has to be able to perceive the environment meaningfully (and not only react to stimuli) in order to adjust the behavior and beliefs to the world in a rational way. Without recognizing the cognitive feature of

33. I am using the term "representation" in the general sense pointed out in footnote 19.

34. This point can be put in relation to Campbell's thesis according to which experience of objects explains our ability to demonstratively think about perceived objects. Cf. Campbell, 2002.

35. Mere conditioning (as something that does not involve perception) is not enough here because the child has to note under which circumstances her answers are correct and incorrect, and which the consequences of her actions are. It is difficult to imagine how those things could be realized without perception.

perceptual experience, it would be difficult to explain in what sense the speaker is an agent at all.

Holding that the linkage between beliefs and the world is only causal, Davidson forgets the normative (or rational) relation that exists between beliefs and things. For instance, in order to make the speaker's behavior intelligible, the radical interpreter needs to be able to say, not only that the fact that a rabbit appeared *caused* the speaker's assent to the utterance "There is a rabbit over there", but also that the speaker assents to that utterance *because* she perceives that there is a rabbit over there. The mere fact that the speaker responds discriminatively to stimuli from the environment does not make the speaker's behavior rational. That is why I have emphasized in the previous paragraph that perceiving the world meaningfully is essential for the interpreter as well as for the speaker.

From the point of view that I am suggesting, the correct application of a concept (or term) to an object, property or event, and the truth of beliefs about them do not depend only on what our peers say or think, but also on how we perceive what the world is. In Davidson's account, the possibility of error arises from the interaction between (at least) two subjects interacting with each other in a shared environment. The possibility of error emerges when subjects' responses to a common stimulus do not correlate. It seems that, in Davidson's explanation, the normative character of thought comes exclusively from the intersubjective relations between the subjects. Since, for Davidson, perceptual experience is only a causal intermediary between beliefs and reality, how the world is—what is revealed, in the simplest cases, in perception—cannot be a reason for holding any belief. In Davidson's view, experience provides causes, not reasons for believing. That is why, following McDowell, I think that in Davidson's account there is no place for a normative relation between thought and reality.

Once we adopt a different notion of perceptual experience, a notion according to which perception put us in direct and cognitive contact with the world, it can be said that, besides our peers, what obliges us to correct, in a rational way, our beliefs and uses of concepts (or terms) is the perception of how the world is and how it "resists" to our actions and desires. To know that one has a belief is to know that the belief can be true or false in virtue of how the world is (whatever our peers think). If beliefs are to refer to the world, their empirical content has to depend on how the world is and on our perception of how the world is. That normative dimension between thought and reality is constitutive of the contents of the most basic beliefs because contents are determined by noting—through perceptual experience—in what circumstances the world verifies or falsifies our thoughts.

Consider, for example, the learning of a term such as "tomato". There is no doubt that the role of our peers in the acquisition of that term is very important. But the world—the tomato in this case—is very important too. The child has to be able to perceive some characteristics of the tomato: its weight, its color, its taste. The child has to be able to perceptually recognize the circumstances in which the word "tomato" can be applied correctly to the thing in question. That ability involves the perception of some characteristics of the tomato. It is in virtue of perceiving those characteristics that the child can correctly associate the word "tomato" to the tomato itself. If the child is wrong on one occasion, not only the adults could correct her, but also the world could do so by presenting obstacles to her actions and beliefs. For example, imagine that the child says "tomato" in the direction of a rock, and that she believes that the rock is a tomato. The error can rise when she compares different perceptions of tomatoes with the perceptions of the rock; or when she tries to savor the rock. In these cases, the child should correct her belief and the application of the word "tomato" because of how the world is. It is by perceiving the consequences of her actions and thoughts in the environment that the child can note when she is right or wrong in her use of words and beliefs; and it is in virtue of this noting that the child can apprehend the content of words and beliefs she is learning. There is no doubt that, if the child applied the word "tomato" to whatever there is in the environment, it would mean that she has not learned the meaning of that word (even though the responses were caused by the environment). We learn the meaning of words and the most basic beliefs by connecting—in an *appropriate* way—the words and beliefs which we have to learn with their objects. My suggestion is simply that we would not be able to do so unless we could perceive (in a cognitive way) the objects and events referred to by the linguistic expressions which we have to learn. It is *perceived* objects and events which we have to learn to associate with words and beliefs.

In short, neither Davidson's coherentism nor his externalism manages to explain how beliefs acquire their contents. Against coherentism it can be said that mere logical relations between beliefs (without a cognitive relation with the world) do not warrant an objective reference. Against externalism it can be said that the mere causal link between beliefs and things does not provide the necessary content. In Kantian jargon: coherentism empties beliefs of contents; externalism posits a blind link between beliefs and reality. The source of the problem is the same one: the lack of recognition that we can establish a cognitive contact with the world by means of perceptual experience. Coherentism encapsulates justification into the sphere of beliefs; externalism posits a direct linkage between thought and reality, but it does not recognize the epistemic nature of that linkage. Plausibly, only by recognizing the fact that perceptual experience is a way through which we can apprehend how the world is, and by accepting that (in virtue of that fact) perceptual experience can justify beliefs, can we hope to avoid the problems of coherentism.[36]

4. *Final remarks: Towards rehabilitating empiricism*

I have tried to show several problems in Davidson's coherentism which derive from his conception of experience and justification. I have implicitly conceded that Davidson's objections to the theory of sense data (or usually called "representationalism") are correct. However, from those correct objections, Davidson draws a wrong conclusion: that perceptual experience does not have cognitive content. That is why he rejects the idea according to which experience can justify beliefs. My objections to Davidson's coherentism suggest a different—although certainly not new—conception of perceptual experience. According to this conception, perceptual experience puts us in cognitive contact with the world: when we have a veridical perceptual experience, we directly perceive how the world is. As I suggested in section 2, there is a plausible notion of experience which explains how we can have cognitive access to the world by perception. According to that notion, perceptual experience opens the world to us, we perceive without epistemic intermediaries how the world is. In that view, the perceived objects not only appear in different ways and reveal its characteristic properties depending on our location into space, but also they constitute thereby part of the perceptual content. We perceive objects by perceiving how they appear in virtue of our perspective. Accordingly,

36. It could be objected that to appeal to perceptual experience—as I recommend it—only post-pones the skeptical problem. If this objection were correct, I would not be in a better position than Davidson. But it is not so. My objection to Davidson's coherentism is that it makes skepticism inevitable, that it makes the world unknowable even in a non skeptical scenario. My position does not have that problem, even though I still owe an answer to skepticism.

to perceive is not to judge or to believe. In perception things themselves are presented as being thus and so.

By adopting this conception of perceptual experience, it is possible to explain how experience can justify beliefs. What Davidson sees as impossible becomes possible: to get outside "our skins to find out what is causing the internal happening of which we are aware". We can make a confrontation between beliefs and reality because we have cognitive and direct access, by means of perception, to how the world is. That conception of perceptual experience makes it possible to hold a different notion of justification as well: not only are beliefs able to enter into logical relations with other beliefs, but perceptual experiences can also do that. It can be seen, then, that we have rehabilitated the epistemic dimension of empiricism. Although I have not given any details about exactly how perceptual experience can justify beliefs, the objections to Davidson's coherentism I offered constitute indirect reasons for thinking—I hope—that perceptual experience can have the epistemic role that empiricism used to attribute to it.

References

Anscombe, G. E. M. 1965: "The Intentionality of Sensation: A Grammatical Feature". In: Anscombe, G. E. M., *Metaphysics and The Philosophy of Mind: Collected Philosophical Papers*. Vol. II. Oxford: Blackwell, 3–20.

Austin, John 1962: *Sense and Sensibilia*. Oxford: Oxford University Press.

BonJour, Laurence & Sosa, Ernest 2003: *Epistemic Justification*. Oxford: Blackwell.

Brandom, Robert 1994: *Making It Explicit. Reasoning, Representing and Discursive Commitment*. Cambridge: Harvard University Press.

— 1998: "Perception and Rational Constraint". *Philosophy and Phenomenological Research* 58, 241–259.

— 2000: *Articulating Reasons. An Introduction to Inferencialism*. Cambridge: Harvard University Press.

— 2002: "Non-inferential Knowledge, Perceptual Experience, and Secondary Qualities: Placing McDowell's Empiricism". In: Nicholas Smith (ed.), *Reading McDowell. On Mind and World*. London: Routledge, 92–105.

Brewer, Bill 1999: *Reason and Perception*. Oxford: Oxford University Press.

Campbell, John 2002: *Reference and Consciousness*. Oxford: Clarendon Press.

Davidson, Donald 1984a: "On the Very Idea of a Conceptual Scheme". In:

Donald Davidson, *Inquiries into Truth and Interpretation*. Oxford: Clarendon Press, 183–198.
— 1984b: "The Method of Truth in Metaphysics". In: Donald Davidson, *Inquiries into Truth and Interpretation*. Oxford: Clarendon Press, 199–214.
— 1999a: "Reply to John McDowell". In: Lewis Hahn (ed.), *The Philosophy of Donald Davidson*. Chicago and La Salle: Open Court, 105–108.
— 1999b: "Reply to A. C. Genova". In: Lewis Hahn (ed.), *The Philosophy of Donald Davidson*. Illinois: Open Court, 192–194.
— 1999c: "Reply to Thomas Nagel". In: Lewis Hahn (ed.), *The Philosophy of Donald Davidson*. Illinois: Open Court, 207–210.
— 1999d: "Reply to Barry Stroud". In: Lewis Hahn (ed.), *The Philosophy of Donald Davidson*. Illinois: Open Court, 162–166.
— 2001a: "A Coherence Theory of Truth and Knowledge". In: Donald Davidson, *Subjective, Intersubjective, Objective*. Oxford: Clarendon Press, 137–153.
— 2001b: "Comments on the Karlovy Vary Papers". In: Petr Kotatko, Gabriel Segal & Peter Pagin (eds.), *Interpreting Davidson*. Stanford: CSLI Publications, 285–308.
— 2001c: "Empirical Content". In: Donald Davidson, *Subjective, Intersubjective, Objective*. Oxford: Clarendon Press, 159–176.
— 2001d: "Epistemology Externalized". In: Donald Davidson, *Subjective, Intersubjective, Objective*. Oxford: Clarendon Press, 193–204.
— 2001e: "Three Varieties of Knowledge". In: Donald Davidson, *Subjective, Intersubjective, Objective*. Oxford: Clarendon Press, 205–220.
— 2004a: "The Problem of Objectivity". In: Donald Davidson, *Problems of Rationality*. Oxford: Clarendon Press, 3–18.
— 2004b: "What Thought Requires". In: Donald Davidson, *Problems of Rationality*. Oxford: Clarendon Press, 135–150.
— 2005: "Seeing Through Language". In: Donald Davidson, *Truth, Language, and History*. Oxford: Clarendon Press, 127–142.
Evans, Gareth 1982: *The Varieties of Reference*. Oxford: Clarendon Press.
Génova, A. C 1999: "The Very Idea of Massive Truth". In: Lewis Hahn (ed.), *The Philosophy of Donald Davidson*. Chicago and La Salle: Open Court, 167–191.
Haack, Susan 1993: *Evidence and Inquiry. Towards Reconstruction in Epistemology*. Oxford: Blackwell.
Ludwig, Kirk 1992: "Skepticism and Interpretation". *Philosophy and Phenomenological Research* Vol. LII (2), 317–340.
Martin, Michael 2009: "The Reality of Appearances". In: Alex Byrne & Heather Logue, (eds.), *Disjuntivism*. Cambridge, MA: The MIT Press, 91–116.
McDowell, John 1994: *Mind and World*. Cambridge: Harvard University Press.

— 1998a: “Criteria, Defeasibility, and Knowledge”. In: John McDowell, *Meaning, Knowledge, and Reality*. Cambridge: Harvard University Press, 369–394.
— 1998b: “Singular Thought and the Extent of Inner Space”. In: John McDowell, *Meaning, Knowledge, and Reality*. Cambridge: Harvard University Press, 228–259.
— 2002: “Responses”. In: Nicholas Smith (ed.), *Reading McDowell*. London: Routledge, 269–305.
— 2009b: *Having the World in View*. Cambridge: Harvard University Press.
Noë, Alva 2004. *Action in Perception*. Cambridge, MA: The MIT Press.
Rorty, Richard 1998: *Truth and Progress*. Cambridge: Cambridge University Press.
— 1999: *Philosophy and Social Hope*. London, Penguin.
Searle, John 1983: *Intentionality. An Essay in the Philosophy of Mind*. Cambridge: Cambridge University Press.
Sellars, Wilfrid 1997: *Empiricism and the Philosophy of Mind*. Cambridge: Harvard University Press.
— 1975: “The Structure of Knowledge”. In: Hector Neri-Castañeda (ed.), *Action, Knowledge and Reality*. Indianapolis: The Bobbs-Merrill Company, 295–248.
Snowdon, Paul 1992: “How to Interpret ‘Direct Perception’?”. In: Tim Crane (ed.), *The Contents of Experience*. Cambridge: Cambridge University Press, 48–78.
— 2009: “The Objects of Perceptual Experience”. In: Alex Byrne & Heather Logue (eds.), *Disjuntivism. Contemporary Readings*. Cambridge, MA: The MIT Press, 49–74.
Stroud, Barry 1999: “Radical Interpretation and Philosophical Scepticism”. In: Lewis Hahn, *The Philosophy of Donald Davidson*. Chicago: Open Court, 139–161.

Grazer Philosophische Studien
85 (2012), 237–253.

STRONG, THEREFORE SENSITIVE: MISGIVINGS ABOUT DEROSE'S CONTEXTUALISM

Jon COGBURN & Jeffrey W. ROLAND
Lousiana State University

Summary

According to an influential contextualist solution to skepticism advanced by Keith DeRose, denials of skeptical hypotheses are, in most contexts, *strong* yet *insensitive*. The strength of such denials allows for knowledge of them, thus undermining skepticism, while the insensitivity of such denials explains our intuition that we do not know them. In this paper we argue that, under some well-motivated conditions, a negated skeptical hypothesis is strong only if it is sensitive. We also consider how a natural response on behalf of DeRose appears to be equally available to his primary rival (viz., the sensitivity theorist).

Introduction

Keith DeRose (1995, 2004) has advanced an influential contextualist solution to the skeptical problem. On contextualist accounts of knowledge, there are some contexts where one genuinely doesn't know whether one is (say) a brain in a vat and other contexts where one does know whether one is a brain in a vat. Part of what distinguishes DeRose's contextualist solution is the detailed and *prima facie* plausible story about how these contexts match some of the key intuitions that lead to the skeptical problem in the first place. For DeRose, epistemology must not just explain why the skeptic is wrong, but also why the skeptic sometimes *seems* right. Why is it sometimes so tempting to believe that, for all we know, we really might be BIV?[1]

In brief, DeRose's answer is that skepticism sometimes seems to be right because negated skeptical hypotheses such as 'I am not a BIV' fail to have a certain modal property known as *sensitivity* (characterized below), and sensitivity is nearly correct as a third necessary condition of knowl-

1. We use the standard abbreviation, 'BIV' for 'brain in a vat' and cognate expressions.

edge. As a result, we're tempted to say that we don't know the denials of skeptical hypotheses, and this opens the way for the skeptic's argument. However, for all its seeming rightness, skepticism is wrong according to DeRose because in most contexts negated skeptical hypotheses do have a certain other modal property, known as *strength*. So though for very good reasons it seems like we cannot know negated skeptical hypotheses, in fact we can.

Careful attention to the underlying modal principles produces an argument against this very aspect of DeRose's explanation of skeptical intuitions. Certain sorts of beliefs, arguably among them skeptical beliefs (i.e., beliefs that *p*, where *p* is a skeptical hypothesis[2]), are such that there is no context where they are strong in DeRose's sense without also being sensitive. So DeRose is not entitled to the claim that negated skeptical hypotheses are both strong and insensitive. We begin (§ 1) by briefly setting out DeRose's solution to skepticism and introducing a framework for that solution. We continue (§ 2) with a discussion of how DeRose uses sensitivity to explain skeptical intuitions. Next (§ 3), we formulate and motivate a feature of skeptical beliefs which opens the way to a proof (§ 4) that sensitivity can't do the work DeRose asks of it. Finally (§ 5), we suggest ways in which the sensitivity theorist[3] might be able to help herself to DeRose's deep insights into the role that context plays in epistemic evaluation.

1. *DeRose's contextualist solution*

1.1 *The skeptical argument*

For the duration of this paper, fix *H* to be a generic skeptical hypothesis (e.g., 'An evil demon is deceiving me about nearly everything' or 'I am a BIV') and let *O* be a statement expressing some ordinary contingent proposition that we typically take ourselves to unproblematically know (e.g., that I have hands).[4] Then, as is common nowadays, the skeptical

2. We use lower-case, italicized roman letters (other than '*w*') throughout as variables for declarative statements. We also sometimes blur the distinction between statements and propositions; no problems result from this blurring.

3. Throughout, we use 'sensitivity theorist' (and cognate expressions) as DeRose (2004) uses 'direct sensitivity theorist', roughly as one who takes knowledge to be true, sensitive belief.

4. We use upper-case, italicized roman letters (other than '*C*' and '*S*') throughout to represent particular declarative statements. We also often write as an individual to simplify examples.

problem can be generated by the following *skeptical argument*.

(SA1) I don't know that $\neg H$.
(SA2) If I don't know that $\neg H$, then I don't know that O.
(SAC) Therefore, I don't know that O.

The skeptic affirms all three of these, and so holds (SA1), (SA2), and (SAC). Sensitivity theorists such as Robert Nozick (1981) affirm (SA1) yet try to deny (SA2) in order to deny (SAC), and so hold (SA1), ¬(SA2), ¬(SAC). And Mooreans such as Ernest Sosa (1999, 2000) deny (SA1) as a way to affirm (SA2) while denying (SAC), thus holding ¬(SA1), (SA2), ¬(SAC).[5]

DeRose affirms (SA2) and then holds that epistemic context determines whether the skeptic or Moorean is correct. A strict epistemic context such as the typical philosophy classroom requires a lot from the knower, and for DeRose is one where skepticism is correct. On the other hand, permissive epistemic contexts such as the vast majority of mundane situations are ones where Mooreanism is correct. This is the payoff of DeRose's contextualism: in normal contexts, the Moorean is correct and skepticism is avoided; only in esoteric contexts is skepticism a threat.

1.2 *Strict versus permissive epistemic contexts*

One of the more distinctive components of DeRose's theory is the way in which it characterizes knowledge in terms of "strong enough true belief." In contrast to Sosa's safety account,[6] DeRose writes:

> The main difference is that ... [safety] is troubled only by worlds (especially very nearby worlds) in which one believes that p but p is not the case (ways in which S would easily believe that p without it being the case that p), while strength is disturbed both by that and by the presence of worlds (especially very nearby worlds) in which p is true but S disbelieves it. (DeRose 2004, 34)

Sosa helpfully characterizes DeRose's account in this manner:

5. There are interesting analogies to these positions in the philosophy of vagueness (Keefe 2007). If you consider the sorites paradox as an analogous inference (with many (SA2) steps), then: (1) the skeptic is the nihilist, (2) the sensitivity theorist the super (or sub) valuationist, and (3) the Moorean the epistemicist. Interestingly, one could model the contextualist's knowledge predicate using supervaluations, suggesting already an affinity between sensitivity theory and contextualism.

6. According to Sosa (1999, 2000), S's true belief that p is knowledge for S just in case it is *safe*: were S to believe that p, it would be true that p.

> To know that p is to believe that p with enough "strength," where the strength of one's belief is stronger the more remote are any worlds wherein one first fails to match the fact of the matter as to whether p. That is to say, the strength of one's belief is directly proportional to the remoteness of the least remote worlds wherein it fails to be the case both that $[p \wedge B(p)]$ and that $[\neg p \wedge \neg B(p)]$. (Sosa 2004, 280)

In other words, the strength of *S*'s belief that p is inversely proportional to the closeness of the nearest world where either $p \wedge \neg Bel(S, p)$ or $\neg p \wedge Bel(S, p)$ obtain.[7] So, according to DeRose, my belief that I have hands is strong enough to count as knowledge because the closest worlds where I have hands and don't believe that I have hands, or where I lack hands and believe I have hands, are far distant from the actual world.

Though DeRose might himself find this unnatural, one can beautifully represent the way in which context partially determines strength in terms of counterfactual conditionals by rendering their semantics as linguists are likely to do, rather than according to the more common Lewisian approach favored by philosophers. In the natural language formal semantics and pragmatics literature (Kadmon 2001) it is much more usual to see something like this for counterfactual semantics:

> LING $\Box\!\!\rightarrow$. $p \Box\!\!\rightarrow q$ is true in world w and context C just in case every C-relevant p-world is also a q-world.[8]

Here it is understood that C-relevance imposes a similarity metric such that C-relevant worlds are closer than non C-relevant worlds, and for any such metric the actual world is the closest world. This approach is closely connected to the Lewisian approach to counterfactual semantics more commonly followed by philosophers—where $p \Box\!\!\rightarrow q$ is true in world w and context C just in case the closest C-relevant p-world is also a q-world—in that, if $p \Box\!\!\rightarrow q$ holds in the linguists' sense, it will also hold in the philosophers' sense.

The linguists' approach is particularly well suited to DeRose's concept of strength for a couple of reasons. First, by using it we can define strength without facing what DeRose (2004) has called the "true–true problem" that besets standard explications of safety-based epistemology.[9]

7. We use '$Bel(S, p)$' to abbreviate 'S believes that p'.
8. We follow convention in using $\Box\!\!\rightarrow$ as a connective for the counterfactual conditional.
9 For the true-true problem, see (DeRose 2004).

STRENGTH. S's belief that p is strong in context C if and only if $p \mathrel{\Box\!\!\rightarrow} Bel(S, p)$ and $Bel(S, p) \mathrel{\Box\!\!\rightarrow} p$ are true in C.

This definition avoids the true–true problem because (Ling $\Box\!\!\rightarrow$) makes sense of false counterfactual conditionals with true antecedents, something the Lewisian approach cannot do. This is essential because of course one might have a true belief that fails to be knowledge, but on the Lewisian account the actual truth of S's belief, combined with the fact that S does actually believe that p, trivially renders both $p \mathrel{\Box\!\!\rightarrow} Bel(S, p)$ and $Bel(S, p) \mathrel{\Box\!\!\rightarrow} p$ true.[10]

Second (and this is certainly why they tend to characterize the underlying semantics in this manner), the linguists' semantics allows for much more elegant characterization of the way context selects for varying kinds and strengths of modality.[11] We can use it to define epistemic context C as being more permissive than epistemic context C' (equivalently that C' is stricter than C) if the C-relevant worlds are a proper subset of the C'-relevant worlds. Then for example, p could be strong relative to C and weak relative to C'. The more worlds an epistemic context rules in as relevant for consideration, the stricter that context.

10. One can solve the true-true problem while continuing to utilize Lewisian semantics by giving up strong centering, according to which the actual world is uniquely closest to itself. Weak centering, by contrast, still holds that no world is closer to the actual world than itself, but permits non-actual worlds to be equally close. Such worlds might be ones where a counterfactual conditional with actually true antecedent and consequent has a true antecedent and false consequent. We don't opt for this solution for three reasons.

First, the linguists' semantics presented here avoids the true-true problem without having to sacrifice centering. Second, in (Cogburn and Roland MS) we argue that recent versions of Sosan safety-based accounts should follow linguists more generally and take conditionals as generalized quantifiers over possible worlds. So comparing strength-based and safety-based accounts is much easier if we use an instance of this scheme in modeling DeRose, as we have here. Third, as far as we can tell it's more difficult to logically regiment DeRose's theory using Lewisian semantics with weak centering (and without strong centering). On the weak centering version of Lewisian semantics, the truth or falsity of a counterfactual is still determined by what happens in one set of worlds equidistant from the actual world. But on DeRose's theory the strength of a belief in a context is a function of the truth of two conditionals in a set of non-equidistant worlds picked out by the context, with the subset relation (as opposed to "closer than") on such sets of worlds mirroring levels of strength required by context. As we show below, the linguists' semantics captures this exactly. We thank an anonymous referee for encouraging us to expand on this point.

11. See, e.g., the ground-breaking work in (Kratzer 1977) and (Roberts 1989), respectively on the characterization of conditionals as generalized quantifiers over possible worlds and on how modal subordination is an instance of domain selection more broadly.

This allows us to make clear sense of DeRose's contextualism. Suppose, as is standard in the contemporary epistemology literature, we are not in fact BIV and that by any reasonable epistemic metric worlds where we are BIV are very dissimilar to (distant from) the actual world. A very strict epistemic context C^* would rule in so many worlds that the class of C^*-relevant worlds would include worlds where either I'm not a BIV and I believe that I am (making $\neg H_{biv} \Box\!\!\rightarrow Bel(S, \neg H_{biv})$ false in C^*, where H_{biv} is the hypothesis 'I'm a BIV') or worlds where I believe that I'm not a BIV but I am one (making $Bel(S, \neg H_{biv}) \Box\!\!\rightarrow \neg H_{biv}$ false in C^*).[12] In contexts such as C^*, ordinary beliefs such as that I have hands are just as likely to not be strong (since DeRose accepts the skeptical inference, (SA2)). For DeRose, the philosophy classroom tends to select such contexts. On the other hand, a more permissive epistemic context would rule in far fewer relevant worlds, excluding worlds where competent epistemic agents either fail to believe they have hands or believe they have hands even though they don't. So where O expresses the proposition that S has hands, both $O \Box\!\!\rightarrow Bel(S, O)$ and $Bel(S, O) \Box\!\!\rightarrow O$ are true in permissive contexts. Likewise, given the validity of (SA2) on DeRose's account, S's belief that $\neg H_{biv}$ is also strong in permissive contexts, as there is no world in the restricted set either where S is not a BIV and she fails to believe that she is not or where S believes she is not a BIV and she in fact is. That is, both $\neg H_{biv} \Box\!\!\rightarrow Bel(S, \neg H_{biv})$ and $Bel(S, \neg H_{biv}) \Box\!\!\rightarrow \neg H_{biv}$ are true in permissive contexts.

2. *Sensitivity and insensitivity*

Our use of linguists' account of counterfactual semantics enables us to recapitulate DeRose's contextualism entirely in terms of the strength of a belief. Whether one epistemic context C is stricter than another C' depends entirely on the set of possible worlds which are relevant to assessing the truth of the strength counterfactuals in C being a proper superset of the set of worlds which are relevant to assessing the truth of those counterfactuals

12. Notice that if one understood belief so that an agent believes every consequence of her beliefs, then we would need to add a consistency constraint on belief sets or adopt a paraconsistent logic in order for the first of these counterfactuals to be false in C^*. (See p. 13 for discussion related to the paraconsistency option.) However, we take it that the notion of belief operative in these debates isn't closed under consequence in the way indicated. Thanks to an anonymous referee for raising this issue.

in C'. This is sufficient to model the contextualist's view that skepticism is correct in strict epistemic contexts and incorrect in more permissive ones.

DeRose, however, adds another commitment to this picture. He holds that epistemic contexts strict enough to support (SA1) will be ones where the negated skeptical hypothesis $\neg H$ is insensitive in Nozick's (1981) sense. For the sensitivity theorist, S knows that p just in case S's belief that p is true and sensitive, in that were p false then S would not believe that p. That is, S's belief that p is sensitive just in case $\neg p \mathrel{\Box\!\!\rightarrow} \neg Bel(S, p)$ is true. On the other hand, S's belief that p is insensitive just in case $\neg p \mathrel{\Box\!\!\rightarrow} Bel(S, p)$ is true. That something like sensitivity is a necessary condition for knowledge is a redoubtable intuition in epistemology, arguably going back as far as (Russell 1912/1997). It is so redoubtable that both Sosa (1999, 2000) and DeRose take it to be one of the tasks of epistemology to explain why people are drawn to it. For Sosa, the fact that the counterfactual conditional expressing sensitivity is the contraposition of his own safety condition, plus the fact that contraposition is valid for material conditionals but not counterfactuals (on the Lewisian semantics),[13] explains why people wrongly think sensitivity is a necessary condition on knowledge. But then, for Sosa, this explains why people wrongly think that we do not know the denials of skeptical hypotheses, since (according to DeRose and Sosa) the denial of a skeptical hypothesis is insensitive (i.e., $H \mathrel{\Box\!\!\rightarrow} Bel(S,\neg H)$ is true).

DeRose agrees that it is the insensitivity of negated skeptical hypotheses that leads us to conclude that we do not know skeptical hypotheses to be false.

> Suppose a direct sensitivity theorist uses the phenomena I've cited as reasons for thinking that [the subjunctive conditional account] is on the right track as support for the view knowledge is roughly true, sensitive believe (with perhaps some other necessary conditions for knowledge also thrown in, to taste). Such a theorist may think there is an important property of beliefs, sensitivity*, which is necessary for knowledge, and of which the conditionals in terms of which sensitivity is defined give a good first approximation. (DeRose 2004, 28)

---

13. We should note that the semantics we use in this paper for DeRose's account validates contraposition, but nothing hangs on this. Our use of $\Box\!\!\rightarrow$ here only formalizes DeRose's modal notion of strength. Nothing we say presupposes that sensitivity or safety theorists must use a contraposing notion of counterfactuals. In particular, the relevant sort of sensitivity claim to which we will argue DeRose is committed can be stated using sensitivity theorists' preferred semantics for counterfactuals. See n. 17.

DeRose argues that both he and Sosa should be committed to this.

> We both appeal to certain beliefs' insensitivity to explain why they give an appearance of no-knowledge, but we both do so in an indirect way: Neither of us pictures knowledge as being anything like sensitive true belief, so we both must explain why insensitive beliefs so often seem not to be knowledge even when they have what we claim is the real knowledge-making property (safety, strength) to a high degree. (DeRose 2004, 35)

However, DeRose diverges from Sosa precisely in the explanatory role that contextualism plays here.

> I believe that a contextualist version of the picture of knowledge under consideration can provide a powerful explanation of why insensitive beliefs can so often give an appearance, of no-knowledge, even when they are quite strong, and can thereby provide a powerful solution to the particular skeptical paradox that both Sosa and I wrestle with. (DeRose 2004, 35)

For DeRose, it is precisely because negated skeptical hypotheses are both insensitive and "quite strong" that the skeptical problem is so alluring. But this is precisely where problems arise for his view.

3. *Evidential indistinguishability*

Say that propositions p, p' are *evidentially indistinguishable* just in case there are no experiences one might have such that on the basis of those experiences one could rationally judge p to be true and p' to be false. Propositions are *evidentially distinguishable* just in case they're not evidentially indistinguishable. Lots of propositions are clearly evidentially distinguishable (e.g., the proposition that there is a cat on the mat and the proposition that water boils at room temperature). Moreover, lots of propositions are evidentially distinguishable from their denials. The proposition that there is a cat on the mat is clearly such a proposition. Intuitively, evidentially indistinguishable propositions are equally well confirmed by any experiences relevant to their confirmation.[14] We can make the notion of evidential indistinguishability more precise as follows. Let an agent S's *evidential base* for a proposition p be the sum total of S's experiences that bear on the truth value of p. A proposition is confirmed for an agent S if and only if S's evidential base for p makes it rational for S to judge p to

14. Cf. empirically equivalent theories.

be true. Then propositions p, p' are evidentially indistinguishable if and only if $p \wedge \neg p'$ is not confirmed for any agent.[15]

What's important for present purposes is that one of the signature properties of propositions expressed by skeptical hypotheses such as 'I am a BIV' is that they are by design supposed to be evidentially indistinguishable from their denials. Suppose H is a true skeptical hypothesis. According to the skeptic there is no experience that could warrant the assertion that H, just because all experience is by design systematically misleading on this very score. So there is no confirming evidential base for the proposition that H. This would not be of great moment if it were any easier to confirm the proposition that $\neg H$. For if we could confirm that, then the evidential base favoring it would be a straight refutation of skepticism. So any self-respecting skeptical hypothesis must at least attempt to be one such that if it is false, no experience could warrant the assertion that it is false, i.e., no evidential base confirms the proposition that it's false. Put another way, H and $\neg H$ are evidentially indistinguishable just in case there is no confirming evidential base for $H \wedge \neg\neg H$. But a confirming evidential base for $H \wedge \neg\neg H$ would be a confirming evidential base for H. However, again, there is no confirming evidential base for H by design. So H and $\neg H$ are evidentially indistinguishable.

Now, obviously, just because the skeptic intends skeptical hypotheses to be evidentially indistinguishable from their denials, we don't need to accept that they in fact are. But we should accept that one needs an argument for evidential distinguishability, since skeptical hypotheses certainly do at least give a strong impression of being evidentially indistinguishable from their denials. If no evidence whatsoever could hold in favor of p, it's very hard to see how any evidence whatsoever could hold against p. This entails that if a proposition that is evidentially indistinguishable from its negation is indeed sensitive, then the negation of that proposition should be sensitive as well. This is easiest to see if we back up and consider the intuition that sensitivity is designed to capture, that true beliefs that fail to be knowledge do so precisely because they involve luck. Now suppose that the proposition that it is two o'clock is for Sam evidentially indistinguishable from the proposition that it is not two o'clock. This means Sam

15. This dependence on agents and experiences gives that the evidential indistinguishability of propositions may vary over time or worlds, which seems to us as it should be. What's important for present purposes is that skeptical hypotheses and their denials are stably evidentially indistinguishable. As we'll see in the next section, this very fact underwrites DeRose's claim that skeptical hypotheses and their denials are stably insensitive.

has equal access to information that could confirm either proposition with equal reliability. But then Sam is just as unlikely to be lucky with respect to the belief that it is two o'clock as he is with respect to the belief that it is not two o'clock. This has the consequence that the sensitivity theorist understands the counterfactual 'were it not the case that it is two o'clock, then Sam would not believe it is two o'clock' as entailing 'were it not the case that it is not two o'clock, then Sam would not believe that it is not two o'clock'. The absence of luckiness that justifies the claim that his belief that it is two o'clock is sensitive also justifies the sensitivity of his belief that it is not two o'clock (did he in fact believe that). So if a proposition and its negation are evidentially indistinguishable from one another, it follows that the sensitivity of one entails the sensitivity of the other.

We should repeat that we don't take ourselves to have definitively established that skeptical hypotheses are in fact evidentially indistinguishable from their denials (though we argue below that DeRose is committed to this being the case). All we've tried to show is that the view that a given skeptical hypothesis is evidentially distinguishable from its denial requires an argument, and that if skeptical hypotheses are evidentially indistinguishable from their denials, then they and their denials succeed or fail together to be sensitive—either both are sensitive or neither is.

4. *Strength entails sensitivity*

The evidential indistinguishability of skeptical hypotheses and their denials undermines DeRose's contention that the denial of a skeptical hypothesis can be strong yet insensitive. This, in turn, undermines achievement of one of his main goals: the attempt to harness sensitivity theory to explain the intuitive attraction of skepticism. The proof that strength entails sensitivity is a fairly straightforward piece of logic that requires only one additional premise of any potential controversy, which we now introduce. Say that an agent S is a member of the *laity* (to be contrasted with the *Priesthood*, see e.g. (Priest 2002)) with respect to p if and only if, necessarily, if S believes that p is false, then S does not believe that p. We formalize this as

Laity. $\forall p \forall S[La(S, p) \leftrightarrow \Box(Bel(S, \neg p) \rightarrow \neg Bel(S, p))]$

Let H_{biv} be the BIV hypothesis and assume that DeRose is a member of the laity with respect to H_{biv}. As far as we know, no one has defended the view that skeptical hypotheses are the kind of statements one could both disbelieve and believe. If one could motivate dialetheism with respect to skeptical beliefs, then it would not be an obviously crazy position (just as it is not an obviously crazy position with respect to statements such as 'This statement is false'). But even with the background of self-reflexive epistemic paradoxes,[16] dialetheism with respect to skeptical beliefs has not been defended. So we take ourselves to be justified in assuming that DeRose is a member of the laity with respect to H_{biv}.

The proof begins with one half of the definition of DeRosean strength and DeRose's claim that the denial of H_{biv} is strong.

1. $\forall p \forall S[Str(p, S) \rightarrow (p \;\Box\!\!\rightarrow Bel(S, p))]$ (by Strength).

This just says that for all statements p and agents S, if p is strong for S, then $p \;\Box\!\!\rightarrow S$ believes that p holds in the sense of (Ling $\Box\!\!\rightarrow$). Then we have:

2. $Str(\neg H_{biv}, d)$ (DeRose's assumption),

which just says that the denial of H_{biv} is strong for DeRose. These two together entail the following:

3. $Str(\neg H_{biv}, d) \rightarrow (\neg H_{biv} \;\Box\!\!\rightarrow Bel(d, \neg H_{biv}))$ (from 1, by $\forall$ elimination),
4. $\neg H_{biv} \;\Box\!\!\rightarrow Bel(d, \neg H_{biv})$ (from 2 and 3, by $\rightarrow$ elimination).

Then from (Laity) and the claim that DeRose is a member of the laity with respect to H_{biv} we get:

5. $\forall p \forall S[La(S, p) \rightarrow \Box(Bel(S, \neg p) \rightarrow \neg Bel(S, p))]$ (by Laity),
6. $La(d, H_{biv})$ (assumption),
7. $La(d, H_{biv}) \rightarrow \Box(Bel(d, \neg H_{biv}) \rightarrow \neg Bel(d, H_{biv}))$ (from 5, by $\forall$ elimination),
8. $\Box(Bel(d, \neg H_{biv}) \rightarrow \neg Bel(d, H_{biv}))$ (from 6 and 7, by $\rightarrow$ elimination).

16. See (Sundholm 2008), an essay titled "A Novel Paradox?" One of the present authors must respond with an emphatic "No!" The same epistemic paradox is discussed at length in (Cogburn 2004), where in a footnote the author credits himself and Roy Cook with having jointly discovered it in the mid 1990s. Of course, it is not improbable that someone else discovered it prior to that, and all such discussions rest on the shoulders of others (Kaplan and Montague 1960). But it is notable that the paradox of Cogburn and Cook does not require epistemic closure.

So on (4) we have that if H_{biv} were false, it would be the case that DeRose believes that it is false, and on (8) we have that necessarily if DeRose believes H_{biv} to be false, then he does not believe H_{biv}. By the following weak transitivity principle for counterfactuals,

WTC. $p \mathrel{\Box\!\!\rightarrow} q, \Box(q \rightarrow r) \vdash p \mathrel{\Box\!\!\rightarrow} r,$

it therefore follows that,

9. $\neg H_{biv} \mathrel{\Box\!\!\rightarrow} \neg Bel(d, H_{biv})$ (from 4 and 8, by WTC).

But this is just to say that H_{biv} is sensitive for DeRose.[17] Hence, our proof shows that if the denial of H_{biv} is strong (as DeRose claims), then H_{biv} itself is sensitive.

The sensitivity of H_{biv} is a problem for DeRose. He likes to characterize the insensitivity of an agent *S*'s belief that *p* as a "block" to judging that *S* knows that *p*, and he argues that the fact that many cases of no-knowledge can be modified so that

(a) they flip from no-knowledge to knowledge cases, and
(b) the relevant belief flips from insensitive to sensitive

is evidence that his account of the attractiveness of skeptical hypotheses is on the right track.

Consider, for instance, a standard lottery case in which (prior to the winning numbers being announced) we judge that we don't know that our ticket will lose. DeRose identifies the insensitivity of our belief that we will lose (i.e., the truth of the counterfactual 'were we to have the winning ticket we would believe that we didn't') as the block to judging that we know we will lose. After the winning numbers have been announced, however, and they indeed aren't the numbers on our ticket, we would like to say that we know we have a losing ticket. But we also would no longer

17. One might object that line (9) of the above proof is not in fact a statement of sensitivity, since sensitivity theorists intended something different by the counterfactual conditional than what we are calling the linguists' account. Such an objection has no purchase however, as $p \mathrel{\Box\!\!\rightarrow} q$ in the sense given in the above proof entails $p \mathrel{\Box\!\!\rightarrow} q$ understood in sensitivity theorists' preferred sense, whether Lewisian or in terms of generalized quantification over *most* of some set of contextually selected worlds, as we argue one should prefer in the context of safety theory (Cogburn and Roland MS).

say that were we to have the winning ticket we would believe that we didn't. Rather, we'd say that a different counterfactual is true, viz., 'were we to have the winning ticket, we would not believe that we have a losing ticket'. But this is just to say that the belief that we have a losing ticket is sensitive. The block has been removed. From this example and others similar to it, DeRose concludes:

> Again and again, [my account] posits a certain block to our judging that we know, and the changes that would clear the way for our judging that we know also remove this block. This makes it difficult not to believe that [my account] is at least roughly correct. (DeRose 1995, 25)

The situation is different for skeptical hypotheses.

> In the case of the BIV hypothesis, it's hard to test [my account] in this way, for it's difficult to imagine a situation in which it seems a subject does know that she's not a BIV. But this only confirms [my account]: While it's difficult to imagine a situation in which one seems to know that one's not a BIV, it's likewise difficult to imagine circumstances in which the block [my account] posits is removed. It's difficult, that is, to imagine a situation in which someone believes they're not a BIV but in which the conditional *If S were a BIV, then S would believe she wasn't a BIV* isn't true. For, as the BIV hypothesis is formulated, one's brain is electrochemically stimulated so that one has precisely those sensory experiences one actually has had. But wouldn't one then have formed precisely those beliefs that one actually has formed, including the belief that one's not a BIV? (DeRose 1995, 25f., original emphasis)

DeRose's point here is that, in contrast to non-skeptical cases, it appears that cases involving skeptical hypotheses cannot be modified so as to satisfy (a) and (b) above. The result of not being able to satisfy (b) with respect to skeptical cases is that negations of skeptical hypotheses are stable with respect to insensitivity—it appears they can't be flipped from insensitive to sensitive. In particular, *S*'s belief that $\neg H_{biv}$ is stably insensitive. For our purposes, there are two important things to notice about the argument of this passage.

First, the argument of this passage applies equally well to an agent's belief that H_{biv} (were she to have such a belief). Just as it's hard to imagine a situation in which someone believes they're not a BIV but in which the conditional 'If *S* were a BIV, then *S* would believe she wasn't a BIV' isn't true, it's hard to imagine a situation in which *S* believes she is a BIV (i.e., $Bel(S, H_{biv})$) but in which the conditional 'If *S* weren't a BIV, then *S* would believe she was a BIV' isn't true—and for the very same reasons.

By design, H_{biv} guarantees that S will have had the same sensory experiences as she has actually had at every C-relevant world, and so it seems she should also have the same beliefs as she actually has, including the belief that she's a BIV. So in every C-relevant world, the consequent of $\neg H_{biv} \;\Box\!\!\rightarrow Bel(S, H_{biv})$ is true, and thus so is the counterfactual itself. It follows that H_{biv} is stably insensitive. But then H_{biv} is not sensitive, contrary to (9). DeRose appears to be caught in a contradiction.

Moreover, even if this contradiction is merely apparent (and this is the second thing to notice), the reasoning DeRose deploys in support of the stable insensitivity of S's belief that $\neg H_{biv}$ relies on the evidential indistinguishability of H_{biv} and $\neg H_{biv}$. If S has precisely the same sensory experiences in BIV worlds and non-BIV worlds, as DeRose contends, then no sensory experience will allow S to distinguish between BIV and non-BIV worlds. Thus DeRose himself countenances the evidential indistinguishability of H_{biv} and $\neg H_{biv}$. But then we have:

10. $\neg\neg H_{biv} \;\Box\!\!\rightarrow \neg Bel(d, \neg H_{biv})$ (from 9, evidential indistinguishability of H_{biv} and $\neg H_{biv}$).

Q.E.D. Hence, if $\neg H_{biv}$ is strong, then it is also sensitive. But there's nothing special here about $\neg H_{biv}$. So this result generalizes to all skeptical hypotheses.[18]

So, unless one has a compelling story about why skeptical hypotheses fail to be evidentially indistinguishable from their denials (and not only have we not seen anyone argue for anything that would entail the relevant distinguishability, DeRose appears unable to avail of himself of such an argument were it to be given), or unless one defends dialetheism with respect to skeptical hypotheses (again, we haven't seen such a defense) one cannot hold, as DeRose does, that the denial of skeptical hypotheses are simultaneously strong and insensitive. Something's got to give.

5. *Concluding remarks*

The easiest way for DeRose to respond would be to simply abandon the idea that the insensitivity of negated skeptical hypotheses plays an impor-

18. Thanks to an anonymous referee for pressing us on the move from (9) to (10) and its connection to evidential indistinguishability, as well as on the status of (9) itself.

tant role in explaining skeptical intuitions. This seems promising to us. Why can't the contextualist idea that in some epistemic contexts skeptical hypotheses are known to be false and in some contexts they're not be enough of an explanation for the perennial draw of skepticism? And, as is clear from our discussion in § 1, DeRose's picture of how this works suggests a formal model that is clear and intuitively compelling.

Sorting this out would take some work though. For if one accepts that negated skeptical hypotheses are sensitive, then one has to accept that the sensitivity theorist has the resources to combat skepticism. Since the supposed failure of sensitivity theory to combat skepticism serves as one of the principal reasons for preferring both safety theory and DeRose's theory of strength to sensitivity theory, this raises the issue of the extent to which the formal model from § 1 is nothing more than epicycles. Why all the bother with strength and safety if sensitivity theory undermines skepticism in any case?

The contextualist should double down here. Strength is not an absolute notion, and indeed the formal characterization in § 1 starts with the characterization of what it is for one context to be stricter than another in terms of inclusion of relevant sets of possible worlds. The same belief then can be strong in a permissive context and weak in a strict one. But if this is the understanding of the underlying modal semantics, the DeRosean would interpret the formal derivation in the previous section in a very specific manner. The claim that a negated skeptical hypothesis is strong is the claim that certain material conditionals hold across a suitably permissive set of possible worlds, a set selected by the linguistic and epistemic context in which the negated skeptical hypothesis is asserted. But then, to avoid equivocation, the conclusion that the negated skeptical hypothesis is sensitive must also be understood using the same set of contextually determined worlds. So all we've really established is that, for DeRose, any context in which a non-Priestly belief that is also evidentially indistinguishable from its negation[19] is strong is also a context in which that belief is sensitive. But this is consistent with there being other contexts where the same belief is insensitive.

By the proof of § 4, contexts in which a negated skeptical hypothesis is insensitive will also be contexts where it is not strong. So, again, the claim that a belief is strong yet insensitive still cannot be used by DeRose to explain skeptical intuitions. For DeRose it really is the relativization of

19. Beliefs are evidentially indistinguishable just in case their propositional contents are.

the notion of strength to context, such that a belief can be strong in some contexts and not in others, that ends up doing all of the work. He should say that skepticism often seems right because in some (strict) contexts we do not know skeptical hypotheses to be false, and then use strength to explain how this works.

This being said, we would be remiss not to conclude with a *mihi quoque* argument on behalf of the sensitivity theorist. If the characterization of strictness of context in terms of nested sets of possible worlds is doing all of the work for DeRose, then there is no reason the sensitivity theorist could not adopt the same contextualist modal semantics for her own account of knowledge. But then, the alleged problems with sensitivity theory weren't something to be fixed by replacing it with some cognate notion like safety or strength, but rather by accepting the contextualism of some defenders of these two notions. Of course, fully defending such a position would be another endeavor, but the fact that the sensitivity theorist can now argue that the strength theorist is already committed to a contextualist sensitivity theory with respect to skeptical hypotheses (in that negated skeptical hypotheses are sensitive in some contexts and not in others, exactly corresponding to whether they are strong or not strong in those very same contexts) is already some evidence in its favor. For the sensitivity theorist is in an excellent position to avail herself of DeRose's insights into the nature of epistemic context without what should now appear to her to be DeRose's and Sosa's modal epicycles.[20]

References

Cogburn, Jon 2004: "Paradox Lost". *Canadian Journal of Philosophy* 34, 195–216.

Cogburn, Jon & Roland, Jeffrey W. (unpublished): "Safety and the True–True Problem". *Unpublished manuscript.*

DeRose, Keith 1995: "Solving the Skeptical Problem". *Philosophical Review* 104, 1–52.

— 2004: "Sosa, Safety, and Skeptical Hypotheses". In: John Greco (ed.), *Sosa and His Critics*. Oxford: Blackwell Publishers, 22–41.

Kadmon, Nirit 2001: *Formal Pragmatics*. Malden: Wiley-Blackwell.

Kaplan, David & Montague, Richard 1960: "A Paradox Regained". *Notre Dame Journal of Formal Logic* 1, 79–90.

20. Thanks to two anonymous referees for this journal for valuable comments.

Keefe, Rosanna 2007: *Theories of Vagueness*. Cambridge: Cambridge University Press.

Kratzer, Angelika 1977: "What 'Must' and 'Can' Must and Can Mean". *Linguistics and Philosophy* 1, 337–355.

Nozick, Robert 1981: *Philosophical Explanations*. Cambridge, MA: Harvard University Press.

Priest, Graham 2002: *In Contradiction: A Study of the Transconsistent*. Oxford: Oxford University Press.

Roberts, Craige 1989: "Modal Subordination and Pronominal Anaphora in Discourse". *Linguistics and Philosophy* 12, 683–721.

Russell, Bertrand 1912/1997: *The Problems of Philosophy*. Oxford: Oxford University Press.

Sosa, Ernest 1999: "How to Defeat Opposition to Moore". *Philosophical Perspectives* 13, 141–153.

— 2000: "Skepticism and Contextualism". *Philosophical Issues* 10, 1–18.

— 2004: "Reply to Keith DeRose". In: John Greco (ed.), *Sosa and His Critics*. Oxford: Blackwell Publishers, 276–282.

Sundholm, Göran 2008: "A Novel Paradox?" In: Cedric Dégremont, Laurent Kieff & Helge Rückert (eds.), *Dialogues, Logics and Other Strange Things: Essays in Honour of Shahid Rahman*. London: College Publications, 375–378.

Grazer Philosophische Studien
85 (2012), 255–277.

BREWER'S SWITCHING ARGUMENT

Andre ABATH
University of Minas Gerais

Summary

In his *Perception and Reason*, Bill Brewer argues that one can only have empirical beliefs if one's perceptual experiences serve as reasons for such beliefs. His argument for this idea relies on a premise according to which in order for the relations with perceptual experience to determine the contents of empirical beliefs, these relations must be reason-giving. He offers an argument for this premise, the so-called Switching Argument. In this paper, I show that the Switching Argument does not work, in which case we have no reason to accept the premise the argument is supposed to establish, and hence no reason to accept Brewer's argument in favour of the idea that one can only have empirical beliefs if one's perceptual experiences serve as reasons for such beliefs.

1. *Introduction*

In his *Perception and Reason* (1999), Bill Brewer offers an argument in favour of the view that one can only have empirical beliefs—i.e., beliefs that are about the external world around the subject—if one's perceptual experiences serve as reasons for such beliefs.

The argument, in his words, is the following:

(1) The most basic beliefs about the spatial world have their contents only in virtue of their standing in certain relations with perceptual experiences.
(2) Only reason-giving relations could possibly serve the content-determining role required by (1).
(3) The hypothesis that perceptual experiences do not provide reasons for empirical beliefs rules out the possibility of beliefs about a mind-independent spatial world. (Brewer 1999, xiv)

Put in this way, it is not clear what, exactly, the argument is. But I take it that Brewer has the following in mind. According to (1), relations with perceptual experiences determine (or give) the content of empirical beliefs. The idea, then, is that empirical beliefs come to be about certain objects and/or properties in the external world thanks to the relations between such beliefs and perceptual experiences. According to (2), in order for the relations with perceptual experiences to determine the contents of empirical beliefs, these relations must be reason-giving. That is, the relations between perceptual experiences and empirical beliefs must be such that perceptual experiences serve as reasons for empirical beliefs, if the relations with perceptual experiences are to give the content of empirical beliefs. The conclusion, then, clearly put, is that, in order for one to have empirical beliefs, perceptual experiences must serve as reasons for such beliefs.[1]

Now, the most controversial premise in this argument is clearly (2), according to which in order for the relations with perceptual experiences to determine the contents of empirical beliefs, these relations must be reason-giving. As we will see in detail later on, Brewer's understanding of reasons for belief is clearly internalist, in the sense that, in order for perceptual experiences to be in reason-giving relations with empirical beliefs, they must be the subject's *own reasons* to form empirical beliefs, that is, they must be reasons for empirical beliefs from the subject's point of view (Brewer 1999, 151). With the idea of reason-giving relations between perceptual experiences and empirical beliefs so understood, premise (2) goes against most theories of content in the market. It goes against causal theories of content, according to which a belief gets to have a content in virtue of causal relations between the belief and the world (Putnam 1982, Davidson 1989). It also goes against covariation theories of content, according to which the content of our beliefs is determined in terms of a lawful covariation between beliefs and items in the world (Dretske 1981, Fodor 1987). And it also goes against teleological theories of content, according to which a belief is about an item in the world because it is its function to be about such an item (Millikan 1984, Papineau 1987). In all of these theories the contents of empirical beliefs are determined independently of reason-giving relations (understood in internalist terms) between perceptual experiences and empirical beliefs. If they presuppose relations between perceptual experiences and empirical beliefs, these relations need only be causal. So, (2) goes against all of them. Brewer's case for (2) should then be particularly strong, since it is supposed to rule out

1. For different arguments in favour of this idea, see McDowell (1994).

most of the dominant theories of content in the market. His argument for (2) is the so-called Switching Argument. In what follows, I will show that the Switching Argument does not work; in that case, we have no reason to accept (2), and, consequently, no reason to accept Brewer's argument in favour of the idea that one can only have empirical beliefs if one's perceptual experiences serve as reasons for such beliefs.

2. *The Switching Argument*

Brewer's Switching Argument is firstly presented at the beginning of *Perception and Reason*, when he writes:

> The Switching Argument establishes that non-reason-giving relations between perceptual experiences and such basic empirical beliefs could not possibly serve this content-determining role [the role stated in (1) above]. For such relations would necessarily leave the subject quite ignorant of which mind-independent object her belief is supposed to be about, in a way which is incompatible with her having the understanding required for this to be a belief *of hers*, about just that thing, at all. (1999, xv/Brewer's italics)

As introduced here, the Switching Argument can be schematized in the following way:

SWITCHING ARGUMENT

(S1) Suppose that a subject has a purported empirical belief that P, and that the subject's perceptual experiences are in non-reason-giving relations with this purported belief.
(S2) If (S1), then the subject will necessarily not know what his purported empirical belief is about.
(S3) The subject then does not have the understanding required to have the belief that P at all.
(S4) So, in order for the relations with perceptual experiences to determine the contents of empirical beliefs, these relations must be reason-giving.[2]

2. (S4) is, of course, (2) above, phrased in a different way.

What I want to question in this argument is (S2). This premise is, at first sight, far from obvious. Why should the fact that a subject's perceptual experiences are in non-reason-giving relations with his purported empirical belief that P imply that he does not know what his purported empirical belief is about? But before we raise objections to the premise, we need to understand in some detail why Brewer takes it to be true. Once this is clarified, the remainder of the argument will also be clarified, and we will be in a position to discuss it critically.

First of all, we need to understand what it means, according to Brewer, for perceptual experiences to be in reason-giving relations with empirical beliefs. I will approach this question by explaining what it means, for Brewer, to say that something is a reason for a belief.

According to him, reasons for belief must have a double role: one is of rational support, the other is causally explanatory. Regarding the rational supporting role, he writes:

> To give a reason in this context is to identify a feature of the subject's situation which makes the relevant judgement or belief (or perhaps action) appropriate, or intelligible, from the point of view of rationality. (Brewer 1999, 151)

The idea here seems to be one familiar from the philosophy of action, and can be made clearer if we consider reasons for action for a minute, before moving back to reasons for belief. To talk about reasons for action is not—not only, at least—to talk about considerations that causally explain why the subject acted as he did. Someone might, for instance, go for a swim in the Bering Sea. And there might be considerations that causally explain why he did so—a sudden desire to challenge oneself, for instance, and the belief that going for a swim in dangerous conditions would be a way to do it. Still, it might be the case that, in a certain sense, he was acting for no reason (that is, no good reason)—in that there were no considerations counting in favour of acting in such a way; or, to put it differently and closer to Brewer's terms, there were no considerations that made the action a rational one ('appropriate, intelligible, from the point of view of rationality'). The idea, then, is that reasons should provide rational support to actions, and not only causally explain them.[3] The same applies

3. In the philosophy of action, this distinction is sometimes taken to be one between explanatory reasons and justificatory reasons. In the words of Stephen Darwall, reasons for action can be '(a) Reasons that are purely explanatory, including anything that serves to explain an action; (b) Reasons that are purely justificatory in that they include any consideration that

to beliefs: reasons should provide rational support to them, in the sense that a given reason should make a given belief a rational one for a subject to have.

However, Brewer also wants reasons to have a causally explanatory role in relation to belief and action. Later on in the book he says that he is also talking about reasons that are 'actually motivating' (1999, 152) and that '... reason giving of this kind is only appropriate when the stated reason's status *as a reason* is causally explanatory ...' (1999, 163/Brewer's italics). This is the second role that reasons must have for Brewer. They must give rational support to belief or action, but they must also be motivational; that is, they must also causally explain why the subject acted or believed as he did. In the example above, the desire to challenge oneself and the belief that going for a swim in dangerous conditions would be a way do so might causally explain why the subject acted as he did, even though this does not give rational support to the action.

According to Brewer, then, reasons for belief or action—genuine reasons —must do both jobs: they must rationally support and causally explain beliefs or actions. Importantly, however, they must do so from the subject's point of view. After all, Brewer is interested

> in reasons *for the subject* to do these things, to take things to be actually the way he believes them to be. These must be the subject's *own* reasons, which figure as such *from his point of view.* (Brewer 1999, 151/Brewer's italics)

Let us now focus on reasons for belief. So, according to Brewer, in order for something to be a reason for a belief, it must rationally support and causally explain the belief from the subject's point of view. Brewer further develops this idea by saying that 'a person's own reasons for believing or doing what she does must, in some sense, be *recognizable* by her as such' (1999, 163/Brewer's italics). What this means, then, is that in order for something to be a reason for a given belief of a subject, it must be recognizable by the subject as giving rational support and causally explaining the belief. Applying the idea to the relation of perceptual experiences and empirical beliefs, we have the following: in order for perceptual experiences to be in reason-giving relations with the empirical beliefs of a subject, the perceptual experiences must be recognizable by the subject as being reasons for such beliefs. In other words, the subject must be able to recognize

---

serves to render an alternative choiceworthy ...' (1983, 34). Many thanks to Leonardo Ribeiro for calling my attention to this.

perceptual experiences as giving rational support and causally explaining empirical beliefs.[4]

As should be clear by now, Brewer's account of reasons for belief favours epistemological internalism. An externalist account of reasons—in which, say, perceptual experiences might be taken as being reasons for beliefs given the fact that they are a reliable mechanism by means of which beliefs are produced—is not considered as providing genuine reasons.

With such an account of reasons in hand, in (S2) of the Switching Argument Brewer is in fact saying that if the perceptual experiences of a subject are not recognized by himself as being reasons for his purported empirical belief that P, then the subject will not know what his empirical belief is about. But now we should ask: why is that so?

In order to understand why Brewer takes (S2) to be true we need to consider his discussion of an account according to which empirical beliefs come to have a content in virtue of causal relations with perceptual experiences, and not reason-giving ones. His aim is to show that such an account will have unacceptable consequences, so that we have to take the relations between perceptual experiences and empirical beliefs as being reason-giving ones, if empirical beliefs are to have a content at all. Following Brewer's terminology, let us call this the 'Familiar Account'. And let us concentrate, like Brewer, on the example of beliefs about colours, and, more specifically, on beliefs about red.

According to Brewer's reading of the Familiar Account, red objects are 'those which have the (perhaps massively disjunctive) microphysical property which normally produces red experiences: this defines what redness in the world actually is' (1999, 55). The Familiar Account then goes on in the following way:

> Red' experiences[5] *indicate* the presence of red objects in virtue of the reliable relations between them. Beliefs that certain mind-independent things are

4. At some points, Brewer seems to be saying that subjects must not only *be able* to recognize reasons as such, but must in fact *do so*. He says, for instance, that 'genuinely reason-giving explanations cite reasons which are necessarily recognized as such in some sense by the subject' (1999, 166). However, what I take him to be saying here is something perfectly compatible with what he says elsewhere in the book: if certain considerations are presented as reasons for a belief, they will only be genuine reasons if recognized by the relevant subject as being his reasons for the belief. Reasons need not be presented every time a belief is formed, however. It suffices that the subject *is able* to recognize something as rationally supporting and causally explaining the belief.

5. When Brewer is talking about red experiences, and not about red as an item in the world, he writes 'red' experiences'.

red acquire their content, in turn, as beliefs that those things have just that colour—that is to say, that microphysical property—on the basis of the relations with red' experiences. (1999, 55/Brewer's italics)

The idea is that a certain microphysical property normally causes red experiences in the subject. These experiences, in turn, indicate (or are about) this property. Red experiences, then, cause certain beliefs in the subject. Since the experiences are about red, the beliefs that they cause are also about red, about a certain microphysical property in the world.

Now, according to the Familiar Account, our perceptual experiences of red are about the property in the world that normally causes them, and such experiences cause beliefs in the subject, which are also about this property in the world. Now, a consequence of this account is that

> had it been a quite different property in the world—a quite different microphysical structure, that is—which happened to be the normal cause of red' experiences, then his beliefs about redness would, on this account, have been beliefs about that instead ... (1999, 56)

The idea, then, is that, if red experiences had normally been caused by a different property in the world, then the experiences would have been about this property. And, that being the case, the beliefs that they cause would also have been about this property.

This point should sound familiar – it is typically made in causal theories of content. Our beliefs about water are about H_2O, since H_2O is the normal cause of such beliefs. But had it been XYZ—a substance perceptually identical to water, but with a different chemical structure—that caused such beliefs in us, they would have been about XYZ.

Now, what is supposed to be the problem with this? For Brewer, it is the following:

> The putative source of empirical content of a person's beliefs about redness in the world is, *as far as the subject himself is concerned*, entirely neutral on which property of things their redness actually turns out to be. In believing that there is something red in front of him, a person is bound to be believing that there is something which is one way or other which things can be, and sometimes are out there. Furthermore, given that redness is defined as the *normal* cause of red' experiences, he will normally be right that there is something that way in front of him. Yet he has not the slightest idea *which* way this is, in the sense that had it been a quite different way, he would be none the wiser. Although it appears to succeed in assigning determinate worldly truth-conditions to a person's beliefs about colours of things, the account on offer does so in a way

which guarantees the thinker's ignorance about which of a range of alternative such truth-conditions these actually are, in this particular case, his ignorance about which of range of alternative microphysical properties redness actually is. (1999, 57/Brewer's italics)

So, what Brewer is saying here is that, if red experiences had been caused by a different property in the world—a different microphysical structure —, nothing would have changed from the point of view of the subject. He would be 'none the wiser'. For the experiences themselves would be phenomenologically identical. So, from the point of view of the subject, red experiences caused by different properties are identical. Now, if that is the case, then the experiences themselves do not reveal to the subject which property they are about. They are 'entirely neutral' regarding this. In that case, when, based on such experiences, the subject employs the concept RED in belief, he does not actually know to what RED refers. He does not know to which microphysical property in the world the concept refers. As Brewer puts it, the subject is 'in a sense ignorant of the semantic values of the secondary-quality concepts which she applies to the things around her' (1999, 67f.).[6] Since, according to Brewer, for a subject to know what his purported empirical beliefs are about is for him to know the reference of the concepts employed in the belief, then, given the Familiar Account, the subject does not in fact know what his purported empirical belief is about.

Thus the general idea is that 'non-reason-giving content determining relations between perceptual experience and beliefs ... lead to this kind of ignorance of semantic value with respect to the relevant empirical concepts' (1999, 68). That is, if the relations between perceptual experiences and empirical beliefs are non-reason-giving, subjects will not know what the concepts they employ in their purported empirical beliefs refer to. And this should be the case in every circumstance. Thus, for any given subject *S*, he will necessarily not know what the concepts he employs in his purported empirical beliefs refer to. If this is the case, then *S* will necessarily not know what his purported empirical beliefs are about. And, in turn, that means that *S* does not fully understand the content of his purported beliefs. As

6. Brewer takes himself as being using the expression 'semantic value' in the way used by Evans (1982) (Brewer 1999, 58ff.). And Evans uses 'semantic value' as a translation for the German word 'Bedeutung' (1982, 7ff.), used by Frege. The most common translation of 'Bedeutung' is 'reference', which is the term I will be using throughout this paper to talk about what Brewer calls the semantic value of a concept.

Brewer puts it, he has 'no real understanding of the way he believes the world to be' (1999, 69). Why is that? Well, because if *S* does not know what the reference of RED, for instance, is, and what his purported beliefs about red are in fact about, then *S* does not actually know what red is. The subject is left with no 'genuine conception of which property redness is ...' (1999, 70). The upshot is that, according the Brewer, given the Familiar Account and the present example, the subject does not have a genuine belief about red at all.

This is Brewer's case for (S2), and for the move from (S2) to (S3). In what follows, I would like to defend the Familiar Account, and argue that, even if perceptual experiences and empirical beliefs are in non-reason-giving relations (assuming Brewer's internalist understanding of reasons for belief), subjects might still know what their empirical beliefs are about. By showing this, I intend, of course, to argue against (S2). However, before doing so, we need to consider a final preliminary question: what is Brewer taking as being required in order for one to know what one's belief is about? Since to know what one's belief is about is to know the reference of the concepts employed in the belief, the question, in fact, is: what is Brewer taking as being required in order for one to know what the reference of a concept is? Only having answered this question can we adequately discuss whether (S2) is true.

3. *Knowing the reference of a concept*

Unfortunately, it is not entirely clear what Brewer is requiring in order for one to know the reference of a concept. One option that he clearly avoids is the one according to which in order to know the reference of a concept such as RED, one must provide a characterization, a description of which property redness is. He says, for instance, that his point is not that 'a person does not know which property redness is ... unless he can give an explicit characterization of which microphysical structural property redness in the world is supposed to be' (1999, 67). So, what then is his point? What is required in order for one to know the reference of a concept?

What appears to be Brewer's requirement for one to know the reference of a concept comes to light in the middle of a discussion about whether we know the reference of natural-kind concepts, such as GOLD. Brewer argues that we can be said to know the reference of such concepts, given the requirement he takes to be in play. The requirement, in his words, is this:

> A person's understanding of her own natural-kind beliefs requires that the word 'gold' has different semantic values on two speaker's lips *only if* they have different conceptions of what gold is, in some sense; where a person's conception, or, as I shall often say, *subjective conception* of what gold—say—is, is what she knows about the semantic value of 'gold' on the basis of those experiences—both perceptual and testimonial—in virtue of her possession of which she associates with that word a concept with just that semantic value. Put slightly differently, a person's subjective conception of gold is constituted by all that she knows of the form 'gold is …' (1999, 70/Brewer's italics)

Brewer is writing in the context of Twin Earth experiments. So the two subjects mentioned here are one on Earth and his molecule for molecule duplicate on Twin Earth (1999, 72). Let us call the subject on Earth subject A, and his counterpart subject Twin-A. So, the idea is that A can be said to know the reference of a natural-kind concept, such as GOLD (and thus can be said to understand his beliefs involving this concept), only if, when the word 'gold' has a different reference on Twin Earth, A's subjective conception of what gold is will be different from the subjective conception of his counterpart on Twin Earth, Twin-A.

Given these remarks, Brewer's requirement for one to know the reference of a concept (let us call it the Knowledge of Reference Requirement, or simply KR Requirement) can be put in the following way:[7]

> KNOWLEDGE OF REFERENCE (KR) REQUIREMENT
> An individual knows the reference of a concept expressed by a given word only if a difference in the reference of the word would necessarily be accompanied by a difference in the individual's (or his molecule for molecule counterpart's) subjective conception associated with that word.

Of course, to properly understand what this amounts to we need to understand what Brewer means by a 'subjective conception' of an object or property, and what makes two subjective conceptions differ. As Brewer says in the passage above, a person's subjective conception of X (say, a given object) is what the person knows about X, where this knowledge is acquired by perception or testimony; or to put in other terms, a person's subjective

7. I will not be considering only natural-kind concepts, as Brewer considers extending the requirement in play here for concepts of secondary qualities, the ones we have been dealing with. See Brewer (1999, 72f.).

conception of X is all of a person's knowledge of the form 'X is …'.[8] Now, what makes two subjective conceptions differ? The requirement, after all, is that a difference in the reference of a word should be accompanied by a difference in the individual's (or his molecule for molecule counterpart's) subjective conception associated with that word. Without knowing what makes two subjective conceptions differ we cannot properly understand the requirement. In one of the few passages in which Brewer deals with this question, he writes:

> The conception itself [of gold] is likely to be essentially demonstrative, 'Gold is *that stuff*', or 'Gold is *that metal*', say: a legitimate subjective conception of which stuff is correctly called 'gold', nevertheless, which differs, as required, between earthlings and their Twin Earth counterparts. This is a perfectly consistent position. Its key component is a strong externalist account of the thinkers' subjective conceptions of semantic values: these features of a person's mental life are themselves world-involving demonstrative-recognitional capacities …(1999, 72/Brewer's italics)

So, Brewer takes our subjective conceptions likely to be essentially demonstrative. For instance, we take gold to be *that metal*, or *that stuff*. Brewer's account of subjective conceptions is also, according to himself, *strongly externalist*, in the sense that subjective conceptions are *world-involving*. But what does that mean? Well, following John McDowell, Brewer takes perceptual experiences to be 'the world's direct impact upon a person's mind' (Brewer 1999, 9). That is, in perceptual experience (when it is veridical) we are presented with the world itself, or, to put it in terms closer to McDowell and Brewer's, we are presented with how things are, with *facts*. As McDowell puts it,

> In a particular experience in which one is not misled, what one takes in is *that things are thus and so* … *That things are thus and so* is the conceptual content of an experience, but if the subject of experience is not misled, that very same thing, *that things are thus and so*, is also a perceptible fact, an aspect of the perceptible world. (1994, 26/McDowell's italics)

8. One might justifiably wonder why a subjective conception must involve *knowledge* of the form 'X is …' and not merely belief. Can I not have a subjective conception of say, gold, based on false beliefs about gold? Intuitively I can, but Brewer rejects such a possibility. According to him, a subjective conception of X must allow one to refer to X in expressing it, and he takes it that genuine reference to an object X requires the subject to have knowledge about X (1999, 251–262). This is, of course, a very strong requirement for one to refer to objects—especially if the requirements for knowledge are internalist, as Brewer wants them to be—and one that can certainly be questioned. We cannot do so here, however.

Brewer agrees, by saying, for instance, that, in perception, 'things just strike him [the subject] as thus and so ...'(1999, 185). So, the idea is that, in veridical perceptual experience, we are presented with the world itself, in the sense that we are presented with facts. Now, as Brewer says in a passage quoted above, subjective conceptions are formed via perception or testimony. Let us stick to cases involving perception. If, for instance, my subjective conception of gold as *that stuff*, or *that metal*, is formed via perception, and if, in perception, we are presented with facts, then *that stuff*, or *that metal* are among such facts. And these facts themselves are then involved in our subjective conception of gold. This is, then, the sense in which demonstrative subjective conceptions are world-involving: they involve facts, grasped, for instance, via perception.

Having clarified this, we can finally understand what makes two subjective conceptions differ. If two subjective conceptions involve different facts, the conceptions themselves are different.

That being so, it can be said, as Brewer in fact says in the passage above, that we do know the reference of the concept GOLD, given the KR Requirement. Suppose we do take gold to be *that metal*, or *that stuff*. Suppose, further, that these are perceptually grasped facts. Now, if subjective conceptions of gold involve such facts, a change in the reference of the word 'gold' would necessarily be accompanied by a change in our subjective conception. After all, our subjective conception of gold in this world involves a given fact: *that metal* (where this is the metal with atomic number 79). Now, suppose that, on Twin Earth, the thing referred by 'gold' is a different chemical element, with a different atomic number. Had our subjective conception of gold been formed on Twin Earth, then it would involve the fact *that metal* (the metal with the atomic number the things referred by 'gold' have on Twin Earth).[9] So, a difference in the reference of the word 'gold' would necessarily be accompanied by a difference in our (or our molecule for molecule counterpart's) subjective conception of what gold is, given that different facts are involved in the subjective

9. Brewer presents a similar case involving proper names, and, more specifically, the name 'Perlman', taken as referring, in this world, to the Israeli violinist. Suppose that, on a possible world, there is a violinist named Zuckerman; in that world, he is responsible for all the musical feats which are attributed to Perlman. Suppose, moreover, that 'Perlman', in that world, refers to Zuckerman. If that is the case, then, says Brewer, 'the two subjective conceptions of which person 'Perlman' names are in fact distinct: in the actual world the subject knows that 'Perlman' names *that man* (Perlman), whereas, in the counterfactual world, she knows that 'Perlman' names *that man* (Zuckerman)' (1999, 85/Brewer's italics).

conceptions. We thus satisfy the KR Requirement. We can then be said to know the reference of GOLD.

4. *Questioning (S2)*

Now that we have identified Brewer's requirement for one to know the reference of a concept (and thus of what one's belief is about) we are in a position to critically discuss (S2). I will argue that the premise is false, even if we adopt Brewer's own requirement for one to know the reference of a concept. To do so, I need to go back to Brewer's target, the Familiar Account, in which the relations between perceptual experiences of colours and empirical beliefs about colours are taken as being causal, and non-reason-giving. According to Brewer, if this is the case, subjects will necessarily not know the reference of the concepts they employ in belief. But is this true if the KR Requirement is in play? I don't think it is. I think subjects will satisfy the KR Requirement even if the relations between empirical belief and perceptual experience are taken as being non-reason-giving. Next I show why.

Brewer takes our subjective conceptions of natural-kinds likely to be essentially demonstrative. We take gold to be *that stuff*. Brewer also takes our subjective conceptions of secondary qualities, such as colours, as likely to be essentially demonstrative. We take red to be *that colour*. Now, suppose that the Familiar Account is true. Suppose that beliefs about red get to be about a certain microphysical property in the world in virtue of causal relations between experiences which are caused by such property and the beliefs in case. Experiences caused by a property M are about M, and beliefs that are, in turn, caused by such experiences will also be about M. In this story, the relations between perceptual experiences and empirical beliefs are purely causal, and non-reason-giving. Now, consider the possibility that a subject A, on Earth, has her perceptual experiences of red and beliefs about red caused by a microphysical property M, and that her molecule for molecule counterpart on Twin Earth, Twin-A, has her perceptual experiences of red and beliefs about red caused by a microphysical property N. If this is the case, then, given the Familiar Account, A's perceptual experiences of red and beliefs about red are about M. And Twin-A's beliefs about red and perceptual experiences of red are about N. But the important point is that, given that subjective conceptions of red involve facts concerning what redness is, A's subjective conception of what red is differs from Twin-A's subjective conception of what red is. According to A's subjective conception, red is *that colour* (property M). However, according to Twin-A's subjective conception, red is *that colour* (property N). The facts involved in the subjective conceptions differ—*that colour*, as property M, differs from *that colour*, as property N. Since the facts involved in the subjective conceptions differ, the subjective conceptions themselves differ. And necessarily so.

If that is the case, subject A satisfies the KR Requirement to know the reference of the concept RED. A difference in the reference of the word 'red' would, after all, necessarily imply that Twin-A's subjective conception of what red is differs from A's subjective conception of what red is. And in that case, subject A can be said to know what her beliefs about red are in fact about, *even though the relations between perceptual experiences and empirical beliefs are taken as being non-reason-giving*. After all, we are supposing that the Familiar Account is true, according to which the relations between perceptual experiences and empirical beliefs are non-reason-giving.

So, given the KR Requirement, the fact that the relations between perceptual experiences and empirical beliefs are non-reason-giving *does not imply*, as (S2) of the Switching Argument states, that the subject will necessarily not know what his purported empirical belief is about. Thus, (S2) of the Switching Argument ends up being false.

5. *Brewer's reply*

Now, Brewer is well aware of this objection, and responds to it in the following way:

> The problem with this move, though, given the overall dialectical position, is that this revised 'familiar account' is no longer one on which the crucial content-determining relations between secondary quality experiences and beliefs are *non-reason-giving*. Thus, it is no objection, from my point of view, that the position is immune to the Switching Argument. Put slightly different, my point is that the revised position is no longer a version of the 'familiar account' ... On the revised position, on which the present strongly externalist bold account of natural kinds is applied to secondary qualities, the subject's actual experiences are supposed to make available a demonstrative subjective conception of which colour in the world redness is, which is *different* from the conception that is made available by colour experiences in the counterfactual case in which redness is a different property of things in

> the world. Red' experiences themselves have to be characterized derivatively, according to the revised position, by reference to the specific external property of redness which they reveal; they are not in themselves neutral on the question of which property this turns out to be. Colour experiences therefore provide reasons for beliefs to the effect that things are M, rather than N, in the actual world. (1999, 73/Brewer's italics)

What we find Brewer saying here is that an account in which subjects have a demonstrative subjective conception of colours, involving facts, cannot be a case in which the relations between perceptual experiences of such colours and empirical beliefs about them are non-reason-giving. Thus, it cannot be a version of the Familiar Account. And, in that case, this does not pose any threat to the Switching Argument, since the argument only concerns cases in which the relations between perceptual experiences and empirical beliefs are non-reason-giving. But why is that, why is such an account necessarily one in which the relations between perceptual experiences and empirical beliefs are reason-giving? According to Brewer, this is the case because, on this account, perceptual experiences make available demonstrative subjective conceptions of the colour in question. In the case of red, red experiences make available a subjective conception of what red is as being *that colour* (say, M). Where phenomenologically identical experiences caused by a different property—say, N—would make available a subjective conception of what red is as being *that colour* (N). If that is so, then, according to Brewer, perceptual experiences are not neutral regarding what property in the world redness is. They reveal such a property. And this is supposed to mean that perceptual experiences are serving as reasons for beliefs.

Does this line of response work? I don't think it does. First of all, it is certainly true that perceptual experience is, in a sense, not neutral here, in that perceptual experiences caused by different properties will be different, will be experiences of different properties, and will cause different beliefs and subjective conceptions in the subjects, since different facts will be grasped. But this was the case all along in the Familiar Account. Perceptual experiences of red caused by a property M are about M, and cause beliefs about M in the subject. Phenomenologically identical experiences caused by a property N are about N, and cause beliefs about N in the subject. In this sense, the perceptual experiences are different, and not neutral. But that, in itself, does not make perceptual experiences reason-giving. For the relations between such experiences and empirical beliefs can, in this story, be purely causal.

Does the fact that subjective conceptions are introduced in the story lead to any difference in this respect, forcing one to say that perceptual experiences are serving as reasons for empirical beliefs? I don't think it does. For a subject can have a subjective conception of, for instance, red as *that colour*, or as *that*, made available by a given perceptual experience, and the subject might employ this subjective conception in empirical belief, while the perceptual experience is only in a causal, non-reason-giving, relation with the empirical belief that employ such conception—at least given Brewer's way of understanding reasons for belief. And, importantly, even in such cases the KR requirement will be satisfied.

Why think, against Brewer, there can be such cases? Well, there are, in fact, empirical reasons for us to believe that such cases can occur, and that they do occur in subjects around us. These subjects are young children. There is empirical evidence suggesting that they have demonstrative subjective conceptions of things, made available by perceptual experiences, and employ such conceptions in empirical beliefs, while their perceptual experiences are in causal, and non-reason-giving, relations with such beliefs. Let us briefly consider such evidence.

First, it is known that demonstratives are among the first words used by children (Diesel 2006, Clark 1978). In a recent paper, Diessel reports that, in examining the speech of four randomly selected 1 and 2 year-old English-speaking children, he found

> 3.857 demonstratives in a corpus of 54.390 words, which means that on the average every 14th word was an instance of *this*, *that*, *here* or *there*. Moreover, the demonstrative *that* was the single most frequent word in the entire corpus (1.794 tokens); the three other demonstratives were less common, but all four demonstratives were among the fifteen most frequent words that children produced. (2006, 483ff.)

There is thus evidence suggesting that demonstratives are used by children in the first stage of language acquisition, and are in fact among the words that are most oftenly used. And these demonstratives are a 'powerful tool that allows the child to make reference to any entity in the surrounding situation without knowing the word for the referent' (2006, 472). So, it seems to be the case that children as young as 2 years old have demonstrative subjective conceptions of things, and employ such conceptions in empirical belief. They take many of the things surrounding them as being just *that*. However, there is empirical evidence suggesting that, typically, until children reach their 4th year, perceptual experiences are not recognizable as being reasons for empirical beliefs. There are, by now, several experiments in the literature that suggest that this so.

Consider, for instance, an experiment conducted by Gopnik & O'Neill (1991), designed to test if young children are able to identify the sources of their beliefs.[10] The subjects in the experiment were twelve 3-year-olds, twelve 4-year-olds and twelve 5-year-olds. They were placed in front of a red tunnel (made out of styrofoam). They were then either shown objects inside the tunnel, requested to feel them, or were told which objects were inside the tunnel by one of the experimenters. Next, they were asked what they had seen, felt, or being told. If this question was answered correctly, they were then asked: 'How do you know what's inside the tunnel?' If they were unable to give a correct answer here, they were then asked whether they had seen the object, felt it or been told about it. The questions were asked separately, so the children only had to answer 'yes' or 'no'. The idea was that if the children were able to identify the sources of their beliefs, they should at least be able to answer these last questions correctly.

The children's performance was analyzed by Gopnik & O'Neill by dividing them into those who committed mistakes in answering the questions and those who did not (this was applied to each age group). The results were the following. While 83% of the 5-year-olds and 75% of the 4-year-olds were able to answer the questions without mistakes, only 17% of the 3-year-olds were able to do so (Gopnik & O'Neill 1991, 392).

I will not take these results as decisive, but they are certainly suggestive. And what they suggest is that, in their 3rd year, most children are not yet capable of identifying the sources of their empirical beliefs. They cannot yet tell if a given belief of theirs was formed as a result of a visual perception, of a tactual perception or of testimony. For most children, this capacity is only present when they reach their 4th year. By then, the experiment suggests, most children have no problem in identifying the sources of their beliefs.

These results are important here for the following reason. They suggest that, for most 3-year-olds, perceptual experiences *are not recognizable as being reasons* for their empirical beliefs. In being unable to identify (recognize) the sources of their empirical beliefs, they are unable to identify (recognize) a given perceptual experience as causing a given empirical belief. Thus, they cannot yet recognize a given perceptual experience as

10. For another experiment along the same lines, see Gopnik & Graf (1988).

causally explaining a given empirical belief. Consequently, they cannot also recognize a given perceptual experience as giving rational support for a given empirical belief. After all, it seems clear that recognizing perceptual experience as causing a belief is necessary for one to recognize it as rationally supporting the belief—at least if the subject's rationality itself is to be preserved. If, for instance, one cannot recognize seeing Mary in the pub as at least partially causing the belief that Mary is not home, one cannot also recognize this perceptual experience as giving rational support to the belief—at least not preserving the subject's rationality. To say 'I'm justified to believe that Mary is not home because I saw her in the pub', and deny that it was this perceptual experience that at least partially caused the belief would be clearly irrational.

That being so, for the majority of 3-year-olds, they cannot recognize perceptual experiences as being reasons for their empirical beliefs. And, according to Brewer's way of understanding reasons for belief, this means that perceptual experiences *are not* in reason-giving relations with their empirical beliefs.

We have seen, however, that we have good reasons to think that 3-year-olds have demonstrative subjective conceptions of things, and empirical beliefs that employ such conceptions. But now we also have good reasons to think that most 3-year-olds are unable to recognize perceptual experiences as being reasons for their empirical beliefs.

So, it might be the case that, in a situation in which perceptual experiences are making available subjective conceptions, the relations between perceptual experiences and the empirical beliefs that employ such conceptions are purely causal, and non-reason-giving—as always, given Brewer's way of understanding reasons for belief. This seems to be typically the case with 3-year-olds.

Still, the KR Requirement would be satisfied in such cases. A 3-year-old might have a subjective conception of red as *that* (property M), where *that* is a perceptually grasped fact. But if this subject (or his counterpart) had been raised in an environment in which experiences of red are caused by a property N, then his subjective conception would be of red as being *that* (property N). So, the 3-year-old's subjective conceptions would be different in the two environments, and necessarily so, since they involve different facts. And the relations between the subject's empirical beliefs that employ such conceptions and his perceptual experiences might be only causal, and non-reason-giving. Still, given the KR Requirement, the subject would know the reference of the concepts employed in empirical belief (and thus what his empirical belief is about), in which case (S2) would be false.

However, at this point, Brewer could object that 3-year-olds, being unable to recognize perceptual experiences as being reasons for their purported empirical beliefs employing subjective conceptions, do not in fact have genuine empirical beliefs at all. So, these subjects would not constitute a problem for his position.

But is such an argumentative strategy really open to Brewer here? I don't think it is. In order for Brewer to argue that 3-year-olds do not have genuine empirical beliefs employing subjective conceptions, he would have to show that these children do not, in fact, know what their purported empirical beliefs are about. If that were the case, then Brewer could safely (following the steps of the Switching Argument) say that 3-year-olds do not have the understanding required to have genuine empirical beliefs employing subjective conceptions.

The reason why it is not open to Brewer to argue along these lines is that, given *his own* requirement for one to know the reference a concept (and thus to know what empirical beliefs are about), the KR Requirement, 3-year-olds *do know* the reference of the demonstrative concepts they employ in their purported empirical beliefs, and thus do know what their purported empirical beliefs are about. After all, as shown just above, the KR Requirement will be satisfied in the case of these children, even if the relations between their perceptual experiences and their empirical beliefs employing subjective conceptions are taken as being non-reason-giving. So, given that the KR Requirement is satisfied in such cases, Brewer cannot argue (at least not following the steps of the Switching Argument) that 3-year-olds do not in fact have genuine empirical beliefs employing subjective conceptions.

Brewer could, however, react in a different way. He could say, as he often does in chapter 6 of *Perception and Reason*, that

> simply in virtue of grasping the content that Φ-thing is thus, he has a reason to believe that that thing is indeed thus; for he necessarily recognizes that his entertaining that content is a response to that thing's actually being thus, given his location and present circumstances. (1999, 205)

The idea here is that the conditions for one to grasp a given content in perception, if satisfied, are sufficient to provide the subjects with a reason to believe in that content. The conditions mentioned here are Kantian in spirit: in order to perceptually grasp the content (the fact) that that is thus, the subject must be able to recognize 'that his current apprehension that things are thus and so is in part due to the very fact they are … to the way things mind-independently are out there' (Brewer 1999, 204). So, perceptually apprehending the content that gold is *that metal* requires recognizing that one is so apprehending because of the fact that gold *is that metal*, because that is how things are in the mind-independent world.

Now, Brewer could then go on to say the following regarding the 3-year-olds we have been considering. If, in perception, they apprehend facts, in a demonstrative way, they must be able to recognize that their apprehension is due, at least in part, to the way things are out there in the world. And, in so recognizing, they would already have a reason to believe that things are as they perceive them to be, even though they are unable to recognize perceptual experiences as reasons for their empirical beliefs. Our objection would then fail, for the cases of 3-year-olds would not be cases of a purely causal, non-reason-giving, relation between perceptual experience and empirical belief.

One way of responding to Brewer here would be to say the following: given that 3-year-olds are typically unable to recognize perceptual experiences as being reasons for empirical beliefs, it is also very unlikely that they are able to recognize that their grasping of demonstrative contents is due to the way things are out there in the world. Still, it seems clear—given that they have beliefs with demonstrative contents—that such contents are grasped. By pursuing this line, we could try to show that Brewer's conditions for one to grasp a content are inadequate. However, doing so would take us too far afield. I will thus end with a simpler reaction to Brewer.

According to him, not only grasping the fact that that is thus in perception gives a reason for a subject to believe in that content—in virtue of the subject's recognition that he is responding to the way things are out there in the world—, but this is also 'itself an account of the way in which such experiences [perceptual experiences] provide reasons for the perceiver's beliefs …' (1999, 205). So, what we have here is an account of how perceptual experiences might provide reasons for empirical beliefs. Notice that, in this account, it is not a requirement for perceptual experiences to provide reasons for empirical beliefs that the subject is able to recognize the experiences as rationally supporting and causally explaining the beliefs. It suffices that the subject recognizes that his grasping of a given fact in perception is a response to the way things are out there in the world. But, if that is so, it seems that there are two different accounts of how perceptual experiences provide reasons for empirical beliefs in *Perception and Reason*: the one presented earlier in the paper, and the one just considered. The problem I want to raise for Brewer is the following: how to reconcile these two accounts—assuming they are reconcilable?

Well, I see only one clear way to do so: to suppose that, among the conditions for a subject to grasp a given content (fact) in perceptual experience, would be one according to which the subject must be able to recognize the experience as being a reason to form an empirical belief with that very content. Thus, the conditions for one to grasp a given content (fact) in perceptual experience would involve not only the Kantian requirement of recognizing that such grasping is due in part to the way things are out there in the world, but would also involve the requirement that one should be able to recognize the perceptual experience as being a reason to form an empirical belief with that very content.

However, if the two accounts are so reconciled, Brewer would be forced to say that most 3-year-old children do not in fact grasp demonstrative contents—since they are unable to recognize perceptual experiences as reasons for empirical beliefs. That would be a difficult line for him to take, given that we have strong reasons to think that 3-year-old children do have demonstrative beliefs—and, therefore, to think that they do grasp demonstrative contents.

So, at a minimum, Brewer faces the challenge of finding a way of reconciling the two accounts of how perceptual experiences provide reasons for empirical beliefs present in *Perception and Reason*, and a way in which he is able to escape the objection we have been pressing—and, at the same time, a way in which he is not forced to say that certain subjects do not grasp certain contents when we have strong reasons to believe they do. Given the arguments presented in this paper, I find the prospects of him being able to do so without changing his views on reasons for belief dim at best. And if he is unable to do so, then, given the KR Requirement, there can be cases in which the relations between perceptual experiences and empirical beliefs are non-reason-giving, and in these cases the requirement will still be satisfied; which means that subjects will still know what the reference of the concepts they employ in belief are, and what their beliefs are about. Thus, (S2) of the Switching Argument is false. The Switching Argument is then unsound.

The Switching Argument, remember, is the argument offered by Brewer in favour of the second premise of the following argument:

(1) The most basic beliefs about the spatial world have their contents only in virtue of their standing in certain relations with perceptual experiences.
(2) Only reason-giving relations could possibly serve the content-determining role required by (1).
(3) The hypothesis that perceptual experiences do not provide reasons for empirical beliefs rules out the possibility of beliefs about a mind-independent spatial world. (Brewer 1999, xiv)

I have tried to show that the Switching Argument does not work. Since Brewer's argument for (2) does not work, and since (2) is not obviously true or intuitive, we have no reason to accept such a premise. Consequently, we have no reason to accept Brewer's argument in favour of the idea that one can only have empirical beliefs if one's perceptual experiences serve as reasons for such beliefs.

References

Brewer, Bill 1999: *Perception and Reason*. Oxford: Oxford University Press.

Clark, Eve 1978: "From Gesture to Word: On the Natural History of Deixis in Language Acquisition". In: Jerome S. Bruner & Alison Agarton (eds.), *Human Growth and Development*. Oxford: Oxford University Press, 85–120.

Darwall, Stephen 1983: *Impartial Reason*. Ithaca, New York: Cornell University Press.

Davidson, Donald 1989: 'The Myth of the Subjective'. In: Michael Krausz (ed.), *Relativism, Interpretation and Confrontation*. Notre Dame: Notre Dame University Press, 159–172.

Diessel, Holger 2006: "Demonstratives, Joint Attention and the Emergence of Grammar". *Cognitive Linguistics* 17, 463–489.

Dretske, Fred 1981: *Knowledge and the Flow of Information*. Cambridge, MA: The MIT Press.

Evans, Gareth 1982: *The Varieties of Reference*. Oxford: Oxford University Press.

Fodor, Jerry 1987: *Psychosemantics*. Cambridge, MA: The MIT Press.

Gopnik, Alison & Graf, Peter 1988: "Knowing How You Know: Young Children's Abilities to Identify and Remember the Sources of Their Beliefs". *Child Development* 59, 1366–1371.

Gopnik, Alison & O'Neill, Daniela 1991: "Young Children's Abilities to Identify the Sources of Their Beliefs". *Developmental Psychology* 27, 390–397.

McDowell, John 1994: *Mind and World.* Cambridge, MA: Harvard University Press.

Millikan, Ruth 1984: *Language, Thought and Other Biological Categories.* Cambridge, MA: The MIT Press.

Papineau, David 1987: *Reality and Representation.* Oxford & Cambridge, MA: Blackwell.

Putnam, Hillary 1982: *Reason, Truth and History.* Cambridge: Cambridge University Press.

Grazer Philosophische Studien
85 (2012), 279–290.

THE VALUE OF FEELINGS FOR DECISION-MAKING

Amadeus MAGRABI
University of Osnabrück

Second prize of the 2011 essay competition for students sponsored by the *Gesellschaft für Analytische Philosophie* (GAP) in cooperation with the *Grazer Philosophische Studien**

1. *Introduction*

The dichotomy between emotional and deliberate decisions is deep-rooted in public belief, philosophy, and science alike. It is reflected in diverse areas, such as poetry (*heart vs. head*), schools of thought (*Romanticism vs. Enlightenment*), and gender stereotypes (*female intuition vs. cold-headed men*). Academic interest in the topic can be traced back to ancient Greece, but still, the debate seems to be flowering more than ever. The question is the following: Should we let our feelings guide decision-making, or should we rather engage in a 'cold' logical analysis of the decision problem? Which of the two strategies leads to better outcomes?

For a long time, it has been the dominant view that reason trumps emotion. Promoted by philosophers like Kant, this view lies in the middle of modern society. People who behave emotionally are thought to break diets, procrastinate important tasks, fail to exercise, take unnecessary risks, or spend money on things they do not need—despite their intention to do otherwise. Regarded in this light, our feelings seem to drive us into decisions we later regret, and distract us from long-term goals.

However, in the past decades, research on emotions has gained a boost of interest, and with it emerged a more positive view of emotions. One of the most famous studies mentioned in this context is the work by neuroscientist Antonio Damasio (1994). Patients in his studies suffered from lesions in prefrontal cortex, which caused them to have a reduced capacity to experience feelings. When confronted with a decision-making task, their performance was severely

* The question of the 2011 competition was: Can we Trust our Feelings? On the Value of Emotions for our Knowledge and Agency. The jury selected from 30 submissions two essays for a second prize and one essay which exceeded the space limit for a special acknowledgement. No first and no third prize was awarded this year.

impaired, although their performance in other cognitive tasks was normal. This phenomenon suggests that feelings play an important role in decision-making, and questions the traditional ideal of 'pure reason'. Still, methodologically speaking, the studies are a rather coarse approach, which does not allow for more specific conclusions about what exactly feelings contribute to the process of making a decision.

In this paper, I want to find out whether we can rely on the guidance of our feelings, or whether we should consciously monitor and disrupt the influence of feelings on decisions, and favor reason. The paper is divided into three parts. In the first part, I will introduce a distinction between *emotions* and *gut feelings*. While this distinction is not necessarily suitable for an ontological characterization of affective space, I do believe that the two phenomena need to be discussed separately with respect to the question at hand. In the second and third part, I will analyze the influence of emotions and gut feelings on decisions in turn. The paper is not an exhaustive overview of all the different theories that are out there; instead, it involves to a large degree my particular view on the debate.

2. *Emotions vs. gut feelings*

There are essentially two separate bodies of literature that are concerned with the influence of feelings on decisions, and it seems like there has been no substantial effort of unification so far.

One body of literature focusses on *emotions*. This research picks out a certain emotion, and tries to find a connection to types of behavioral decisions. These studies suggest that, for example, anger promotes risky decisions (Lerner & Keltner 2001), and guilt promotes pro-social decisions (Nelissen, Dijker, & deVries 2007). Providing a clear definition of emotion has proven to be difficult, but, simplifying somewhat, emotions are frequently characterized in the following way: Emotions are experienced as an episode during which a distinctive conscious feeling persists. This feeling is directed towards something, such as an object or an event (we are afraid *of* something). Emotions are also accompanied by a characteristic pattern of physiological (e.g., heart rate or muscle tension) and cognitive (e.g., attention or memory) changes. As a consequence of these changes, emotions bias certain types of behaviors (Baumeister, Vohs, DeWall, & Zhang 2007). Emotions are usually the result of an appraisal of the current situation (Arnold 1960, Lazarus 1991). On a temporal scale, emotional episodes are longer than impulses, but shorter than moods. In short, emotions set the organism into a stereotyped *mode of processing*, which is triggered when the appropriate pattern is observed (Griffiths 1997). While some emotional responses are rather fixed and primarily based on short-term goals (e.g., fear, anger, or disgust; referred to as *basic emotions*), others are quite flexible and based on more long-term goals (e.g., envy, shame, or guilt; referred to as *higher cognitive emotions*).

The other body of literature focusses on *gut feelings* and *intuitions*, terms which I will use interchangeably. Gut feelings are subtle feelings which are simply positive or negative—liking or disliking. They are typically directed towards beliefs and decision alternatives, and we are usually not aware of their origin. For example, in a quiz show, a candidate might have a positive feeling towards answer C, but he cannot explain where this feeling is coming from. Or imagine someone who feels like he should study medicine, although the arguments speak in favor of economics. Gut feelings do not require effortful thinking; they arise spontaneously and may vanish just as quickly. Experimental research on the influence of gut feelings on decisions is mainly done under the heading of *dual-systems models* (for a recent review, see Evans 2011), which I will explain in a bit more detail in section 4.

Two differences between emotions and gut feelings should be noted: 1) Gut feelings only persist for a short period of time, and they end immediately once the decision has been made. In contrast, emotions often persist for a longer period of time, and may even last long after the causing situation is over. Thus, emotional episodes can influence events that are unrelated to the circumstances towards which the emotion was originally directed. Imagine a husband who gets angry in a fight with his wife. When he goes to work afterwards, his state of anger might make him offend his colleagues, who have nothing to do with his problems with his wife. Emotions can have these 'spill-over effects' (Seo & Barrett 2007), but this does not seem to be the case for gut feelings. 2) Emotions are accompanied by large-scale physiological and cognitive changes of the organism, which, to my knowledge, has not been shown for gut feelings so far. Gut feelings seem to just consist of a feeling along the valence dimension—either positive or negative—which is directed towards a decision alternative. These feelings can also be part of an emotion, but gut feelings can emerge detached of an emotional episode, too.

The distinction between emotions and gut feelings is made in a similar way by Baumeister, DeWall, and Zhang (2007) when they distinguish between 'full-blown, conscious emotions' and 'automatic affect'. However, from my reading of the literature, there seem to be confusions about this distinction in many papers of the philosophical, psychological, and neuroscientific literature alike. It is often not clear whether authors want to talk about the influence of feelings on decisions in general, or whether they are just concerned with either emotions or gut feelings. As I mentioned in the introduction, the distinction is not motivated by the goal of developing an adequate ontology of affectivity. Instead, the distinction is tailored to the question of how feelings influence decisions, since I will argue that the roles emotions and gut feelings play for decision-making are quite different. This discussion will reveal more differences between gut feelings and emotions than I have mentioned at this point.

3. *The influence of emotions on decisions*

3.1 *Nature of the influence*

In this part, I will introduce five general properties of the influence of emotions on decisions. It seems that the literature does not provide an elegant integrative theory on the issue so far.

1) I adopted a view according to which the influence of emotions on decisions is indirect. Emotions set us into certain modes of processing, and via these physiological and cognitive changes, emotions influence the decisions we make (Baumeister, Vohs et al. 2007). For example, disgust has been shown to narrow attention (Gable & Harmon-Jones 2010), which might give us the tendency to overrate the disgusting object and underrate other factors during decision-making. Naturally, this interplay will lead to systematic effects of emotions on decision-making, but there is room for variability. This theory is opposed to views of direct causation, along the lines of 'fear causes flight'. Examples of direct causation are often simply given for emotions like fear, for which the association to a certain behavior is rather straightforward. However, it is unclear what the unique behavior for other emotions, like joy or envy, should be like (Fredrickson 1998). Even the popular example of 'fear causes flight' is oversimplified, because fear can also cause behaviors like freezing, to mention just one example. In experimental research, researchers in fact find that there is a lot of variance concerning the kinds of behaviors that emotions cause (Barrett 2006), which speaks against the view of direct causation.

2) The influence that emotions have on decisions is probably not discrete, but continuous. There seems to be no clear-cut border between emotional and deliberate decisions, but decisions can be mainly based on emotions, or mainly based on deliberate reasoning.

3) It is important to differentiate between *endogenous* and *exogenous* influences of emotions (Seo & Barrett 2007). When the influence is endogenous, the emotional episode is caused by the current situation and relevant to the decision at hand (e.g., anxiety for risky decisions, or sadness for seemingly hopeless decisions). When the influence is exogenous, the emotional episode is unrelated to the current situation, and typically caused by previous circumstances. For example,

imagine someone who is happy that his favorite baseball team won, and then donates money to a homeless person he walks by; someone who is disgusted by the bathroom he has to clean, and does not want to eat lunch afterwards; or someone who is proud after accomplishing a certain task, and then overestimates his ability to perform another task that is completely different. Examples of these 'spill-over effects' occur frequently in everyday life.

4) Emotions appear to have a bias towards rather short-term goals, whereas deliberate reasoning seems more suitable for long-term goals. From an evolutionary perspective, one could argue that emotions have evolved to provide quick stereotyped responses, which require relatively little energy. These responses are absolutely necessary in uncertain environments that require fast decisions. In contrast, our capacity for deliberate reasoning allows for decisions which are more flexible, at the cost of requiring more time and energy.

5) Emotions cannot only influence which option from a given set of decision alternatives we prefer, they might also influence which decision alternatives we even take into account (Hanoch 2002). For instance, one could assume that anger promotes a very restricted perspective on the available decision alternatives, whereas joy makes us broaden our horizon and also consider more creative possibilities. Moreover, emotions can also change the kinds of goals we have, and the kinds of decision problems that we actually tackle (Bagozzi, Baumgartner, Pieters, & Zeelenberg 2000). Hence, emotions could be the solution to what is called the *frame problem* in cognitive science (McCarthy & Hayes 1969). The number of goals and decision alternatives that are logically possible at any given time is virtually infinite, which makes it impossible, or at least utterly impractical, to make decisions solely on the grounds of logical reasoning. Emotions might impose the restrictive pre-selection that is needed. This solution to the frame problem is controversial (Evans 2002), and the theory lacks a specification of how exactly emotions accomplish this task. Still, the basic idea seems to be very fruitful.

In a nutshell, I characterize the influence of emotions on decisions as indirect, continuous, and either endogenous or exogenous. Emotions may fulfill the role of deconstructing decision problems into manageable sizes. Compared to reason, emotions are fast but stereotyped responses, and they seem to have a bias towards more short-term goals.

3.2 *Empirical findings*

Having shed some light on the general kind of influence that emotions have on decisions, I will now introduce empirical work, in which researchers try to find out how particular emotions influence decisions.

The number of published studies on this topic is huge, and the results are very diverse. Some studies appear to suggest that we should not follow our emotions. Typical findings are that, for example, anger leads to overly optimistic choices (Lerner & Keltner 2001), fear leads to egoistic choices (Nelissen et al. 2007), and sadness leads to overly pessimistic choices (Lerner & Keltner 2001). Other studies suggest quite the opposite picture: sadness leads to altruistic choices (Cunningham, Steinberg, & Grev 1980, Manucia, Baumann & Cialdini 1984), and joy leads to variety-seeking choices (Kahn & Isen 1993). I will not try to review and compare all the experiments which have been done; neither will I discuss potential methodological and conceptual shortcomings. My point is merely that these kinds of studies do not allow for a general answer to the question of this paper. No specific emotion is inherently detrimental or beneficial—it all depends on the particular circumstances. While regret might help us to learn from our mistakes in one situation, it might just distract us from more important problems in another situation. Likewise, while fear might make us too hesitant in one situation, it might cause overly rash decisions in another situation. Therefore, the strategy of classifying particular emotions as either helpful or hurtful for decision-making has to fail. The research that I mentioned is of major importance for other areas, but for the question at hand, and at the current state of research, theoretical considerations appear to be more productive.

3.3 *Should we follow our emotions?*

I am now going to examine the normative question of whether we should let emotions guide decision-making. What I have discussed so far suggests the following picture: If it is inevitable that a decision needs to be made quickly, emotional responses are our best estimate of what needs to be done. They provide stereotyped responses to patterns which occurred frequently during evolution, and thereby simplify decision-making for us. But there is a risk of oversimplification, which should not be ignored. If more time is available, and an important decision needs to be made, I would argue that we should not blindly follow our emotions. In particular, we need to be careful if we need to make long-term decisions, or if we experience a very uncommon decision problem, because emotions do not seem to have evolved for these situations. Furthermore, we should be skeptical if we realize that we are experiencing an exogenous emotion. In this case, the emotion arose from a previous situation that is unrelated to the current decision problem. Whether the influence of the emotion is helpful or harmful is then basically just a matter of chance. Thus, in my view, deliberate reasoning should be favored in these cases. It is not a good strategy to blindly follow emotions, but neither is notoriously fighting them. I suggest that we should try to monitor our emotions, and think about what may have caused the emotional episode, such that we can make decisions that benefit from both emotion and reason.

4. *The influence of gut feelings on decisions*

4.1 *Nature of the influence*

Gut feelings are often discussed in the context of dual-systems models (Evans 2008, Kruglanski & Orehek 2007). These models assume that decisions result from interactions between a deliberate and an intuitive system. Whereas the former is slow, effortful, foresightful, and operates consciously, the latter is fast, effortless, unconscious, impulsive, and operates unconsciously. We experience the deliberate system as logical reasoning. Although the intuitive system operates unconsciously, the results enter conscious awareness in the form of gut feelings. The theory of dual-systems models has been the target of some criticism (Gigerenzer & Brighton 2009, Osman 2004), but it seems to provide a quite useful framework to think about gut feelings.

But how exactly does the intuitive system work; where do gut feelings come from? It has been proposed that gut feelings may arise from an unconscious recognition of cues that allow for reliable predictions and have been learned through prolonged experience (Simon 1992). For example, a tennis player might have the feeling that the next shot of his opponent will be placed near the net. This feeling could be based on cues that can be found in certain movement patterns or facial expressions, of which the tennis player is not consciously aware. Facets of this phenomenon have also been studied under the heading of *decision heuristics* (Gigerenzer 2007). For a lot of people, gut feelings still have the connotation of something mysterious and esoteric. But if the above account is right, this 'demystifies' gut feelings, as they simply rely on an unconscious recognition of cues in the environment.

For the development of good intuitions, there has to be 1) an environment with observable regularities, and 2) consistent and reliable feedback, such that effective cues can be learned over time (Kahneman & Klein 2009). In some domains, the feedback is usually straightforward (e.g., for chess players or firefighters). However, in other domains, feedback can be ambiguous and very time-delayed (e.g., for politicians or managers), which makes it hard to develop accurate gut feelings.

Why do we have an 'intuitive system'? One can propose the same evolutionary explanations that I have already mentioned for emotions: Decisions based on gut feelings are fast and effortless, because they systematically simplify the decision problem, and break it down to an analysis of cues. Therefore, gut feelings are especially useful in environments in which every second matters. On the other hand, deliberate reasoning seems to allow for decisions that take more information into account and are therefore are less risky, at the cost of requiring more time and effort.

4.2 *Empirical findings*

Researchers have tried to find out whether deliberate reasoning or trusting one's gut feelings leads to better decisions. In a typical study, researchers may confront participants with a decision task, either with the instruction to rely on their gut feelings, or with the instruction to make a deliberate decision. Afterwards, researchers can compare which of the two instructions led to better performance. In a quite famous study, Dijksterhuis and colleagues (2006) actually found that performance was better when participants relied on their gut feelings. Other researchers have found similar results (Halberstadt & Levine 1999, Harteis & Gruber 2008, Wilson & Schooler 1991). Dijksterhuis and Nordgren (2006) hypothesize that one should only engage in analytical thinking when the decision problem is small—for complex problems, analytical thinking should be avoided. However, many researchers report opposite effects and raise severe criticism against the theories postulated by Dijksterhuis and colleagues (Aczel, Lukacs, Komlos & Aitken 2011, Calvillo & Penaloza 2009, Lassiter, Lindberg, González-Vallejo, Bellezza & Phillips 2009, Newell, Wong, Cheung & Rakow 2009). Empirical studies do not seem to allow for conclusive statements about the value of intuitive judgments at the current state of research.

4.3 *Should we follow our gut feelings?*

Compared to the influences that emotions can have, listening to gut feelings is less 'dangerous'. Gut feelings are very subtle feelings, which, if needed, can be more easily controlled. In addition to that, gut feelings cannot have the exogenous influences that emotions can have, which makes them more trustworthy, as one can be rather sure that they are relevant for the current decision problem.

However, reliable gut feelings can only develop in repetitive environments that provide regular and nonambiguous feedback. If these conditions are not met, gut feelings can lead to terrible decisions. In a famous example of this scenario (Hogarth 2001), an early 20th century physician had intuitions about which patients would soon develop typhoid. To check whether his intuitions are right, he palpated the tongues of these patients. But because he did not wash his hands, he actually caused typhoid, and always confirmed his intuitions even though they were wrong.

These examples show that gut feelings can be misleading, but they obviously do not show that gut feelings are never reliable. There is reason to believe that gut feelings are more frequently beneficial than detrimental, because their development during evolution would have otherwise been unlikely. Gut feelings probably rely on oversimplifications and are less flexible than deliberate reasoning; but when time is short, or when the decision problem is so complex that we are unable to find a solution via deliberate reasoning, gut feelings provide assistance that we could not do without.

5. *Conclusion*

In this paper, I wanted to find out whether decision-making should be guided by feelings or logical deliberations. To answer this question, it is necessary to distinguish between emotions and gut feelings. The main differences between the two concepts are that 1) gut feelings basically just consist of an experience of liking or disliking, whereas emotions are also accompanied by large-scale physiological and cognitive changes, 2) gut feelings can only have endogenous influences on decisions, whereas emotions can have both endogenous and exogenous influences, 3) gut feelings are easier to control than emotions, and 4) gut feelings are more dynamic than emotions, as the conditions that trigger gut feelings change constantly while we gather more knowledge and experience in a certain domain; without a doubt, the conditions that trigger emotions are also subject to learning, but they appear to be more sluggish than gut feelings.

Based on these considerations, I argued that gut feelings are more trustworthy when it comes to making choices: They cannot have exogenous effects, they are easier to control, and more dynamic when it comes to integrating new information. Gut feelings can be considered as a reasonable hint in the decision-making process, whereas we should be more skeptical about emotions.

As a decision-making strategy, deliberate reasoning is more flexible than both emotions and gut feelings, at the cost of requiring more time and energy. However, especially in repetitive environments, flexibility is not always needed, and the 'affective autopilot' usually guides us exceptionally well. But in unfamiliar situations, which we have never experienced before, we should try to carefully monitor and understand our feelings. This way, we can try to find a balance between 1) avoiding the influence of feelings when we find out that they are misleading, and 2) using our feelings as an advisor that helps us to make the right choices.

Aczel, Balazs, Lukacs, Bence, Komlos, Judit, & Aitken, Michael R. F. 2011: "Unconscious Intuition or Conscious Analysis? Critical Questions for the Deliberation-Without-Attention Paradigm". *Judgment and Decision Making* 6, 351–358.

Arnold, Magda B. 1960: *Emotion and Personality*. New York: Columbia University Press.

Bagozzi, Richard P., Baumgartner, Hans, Pieters, Rik, & Zeelenberg, Marcel 2000: "The Role of Emotions in Goal-Directed Behavior". In: S. Ratneshwar, David G. Mick, Cynthia Huffman (eds.), *The Why of Consumption*. London: Routledge, 36–58.

Barrett, Lisa F. 2006: "Are Emotions Natural Kinds?". *Perspectives on Psychological Science* 1, 28–58.

Baumeister, Roy F., DeWall, C. Nathan, & Zhang, Liqing 2007: "Do Emotions Help or Hurt Decision Making?: A Hedgefoxian Perspective". In: Kathleen D. Vohs, Roy F. Baumeister, & George Loewenstein (eds.), *Do Emotions Help or Hurt Decision Making?: A Hedgefoxian Perspective*. New York: Russell Sage Found, 11–16.

Baumeister, Roy F., Vohs, Kathleen D., DeWall, C. Nathan, & Zhang, Liqing 2007: "How Emotion Shapes Behavior: Feedback, Anticipation, and Reflection, rather than Direct Causation". *Personality and Social Psychology Review* 11, 167–203.

Calvillo, Dustin P., & Penaloza, Alan 2009: "Are Complex Decisions Better Left to the Unconscious? Further Failed Replications of the Deliberation-Without-Attention Effect". *Judgment and Decision Making* 4, 509–517.

Cunningham, Michael R., Steinberg, Jeff, & Grev, Rita 1980: "Wanting to and Having to Help: Separate Motivations for Positive Mood and Guilt-Induced Helping". *Journal of Personality and Social Psychology* 38, 181–192.

Damasio, Amtonio 1994: *Descartes Error: Emotion, Reason, and the Human Brain*. New York: Putnam.

Dijksterhuis, Ap, Bos, Maarten W., Nordgren, Loran F., & Van Baaren, Rick B. 2006: "On Making the Right Choice: The Deliberation-Without-Attention Effect". *Science* 311, 1005–1007.

Dijksterhuis, Ap, & Nordgren, Loran F. 2006: "A Theory of Unconscious Thought". *Perspectives on Psychological Science* 1, 95–109.

Evans, Dylan 2002: "The Search Hypothesis of Emotion". *The British Journal for the Philosophy of Science* 53, 497-509.

Evans, Jonathan S. B. T. 2011: "Dual-Process Theories of Reasoning: Contemporary Issues and Developmental Applications". *Developmental Review* 31, 86–102.

— 2008: "Dual-Processing Accounts of Reasoning, Judgment, and Social Cognition". *Annual Review of Psychology* 59, 255–278.

Fredrickson, Barbara L. 1998: "What Good are Positive Emotions?". *Review of General Psychology* 2, 300–319.

Gable, Philip, & Harmon-Jones, Eddie 2010: "The Blues Broaden, but the Nasty Narrows". *Psychological Science* 21, 211–215.

Gigerenzer, Gerd 2007: *Gut Feelings*. New York: Tantor Media.

Gigerenzer, Gerd & Brighton, Henry 2009: "Homo Heuristicus: Why Biased Minds Make Better Inferences". *Topics in Cognitive Science* 1, 107–143.

Griffiths, Paul E. 1997: *What Emotions Really Are: The Problem of Psychological Categories*. Chicago: University of Chicago Press.

Halberstadt, Jamin Brett & Levine, Gary M. 1999: "Effects of Reasons Analysis on the Accuracy of Predicting Basketball Games". *Journal of Applied Social Psychology* 29, 517–530.

Hanoch, Yaniv 2002: ""Neither an Angel nor an Ant": Emotion as an Aid to Bounded Rationality". *Journal of Economic Psychology* 23, 1–25.

Harteis, Christian & Gruber, Hans 2008: "Intuition and Professional Competence: Intuitive versus Rational Forecasting of the Stock Market". *Vocations and Learning* 1, 71–85.

Hogarth, Robin M. 2001: *Educating Intuition*. Chicago: University of Chicago Press.

Kahn, Barbara E., & Isen, Alice M. 1993: "The Influence of Positive Affect on Variety Seeking Among Safe, Enjoyable Products". *Journal of Consumer Research* 20, 257–270.

Kahneman, Daniel & Klein, Gary 2009: "Conditions for Intuitive Expertise: A Failure to Disagree". *American Psychologist* 64, 515–526.

Kruglanski, Arie W. & Orehek, Edward 2007: "Partitioning the Domain of Social Inference: Dual Mode and Systems Models and Their Alternatives". *Annual Review of Psychology* 58, 291–316.

Lassiter, G. Daniel, Lindberg, Matthew J., González-Vallejo, Claudia, Bellezza, Francis S., & Phillips, Nathaniel D. 2009: "The Deliberation-Without-Attention Effect". *Psychological Science* 20, 671–675.

Lazarus, Richard S. 1991: "Cognition and Motivation in Emotion". *American Psychologist* 46, 352–367.

Lerner, Jennifer S., & Keltner, Dacher 2001: "Fear, Anger, and Risk". *Journal of Personality and Social Psychology* 81, 146–159.

Manucia, Gloria K., Baumann, Donald J., & Cialdini, Robert B. 1984: "Mood Influences on Helping: Direct Effects or Side Effects?". *Journal of Personality and Social Psychology* 46, 357–364.

McCarthy, John, & Hayes, Patrick J. 1969: "Some Philosophical Problems from the Standpoint of Artificial Intelligence". *Machine Intelligence* 4, 463–502.

Nelissen, Rob M. A., Dijker, Anton J. M., & de Vries, Nanne K. 2007: "How to Turn a Hawk into a Dove and Vice Versa: Interactions between Emotions and Goals in a Give-Some Dilemma Game". *Journal of Experimental Social Psychology* 42, 280–286.

Newell, Ben R., Wong, Kwan Yao, Cheung, Jeremy C. H., & Rakow, Tim 2009: "Think, Blink or Sleep on It? The Impact of Modes of Thought on Complex Decision Making". *The Quarterly Journal of Experimental Psychology* 62, 707–732.

Osman, Magda 2004: "An Evaluation of Dual-Process Theories of Reasoning". *Psychonomic Bulletin & Review* 11, 988–1010.

Seo, Myeong-Gu, & Barrett, Lisa F. 2007: "Being Emotional During Decision Making—Good or Bad? An Empirical Investigation". *The Academy of Management Journal ARCHIVE* 50, 923–940.

Simon, Herbert A. 1992: "What Is an 'Explanation' of Behavior?". *Psychological Science* 3, 150–161.

Wilson, Timothy D. & Schooler, Jonathan W. 1991: "Thinking Too Much: Introspection Can Reduce the Quality of Preferences and Decisions". *Journal of Personality and Social Psychology* 60, 181–192.

Grazer Philosophische Studien
85 (2012), 291–298.

DO PAIN-ACCOMPANYING EMOTIONS MISLEAD US?—CONSIDERATIONS IN THE LIGHT OF REACTIVE DISSOCIATION PHENOMENA

Stefan REINING
Universitat de Barcelona

Second prize of the 2011 essay competition for students sponsored by the *Gesellschaft für Analytische Philosophie* (GAP) in cooperation with the *Grazer Philosophische Studien**

My aim in this paper is to show that if the reports from so-called dissociation patients are true, then this strongly supports the claim that the emotions accompanying our pain-sensations normally mislead us into believing that these pain-sensations are inherently awful. In order to achieve this, I will first, in section 1, give a short description of the relevant medical phenomena. In section 2, I will deal with Daniel Dennett's argument for the incoherency of the concept of pain—an argument that might appear to initially block my central claim. Finally, in section 3, I will defend my central claim.

1. *The phenomena*

Reflecting on physical pain (henceforth simply called 'pain'), one might think that its sensory aspect can in principle not be separated from its aversive emotional aspect. That is, one might think that if someone perceives herself as having a pain-sensation, then this person thereby also perceives herself as having a sensation that is, taken by itself, awful from her perspective.

This opinion, however, has apparently been falsified by many cases from the (more or less) recent history of medicine. The striking characteristic of all these so-called reactive dissociation phenomena is that the patients, who severely suffered from chronic, intractable pain, reported, after their medical treatment, that

* The question of the 2011 competition was: Can we Trust our Feelings? On the Value of Emotions for our Knowledge and Agency. The jury selected from 30 submissions two essays for a second prize and one essay which exceeded the space limit for a special acknowledgement. No first and no third prize was awarded this year.

the pain was still present, but that they simply did not care anymore about it. Their (allegedly remaining) physical pain had stopped producing any aversive emotions in them.

This kind of result is reported to have occurred regularly in the case of lobotomy patients. In fact, this kind of brain surgery is normally done exactly for the purpose of causing reactive dissociation. Antonio Damasio reports one such case in his book *The Feeling of What Happens*—a severe case of refractory trigeminal neuralgia, "a condition involving the nerve that supplies signals for face sensation in which even innocent stimuli, such as a light touch of the skin of the face or a sudden breeze, trigger an excruciating pain" (Damasio 1999, 74). Damasio describes the patient's condition two days after the operation as follows:

> [H]e had become an entirely different person, relaxed, happily absorbed in a game of cards with a companion in his hospital room. When [asked] about the pain, he looked up and said quite cheerfully that 'the pains were the same,' but that he felt fine now. [...] The operation had done little or nothing to the sensory patterns corresponding to local tissue dysfunction that were being supplied by the trigeminal system. The mental images of that tissue dysfunction were not altered and that is why the patient could report that the pains were the same. And yet the operation had been a success. It had certainly abolished the emotional reactions that the sensory patterns of tissue dysfunction had been engendering. Suffering was gone. The facial expression, the voice, and the general deportment of this man were not those one associates with pain. (Ibid., 74f.)

In another lobotomy case, reported by Paul Brand and documented in Brand & Yancey (1997), a woman suffering from constant, intractable vaginal pain, who was desperate after unsuccessfully having tried different kinds of medication and surgery, finally underwent a lobotomy. Brand reports that

> the woman emerged from surgery completely free of the suffering that had shadowed her for a decade. [...] When I [more than one year later] inquired about the pain, she said, 'Oh yes, it's still there. I just don't worry about it anymore.' She smiled sweetly and chuckled to herself. 'In fact, it's still agonizing. But I don't mind.' (Brand & Yancey 1997, 210)

In the case of cingulotomy, the same pattern is reported to have regularly occurred. In Foltz & White (1962), for example, the case of a patient is described who suffered severe vaginal and perineal pain for more than two years and was reported to show extreme emotional reactions to her pain (cf. Foltz & White 1962, 93f.).

After the operation, her situation changed considerably: "On questioning her, the pain was still present but did not concern her now." (Ibid., 94)

Reactive dissociation phenomena do not only occur after surgical treatment, but also—as, for example, mentioned in Dennett 1978—after the administration of morphine; which is why "some researchers describe the action of morphine (and some barbiturates) as 'reversible pharmacological leucotomy [lobotomy]'." (Dennett 1978, 209).

Nikola Grahek, however, in his 2007 book *Feeling Pain and Being in Pain*, points out that the dissociation phenomena in the kinds of cases considered above only change the patients' emotional reactions to their *chronic*, more or less constant, familiar pains, whereas these patients would still react in a normal, aversive way to new sorts of pain (cf. Grahek 2007, 36). Grahek does not consider these cases as real cases of reactive dissociation; rather, he compares them to those cases in which a person is distressed by her pain because she sees it as a bad sign indicating some sort of life-threatening development. Accordingly, this person would no longer be distressed in this way after having been told that her pain is actually no such sign (cf. ibid., 35f.). What Grahek suggests here is that, in the described cases of lobotomy, cingulotomy, and morphine administration, the patients' aversive emotional reactions before the medical treatment were not primarily caused by the unpleasantness of the pain-sensations, but rather by the pain's presumedly signalling character.

But this suggestion seems to be inadequate, since the patients in the relevant cases already knew *before* their treatment that their pain does not have the signalling character that Grahek describes. And even if they did not know this, the medical treatment that caused their aversive reactions to disappear certainly did not convey any such information to them. However, if one nevertheless shares Grahek's doubts about the cases considered above, one may still—as Grahek himself does—stick to cases of pain asymbolia as real cases of reactive dissociation. Grahek describes them as cases of complete indifference to pain, "while the sensory aspect of pain is fully preserved" (ibid., 39).

For the purposes of the present paper, it does not matter which of the phenomena that I have mentioned best exemplify the dissociation of pain-sensations from the pain's emotional aspect. What matters is merely that there *are* such phenomena, and this is indeed what the empirical data seems to show.

What I wish to claim in the light of these phenomena is that if the patients are right in reporting that the sensory aspect of their pains is still present, then this shows that pain-accompanying emotions—however these emotions are chosen to be called and whatever they consist in[1]—tend to *mislead* us into believing

1. These pain-accompanying emotions might even be conceived as certain bodily perceptions —a view originally developed in James (1884) and Lange (1885), and more recently endorsed

that the pain-sensations that cause our pain-aversive reactions are awful independently of this causal relation. However, since this claim involves the concept of pain, it might initially be blocked by an argument presented by Daniel Dennett, according to which reactive dissociation phenomena indicate that our ordinary concept of pain is incoherent. Before arguing for my claim, I will therefore first discuss Dennett's argument.

2. *Dennett's incoherency argument*

Dennett's strategy to show that our ordinary concept of pain is incoherent consists in formulating two claims which he regards as being conceptual truths about pain, and in demonstrating that the conjunction of these two claims is incompatible with how the reactive dissociation phenomena should be described. The two claims are the following:

1) "[W]e are 'incorrigible' about pain; i.e., if you believe you are in pain, your belief is true; [...]." (Dennett 1978, 226)
2) "[P]ains are awful." (Ibid., 227)

From the conjunction of these two alleged conceptual truths about pain, one can conclude that if one believes that one is in pain, then one is having an awful experience. Now this conclusion does not by itself contain a contradiction. But a contradiction becomes unavoidable as soon as one admits that the patients described in section 1, who are not suffering at all and therefore not experiencing something awful, really have the belief (after their medical treatment) that they are in pain. A contradiction is unavoidable here because one may either claim that the patients' belief is true, which would, together with the supposition that the patients do not experience something awful, contradict (2); or one may claim that the patients' belief is false, which would contradict (1).

Strictly speaking, these remarks do not show that the ordinary concept of pain, as Dennett characterizes it, is, taken by itself, incoherent, because—as mentioned

in Prinz (2004)—, as long as these bodily perceptions do not involve the pain-sensations that the dissociation patients claim to have. In this paper, I will not take any stance regarding the general ontology of emotions. However, in using the term ‚pain-accompanying emotions' I intend to refer to something that is not only present in all usual cases of pain, but also in those horror-cases in which apparently anesthetized surgery patients remain completely conscious despite their lack of ability to use their muscles. (In Dennett 1978, 209, actual cases of this sort are described.) Since it might seem unnatural to say that persons who find themselves in the latter sort of scenario are in certain emotional states, I leave open the possibility that what I call 'pain-accompanying emotions' should actually rather be called 'pain-accompanying feelings'.

above—the incoherency only appears after taking into account our willingness to attribute certain beliefs to the patients. However, Dennett regards the ordinary concept of pain as the culprit within this incoherency-yielding conglomerate, and he therefore regards it as a concept that should be abandoned, in favor of a new, natural-kind concept of pain (cf. ibid., 228).[2]

My first reply to Dennett's argument is that it is quite doubtful whether (1) really is a conceptual truth about pain. Part of Dennett's justification for his claim that the dissociation phenomena function as counter-examples to the conjunction of (1) and (2) is his following remark:

> Surely it is logically possible to be in a dispositional state bearing all the usual earmarks of belief that one is in pain, and yet not be experiencing or undergoing something awful. Not only is this logically possible, it is instanced routinely, by morphine subjects and others. (Ibid., 227)

One may dispute whether the patients' pain-beliefs really bear "all the usual earmarks of belief that one is in pain". As Rick Kaufman points out, normal non-linguistic pain behavior, which the dissociation patients clearly lack, should also be regarded as one of these earmarks (cf. Kaufman 1985, 280f.) Kaufman presents a better example:

> A soldier struck on a battlefield might moan and cry for a medic only to discover that the bullet which knocked him down had merely pierced his canteen, spilling water down his leg, leaving him quite uninjured. It is implausible and arbitrary to say the soldier did not believe he was in pain. (Ibid., 281)

The point that I wish to make here is the following: The fact that Dennett apparently expects his readers to agree to the first sentence of the passage just cited (or at least to the part after the word 'surely'), and—more importantly—the fact that it is plausible to regard Kaufman's description of the soldier as conceptually possible, indicate that (1) is not a conceptual truth about pain. After all, our dispositions to agree or disagree with Dennett's logical-possibility claim (at least if it is also taken as a conceptual-possibility claim), and our dispositions to regard or not regard the soldier-example as conceptually possible, are *criteria* of what our concept of pain is actually like. Therefore, it should be suspected that Dennett's argument is based on at least one false premise.

2. Dennett's proposal to introduce a natural-kind concept of pain, in order to let science discover whether the reactive dissociation patients really are in pain, is very problematic. But I will not elaborate on this any further, especially because it is not necessary for my argumentation.

My second reply to Dennett's argument runs as follows: Even if his argument were sound, the claim that I have introduced at the end of section 1 as the claim that I primarily wish to defend in this paper, namely

(C) If the patients' claim that the sensory aspect of their pains is still present is true, then this shows that the emotions normally accompanying our pain-sensations tend to mislead us into believing that the pain-sensations that cause these emotions are awful for us independently of this causal relation.

can be reformulated (without being essentially modified) so as to avoid being blocked by the alleged incoherency of the concept of pain. The reformulated claim is the following one:

(C*) If the patients' claim that the sensory aspect of what they would call their 'past pains' is still present is true , then this shows that the emotions normally accompanying what we would call 'pain-sensations' tend to mislead us into believing that the sensations that cause these emotions are awful for us independently of this causal relation.

I will now defend C*, using—for the sake of brevity—the expression 'pain-sensation' in order to refer to the sensory aspect of what people normally call 'pain'.

3. *The misleading character of the emotions accompanying our pain-sensations*

The most obvious possible objection against C* runs as follows: One might claim that while *our* pain-sensations are awful for *us*, the sensations that the dissociation patients call their 'pains' are simply not awful for *them*; and this difference regarding awfulness does not merely consist in the fact that our pain-sensations cause aversive emotional reactions in us. Therefore, so the objection goes, the emotions accompanying our pain-sensations do not *mislead* us into believing that our pain-sensations are awful for us independently of their causal relation to these emotions, since our pain-sensations *are* awful for us independently of this causal relation; whereas the patients' sensations are not awful for them at all.

Here is my answer to this objection: If the patients' sensations and our pain-sensations do not differ regarding their *sensory* aspect—a condition that is built into C*—then there is no plausible candidate for what the difference could be apart from the difference regarding the causation of certain emotions. The *inherent* degree of awfulness of the sensations simply cannot differ if the sensations are supposed to have the same sensory character. Claiming the opposite would be analogous to claiming that someone could experience the same sugar-taste as before, but without the sweetness.

Leaving this objection behind, I will now describe what I take to be the misleading character of our pain-sensations. It can hardly be disputed that we normally regard the awfulness of our pain-sensations as not only consisting in the character of the emotions caused by these sensations, but also as at least partly consisting in the inherent character of these sensations. But if the dissociation patients are right in claiming that the sensory quality of what they call their 'pain' is still the same as it was before their medical treatment, then there seems to be no way of convincingly denying that the awfulness of all these qualitatively identical sensations does not consist in anything else than their causal relations to certain emotions (unless one wishes to claim that the patients in fact suffer from their sensations without being able to express their suffering—a claim for which there is no evidence). To be sure, I do not intend to downplay the awfulness of what we call 'pain'. What I claim is that the emotions accompanying our pain-sensations normally mislead us into locating the awfulness in the wrong place. More precisely, what I am pointing to here is the phenomenon of *projection*: The emotions accompanying our pain-sensations normally mislead us into projecting their awfulness into the sensations that cause them.

This phenomenon can be illustrated by looking at examples that show a similar pattern: People tend to be sad after losing a game, and they tend to be angry after being insulted. As in the case of pain-sensations, people tend to project the (relatively mild) awfulness of these emotions of sadness and anger into something else, namely into the inherent characteristics of the state of having lost a game and the state of having been insulted, respectively.

There is, however, a significant difference between these two cases and the case of pain-sensation: In the former cases, the tendency to project is clearly weaker than in the latter case; and if, in the former cases, a projection occurs, then it is much more transparent to us than in the latter case. It often happens, for example, that people who are sad after losing a game do not really wish that they had won the game, but instead wish not to be the kind of person who is sad after losing a game. And something analogous can be said regarding the case of being angry after having been insulted. In spite of being emotionally affected in the way described, the people in these examples have overcome the tendency to *project* the respective emotions. In contrast to that, our tendency to project in the case of pain-sensations is so deeply entrenched that it seems hardly possible in non-pathological cases for people to overcome this tendency.

One might claim that this disanalogy regarding the degree of transparency speaks against there being any projection at all in the case of pain-sensations.

However, this disanalogy can be explained by the fact that people normally have a better grasp of the causal chain between, for example, their losing a game and their being sad than of the causal chain between their pain-sensations and the accompanying emotions. In fact, people normally do not realize that there is any such principally interruptible chain at all between their pain-sensations and the accompanying emotions—the presence of such a chain only occurs to us after, for example, hearing about dissociation phenomena. So even if the projection-thesis regarding pain-sensations might seem to be prima facie unbelievable, this unbelievability can at least be explained.

It might still be that the antecedent of C* is false, in which case C* would be true even if there were no projection in the case of pain-sensations. It might, for example, be that the patients' medical treatment distorts their memories of their past pain-sensations and that they therefore make false similarity-claims regarding their past and present sensations. But as long as there is no evidence for the occurrence of this kind of distortion, it seems plausible that if C* is true —which I have argued for—, then the projection-thesis is true as well.

References

Brand, Paul & Yancey, Philip 1997: *The Gift of Pain. Why We Hurt and What We Can Do About It.* Grand Rapids, MI: Zondervan Publishing House.

Damasio, Antonio 1999: *The Feeling of What Happens. Body, Emotion and the Making of Consciousness.* London: Random House.

Dennett, Daniel C. 1978: "Why You Can't Make a Computer that Feels Pain". In: Daniel C Dennett, *Brainstorms. Philosophical Essays on Mind and Psychology.* Cambridge, MA: MIT Press, 190–229.

Foltz, Eldon L. & White, Lowell E. 1962: "Pain 'Relief' by Frontal Cingulotomy". *Journal of Neurosurgery* 19, 89–100.

Grahek, Nikola 2007: *Feeling Pain and Being in Pain* (2nd ed.). Cambridge, MA: MIT Press.

James, William 1884: "What Is an Emotion?". *Mind* 9, 188–205.

Kaufman, Rick 1985: "Is the Concept of Pain Incoherent?". *Southern Journal of Philosophy* 23, 279–283.

Lange, Carl G. 1885: *Om Sindsbevaegelser. Et psyko-fysiologisk Studie.* Copenhagen.

Prinz, Jesse J. 2004: *Gut Reactions. A Perceptual Theory of Emotion.* New York: Oxford University Press.

Grazer Philosophische Studien
85 (2012), 299–309.

PATRIOTISM
A CASE STUDY IN THE PHILOSOPHY OF EMOTIONS

Peter KÖNIGS
Universität Tübingen

Special acknowledgement of the 2011 essay competition for students sponsored by the *Gesellschaft für Analytische Philosophie* (GAP) in cooperation with the *Grazer Philosophische Studien**

Ever since the cognitive turn in the philosophy of emotions, it has been a commonplace that emotions are subject to standards of appropriateness. However, once emotions can be appropriate, they may also fail to be so. In the present paper, I would like to draw on these insights applying them to one particular case, namely the case of patriotism. The appropriateness of patriotism, I will argue, varies according to its formal object. We must distinguish between patriotism as 'love of one's country' and patriotism as 'pride of one's country'. I will show that these two different patriotic emotions must be treated differently, but also how they are related to each other. Moreover, I will argue that the debate about the merits and demerits of patriotism is misled or incomplete if it is only concerned with the consequences of group favouritism rather than with the emotional appropriateness of patriotism. I will conclude with some reflections on how this 'emotional' approach to patriotism can teach us about what is really at stake in the debate about patriotism.

Introduction

David Hume points out that pride can be linked with all kinds of things: "Our country, family, children, relations, riches, houses, gardens, horses, dogs, cloaths; any of these may become a cause either of pride or of humility." (Hume 2003, 199) Hume is doubtless right in this observation. However, Hume is not con-

* The question of the 2011 competition was: Can we Trust our Feelings? On the Value of Emotions for our Knowledge and Agency. The jury selected from 30 submissions two essays for a second prize and one essay which exceeded the space limit for a special acknowledgement. No first and no third prize was awarded this year.

cerned with the more important question about pride: Do these objects *deserve* our pride? Hume's theory of emotion is unfit to answer this question. Hume famously conceives of emotions, or passions to use his term, as affects that have a phenomenological quality and motivational force while being devoid of any cognitive component. They cannot be rationally criticised.

However, ever since the cognitive turn in the philosophy of emotions it has been a commonplace that there is a cognitive dimension to emotions (e.g. de Sousa 1987). Emotions do not only have a motivational and a phenomenological dimension, they are also directed at states in the world and represent them as being in certain way. Fear of a barking dog represents the dog as being dangerous. Admiration for Beethoven represents Beethoven as being extremely capable and gifted, as admirable. As a corollary, emotions can represent the world correctly and incorrectly. It would be inappropriate to fear Beethoven while showing admiration for the barking dog.

In recourse to scholastic parlance, this quality of the intentional object of an emotion has been dubbed its formal object; 'dangerous' is the formal object of fear. Emotions can be individuated from each other in terms of their formal objects.

Emotions having a cognitive dimension, it is one task of the philosophy of emotions to produce a general theory of the standards of appropriateness of emotions and thus of their rationality (e. g. de Sousa 1987).[1] The purpose of my essay is more modest in scope, however. I will be concerned with the appropriateness of emotions in only one particular case: that of patriotism. This paper is a case study in the philosophy of emotions applying the findings of the philosophy of emotions to the issue of patriotism. The argument will be that there are two distinct emotions both of which qualify as patriotism. Since they have two different formal objects, the appropriateness of patriotism depends on which of either patriotic emotion we look at. Nonetheless, those two distinct emotions are linked in a significant way that I will also discuss in the course of the essay.

However, before I specify these thoughts, let me make some remarks on the issue of patriotism.

The topic has only recently received extensive philosophical attention. Much of the philosophising about patriotism takes place within the larger context of the debate between liberals and communitarians. Therefore, one common approach to the issue of patriotism is to examine to what degree the somewhat parochial concept of patriotism is compatible with liberal universalism and—if not—which of either concepts should trump.

1. A different question is whether it is *pragmatically* rational to rely on our emotions. (See Arpaly 2000, Jones 2003)

This, however, is not the approach that I would like to take. These certainly important questions have been discussed at length elsewhere, and I will touch upon them as I go. Yet, the focus of the present paper will be a different one: I want to start with assessing the appropriateness of patriotism as an emotion. It is one question whether we should see in patriotism a communitarian value (MacIntyre 2002) or a threat to liberal impartiality (e. g. Gomberg 1990); it is quite another whether patriotism as an emotion is appropriate or not. I will say more about this at a later stage.

Two patriotic emotions

As adumbrated above, I think that in order to make sense of the idea of patriotism we need to distinguish between two kinds of patriotic emotions: Patriotism in the sense of love of one's country and patriotism in the sense of pride of one's country. This difference is only rarely made explicit, even though it will prove crucial when assessing the emotional appropriateness of patriotism.

The main point that I will make in what follows is that love of one's country can be a perfectly appropriate emotion whereas pride of one's country rarely ever is. To see this, let us first look at how these two emotions differ. Although we have an intuitive grasp of their difference, this difference should be spelled out in terms of their differing formal objects.

The formal object of love[2], roughly, implies an extraordinary personal value of the object. This is what makes the object deserving of one's love. Love represents its intentional object as being of extreme personal value. The formal object of pride, on the contrary, is that of being a personal achievement. Pride represents its intentional object as a personal achievement. To be proud means to be pleased with one's own achievement.

As love and pride have two distinct formal objects, their standards of appropriateness differ, too. Let me first assess the appropriateness of love of one's country.

Love of one's country represents one's country as being extremely valuable. It seems, however, that the value of a country can be controversial. Some countries have more beautiful a landscape than others, some countries' cultural heritage is more formidable than others', and some countries have a just and fair political system while others do not. Certainly, the qualities of countries differ. Does that mean that some people are more justified in loving their country than others? I do not think so. It has been pointed out before that the value in question is an agent-relative value. That means that the value is constituted by the special

2. Love is here used in a very generic way, more or less equivalent with strong affection. A more detailed analysis of the concept is not needed for the purposes of this essay.

personal relation that we have to the object of our love: People love their home town because it is the place of their childhood memories. They love the nature of where they have grown up because they are so well acquainted with it in all its manifold facets. And one might love the people of one's country because they share the same language, humour and habits.

This also implies that patriotism need not entail the belief that one's country is objectively the best, since love of one's country is not about objective but about subjective value. Although this is often confused, the confusion is easily debunked.[3]

Of course, I am not saying that this love must be exclusive. It seems perfectly reasonably to love one's country all the while loving, say, France, too. Also, I am not saying that one *has* to love one's country or that love for one's country is never defeasible. Clearly, no such duty to love one's country can be established. What I do want to establish is that one's having a strong personal relation with one's country can make one's love for it appropriate, the value of one's country being a subjective and agent-relative value.

I do not think that this is a controversial claim to make. Most people will agree that one can be justified in loving one's hometown or in loving a country other than one's own. How many people love Italy, and rightly so? I do not see why love of one's own country should not be justified *a fortiori*.

What I do think is unjustified is pride of one's country. To be proud of something means—according to its formal object—to be pleased with one's contribution to a laudable achievement. And while a country and its people may have brought about a whole range of laudable achievements, we, as single individuals, rarely ever make any significant contribution to them. This is evidently true in cases where the achievements in question are achievements of the past. I cannot possibly be proud of, say, the cultural achievements of Italian Renaissance since I have not contributed to it. But even if we do contribute to the achievement, the contribution of a single individual is usually too small to justify pride to any meaningful degree.

There is one obvious rejoinder to my twofold account of patriotism: One could object that I have misdescribed the formal object of pride. Pride, the objection goes, need not be directed at one's own achievements. It may be perfectly appropriate to be proud of somebody else's achievements to which one has only little contributed, or not at all. One may, for instance, be proud of an achievement of a close friend or a relative, or one may be proud that the football club that one supports has won the championship. In such cases, our contribution to

3. "Nationalism" might be the correct term for the belief that one's own country is objectively the best, although I am not sure whether this difference between patriotism and nationalism is really captured by natural language.

the actual achievement is too small to justify our being proud to any significant degree. So it seems that our contributing to the achievement is not a necessary requisite for legitimate pride. And if this is so, then the case against national pride collapses.

To me, however, this objection seems flawed. Although it is indeed very common to display pride in cases like the ones just described, I do not think that this emotional response is warranted. Rather, we seem to fall victim to some kind of emotional fallacy, as it were. I believe that we all have an intuitive grasp of the meaning of pride and that we are therefore, upon reflection, prepared to concede that pride in such cases is unwarranted. However, I do not want to content myself with this intuitive argument. In order to make my claim more compelling, I will endeavour to spell out how this emotional fallacy might originate.

This is where I want to demonstrate that love and pride, although two distinct emotions, are intimately linked. Such inappropriate instances of pride as the ones just described are due to a cognitive delusion that results from one's love for the person (or entity) that really brought about the achievement in question. Or so at least I will argue:

Love does not only imply that one values the intentional object. It also leads one to identify with the object of one's love. This applies, above all, to the interests of the beloved person. There is no longer a strict demarcation between one's own interests and those of the object of one's love. This has been pointed out by many authors; to quote only some of them: Roger Scruton writes: Love means that "all distinction between my interests and your interests is overcome" (1986, 230). Nozick concurs: "What is common to all love is this: Your own well-being is tied up with that of someone (or something) you love [...]. When something bad happens to one you love [...] something bad happens *to you*." (1991, 417) A similar point, though with respect to state actors, has been pointed out by Alexander Wendt. States, Wendt argues, identify with one another and with one another's interests if they come to conceive of each other as friends (Wendt 1992).[4] It seems, then, that love or strong affection leads to a kind of meshing of identities.

One caveat, however, remains to be noted: Mutual identification here seems to be restricted to the identification with one another's interests. I concede that more needs to be said on the psychological mechanisms that lead to the identification with one another's achievements. However, the idea of such a more extended identification is anything but far-fetched. According to Robert Solomon, for instance, love implies "the urge for *shared identity*" (1991, 511) and "the concentration and the intensive focus of mutual definition on a single

4. To be sure, friendship is not love. But if friendship leads to a meshing of identities, then love will do *a fortiori*.

individual, subjecting virtually every personal aspect of one's self to this process." (1988, 197) And Mark Fisher, in much the same fashion, contends that loving people "perceive, feel and act as a single person" (1990, 28). On their accounts, identification goes beyond the identification with each other's interests extending also to roles, perceptions and much more. Therefore, explaining the 'false pride'-fallacy in terms of mutual identification has a high initial plausibility. Thus, the phenomenon of mutual identification among sympathising agents adds to the cogency of my claim that pride of another person's achievement is an emotional failure rather than an adequate emotional response.[5,6]

Before I move on, let me summarize my findings thus far: I argued that the issue of patriotism must be tackled by looking at the appropriateness of patriotic emotions. Although love and pride are two distinct emotions, which must therefore be treated differently, I hypothesised that there is a psychological link between them: it is the very love of one's country that produces 'false pride' of one's country. I have contended that love of one's country can be a perfectly appropriate emotion. This is an important finding that needs to be kept in mind when discussing the issue of patriotism. Even if we reject patriotic favouritism on moral or political grounds, it must be acknowledged that patriotism—when meant as love for one's country—can be defended on emotional grounds, as it were. In the remainder of this essay I will use 'patriotism' and 'love of one's country' interchangeably.[7]

Two ways of looking at patriotism

I have approached the issue of patriotism by looking at the appropriateness of patriotic emotions. I think that we need to take this approach since the question of whether love or pride of one's country are legitimate emotions is an important question in its own right. In this respect, my approach differs from the purely

5. There is another observation that corroborates the thesis that it is love or sympathy that leads one to identify with another person's achievement: As a matter of fact, we are proud of somebody else's achievement only if we sympathise with the person who brought about the achievement. In all other cases we may envy her or admire her or just be indifferent about her achievements – but we will not be proud.

6. One might be tempted to argue that pride is an appropriate emotion for the very reason of this identification: Since this identification leads to the meshing of identities one can be proud of somebody else's achievement since it no longer really is somebody *else's*. I think, however, that such a strong interpretation of the fusion of identities is strongly counter-intuitive.

7. I am inclined to think that love of one's country is essential to patriotism while pride is not. I would assume that one cannot be a patriot without loving one's country in one way or another, while I do not think that one has to be proud.

moral or political approach to patriotism of philosophers who investigate the merits and demerits of patriotic favouritism. In what follows, I want to give three reasons why I think we should disentangle these two dimensions of patriotism, i. e. the appropriateness of love for one's country on the one hand and the pros and cons of group favouritism on the other hand.

First, I do not think that patriotism implies favouritism by definition. McCabe rejects patriotism on the grounds that it "stresses partiality, particularity, and showing special concern for one's co-nationals" (McCabe 1997, 203). He sees no way to reconcile patriotism with liberal impartiality the latter of which McCabe thinks ought to trump. What is noteworthy about McCabe's account of patriotism is that he defines patriotism in terms of the moral obligations it entails. Patriotism essentially *means* "that co-nationality is a morally significant relationship that may impose special duties and sanction special treatment" (ibid.). It is, to my mind, a flawed definition of patriotism. Contrary to McCabe and other theorists of patriotism (e. g. Nathanson 1989, MacIntyre 2002), I do not think it correct to include in the definition of patriotism the preparedness to "promote its [scil. one's country's] well-being even if that can only be done at the cost of diminishing the well-being of other countries" (Nathanson 1989, 536) or even "a willingness to go to war on one's community's behalf" (MacIntyre 2002, 289). To be a patriot means, first and foremost, to love one's country, and we should not, by way of definition, equate patriotism with a certain pattern of behaviour.

One cannot, on pain of contradiction, claim to be a patriot without loving one's country in one way or another. But there is no such contradiction in being a patriot without being willing to benefit one's own country at the cost of others. This confusion arises, it seems, because if one loves someone or something one is at least strongly disposed to favour the object of love. I concur with David Velleman, however, who points out: "The account of love offered by many philosophers sounds to me less like an analysis of the emotion itself than an inventory of the desires and preferences that tend to arise in loving relationships of the most familiar kinds." (1999, 342) Patriotism is essentially love of one's country, and it is only contingently linked with favouritism and the like.

One might object that, even if we grant that favouritism is not part of the definition of patriotism, love for one's country still morally obliges one to favouritism and partiality. I do not think it does. This is the second reason why I think we should disentangle these two dimensions of patriotism. The link between the appropriateness of an emotion and its normative force is contentious. *Prima facie*, it seems perfectly reasonable to recognize the personal value of some object all the while thinking that one ought not give this object undue preference. Many utilitarians, to give an obvious example, will agree that one's love for a particular person may be an appropriate emotion all the while holding that, nonetheless, all sentient beings must be treated equally.[8] And even philosophers who stress that evaluative emotions can point to normative reasons for action concede that other reasons may be overriding. Emotions equip us at best "with a *prima facie* reason for action" (Döring 2007, 391; emphasis P. K.). Emotions, being cognitions, may reveal or possibly co-constitute the value of the emotions' intentional objects.[9] But a straightforward inference from the axiological to the deontic is not warranted. Thus, if one wants to establish that love of one's country justifies or obliges one to partial treatment, this is a claim that must be independently argued for.

That means that we *can* discuss the appropriateness of patriotism independently from alleged moral implications.

Third, and relatedly, if we do not disentangle these two aspects our inquiry will yield few new insights. Once we define patriotism as justifying partiality, pointing out that patriotic partiality is incompatible with liberal impartiality amounts to little more than a truism.[10] And, most importantly, such a limited approach leaves an important question open: Of course, there is the question of whether love of one's country is valuable (for communitarian considerations) or harmful (for liberal considerations). But there is also the second and distinct question of whether love of one's country is appropriate or inappropriate *tout court*. Although answering the first is certainly a worthwhile and necessary project, the second question is a legitimate question in its own right. Irrespective of whether patriotism implies partiality or not—I argued that no such link need to be assumed—, we just cannot settle the second question by pointing to the unpleasant illiberal implications of patriotism. And this is because emotions have an independent standard of appropriateness. Therefore, if we tackle the issue of patriotism from a merely moral, somewhat consequentialist perspective, the issue will be dealt with without even touching on the question in dispute, namely whether we can be justified in loving our country. If we assess patriotism only in terms of its potential moral implications or political consequences, we throw out the baby with the bath water.

By the same token, pointing out that patriotism fosters enmity and war does not settle the question. I am inclined to agree with critics of patriotism who, seconding Leo Tolstoy (1987), argue that this kind of group attachment is a threat to peace. "Certainly, patriotism may be mobilized for a good cause. But much

---

8. Of course, on the utilitarian account, partial treatment as caused by love may turn out to be *contingently* optimific.

9. This is roughly the position of David Wiggins' (1987) and John McDowell's (1998) Sensibility Theory; see D'Arms/Jacobson (2006).

10. Although it has been tried to reconcile patriotic partiality with liberalism (see e. g. Nathanson 1989).

more easily, it may be mobilized for an unjust one", Kateb (2000, 912) argues convincingly, pointing out that "patriotism is inherently disposed to disregard morality" (909). It has to be stressed, however, that this is only a disposition. As I argued before, love of one's country does not necessarily morally oblige one to acts deemed immoral by liberals. Of course, the negative consequences of patriotism must be taken seriously, but so must the emotional appropriateness of love of one's country.[11] Rather than to delegitimise patriotism altogether by pointing to its potential immoral consequences, we should embrace the dilemma that is posed by patriotism: It is an appropriate emotion which at times, deplorably enough, motivates patriots to ghastly deeds.

To conclude: not only *can* we discuss the appropriateness of patriotism independently from potential moral implications, but we *have* to if we want to take the issue of patriotism seriously.

Patriotism and recognition

In the remainder of my paper, I would like to point to one further insight that we may draw from this more nuanced approach to the issue of patriotism: My more balanced account of patriotism can provide resources to explain the aggressiveness and relentlessness that is often present in debates about patriotism. Usually, the focus is on whether the consequences of patriotism are desirable or not. However, it seems likely that advocates of patriotism will feel misunderstood when debates about patriotism are reduced to its pros and cons in such a fashion. As I have stressed in my paper, there is more to the question of patriotism than this: we also have to look at the appropriateness of patriotism as an emotion. Now, it is plausible to assume that it is the appropriateness of their love of their country that is of paramount importance to advocates of patriotism, while they might even share liberal worries as to illiberal drawbacks of patriotism. Arguably, it is the lack of recognition for their love of their country that troubles advocates of patriotism the most. The importance of recognition has been a major issue in political philosophy recently (e. g. Taylor 1994, Honneth 1996). Although recognition has not been linked with emotions yet, extending recognition to the

11. In a way, my critique is reminiscent of D'Arms' and Jacobson's criticism of the "moralistic fallacy" (D'Arms/Jacobson 2000). This fallacy is committed whenever one denies "that an emotion can be fitting despite being wrong (or inexpedient) to feel" (D'Arms/Jacobson 2000, 69). That means that it can be appropriate to feel an emotion (e. g. envy) which would be imprudent or morally objectionable to feel. My point, however, is that it can be appropriate to feel a (positively evaluative) emotion which is not itself morally objectionable but which *motivates* one to do morally objectionable deeds.

appropriateness of emotions seems only natural. The importance of recognition in this context is obvious: Recognizing the appropriateness of love means recognizing the value of the object of love.[12] By dismissing patriotic love altogether, we dispute that one's country is worthy of love. And given that love of one's country is a very personal emotion, as pointed out at the outset of this paper, disputing the appropriateness of this emotion is a somewhat bold and harmful statement to make. Debates about patriotism are often very emotional and personal because there is always a hidden thoroughly emotional and personal dimension to the issue of patriotism disregard of which is not only negligent but hurtful.

References

Arpaly, Nomy 2000: "On Acting Rationally Against One's Best Judgement". *Ethics* 110, 488–513.

D'Arms, Justin/Jacobson Daniel 2000: "The Moralistic Fallacy: On the 'Appropriateness' of Emotions". *Philosophy and Phenomenological Research* 61, 65–90.

— 2006: "Sensibility Theory and Projectivism". In: David Copp (ed.), *The Oxford Handbook of Ethical Theory.* Oxford: Oxford University Press, 186–218.

De Sousa, Ronald 1987: *The Rationality of Emotion*. Cambridge, MA: The MIT Press.

Döring, Sabine A. 2007: "Seeing What to Do: Affective Perception and Rational Motivation". *Dialectica* 61, 363–394.

Fisher, Mark 1990: *Personal Love*. London: Duckworth.

Gomberg, Paul 1990: "Patriotism Is Like Racism". *Ethics* 101, 144–150.

Honneth, Axel 1996: *The Struggle for Recognition: The Moral Grammar of Social Conflicts*. Cambridge: Polity Press.

Hume, David 2003: *A Treatise of Human Nature.* Mineola: Dover.

Jones, Karen 2003: "Emotion, Weakness of Will, and the Normative Conception of Agency". In: Anthony Hatzimoysis (ed.), *Philosophy and the Emotions.* Cambridge: Cambridge University Press, 181–200.

Kateb, George 2000: "Is Patriotism a Mistake?". *Social Research* 67, 901–924.

MacIntyre, Alasdair 2002: "Is Patriotism a Virtue?". In: Derek Matravers & Jon Pike (eds.), *Debates in Contemporary Philosophy*. London/New York: Routledge, 286–300.

McCabe, David 1997: "Patriotic Gore, Again". *The Southern Journal of Philosophy* 35, 203–223.

12. Of course, bearing in mind the relativity of love, this does not mean that we must ourselves love the object. But we should acknowledge that somebody else's love can be appropriate.

McDowell, John 1998: "Values and Secondary Qualities". In: Idem, *Mind, Value, and Reality*. Cambridge, MA: Harvard University Press, 131–150.
Nathanson, Stephen 1989: "In Defense of 'Moderate Patriotism'". *Ethics* 99, 535–552.
Nozick, Robert 1991: "Love's Bond". In: Kathleen Higgins (ed.), *The Philosophy of (Erotic) Love*. Lawrence: Kansas University Press, 417–432.
Scruton, Roger 1986: *Sexual Desire*. London: Weidenfeld and Nicolson.
Solomon, Robert 1988: *About Love: Reinventing Romance for Our Times*. New York: Simon & Schuster.
— 1991: "The Virtue of (Erotic) Love". In: Kathleen Higgins (ed.), *The Philosophy of (Erotic) Love*. Lawrence: Kansas University Press, 492–518.
Taylor, Charles 1994: "The Politics of Recognition". In: Charles Taylor et al. (eds.), *Multiculturalism*. Princeton: Princeton University Press, 25–73
Tolstoy, Leo 1987: "Patriotism, or Peace?". In: Idem, *Writings on Civil Disobedience and Nonviolence*. Philadelphia: New Society Publishers, 137–147.
Velleman, David J. 1999: "Love as a Moral Emotion". *Ethics* 109, 338–374.
Wendt, Alexander 1992: "Anarchy Is What States Make of it: The Social Construction of Power Politics". *International Organization* 46, 391–425.
Wiggins, David 1987: "A Sensible Subjectivism?". In: Idem, *Needs, Values, Truth: Essays in the Philosophy of Value*. Oxford: Blackwell, 185–214.

Grazer Philosophische Studien
85 (2012), 311–323.

WHAT DO YOUR SENSES SAY? ON BURGE'S THEORY OF PERCEPTION[1]

Christopher GAUKER
University of Cincinnati

According to Tyler Burge, perceptions represent particulars and represent them as belonging to kinds.

> Perception makes reference to particulars in the physical environment. Perception attributes physical properties, kinds, and relations to those particulars—categorizes or groups them. (24)

He really does mean that perceptions represent groupings. He writes:

> A perceptual attributive is an aspect of a perceptual representational content that functions to indicate a repeatable type and to group or characterize purported particulars as of that type. (380; see also 198)

Burge denies that perceptions can represent such things as "elementary particles, teacups, pianos, and recessions" (101; see also 546). But he thinks that objects can be perceived *as bodies* (258) and *as having certain shapes* (323) and *as moving* (24). My primary objective in this review will be to show that Burge has not given us any good reason to think that perceptions group things in any way.[2]

A sentence may consist of a singular term and a predicate. The singular term may denote a particular, and the predicate may represent an entire kind of thing. But a perception is not like a sentence (as Burge acknowledges, 540). It does not contain a predicate. So it is a question in what sense a perception can represent a general kind of thing. If a dog perceives a piece of bone as an *edible*

1. Review essay on: Tyler BURGE, *Origins of Objectivity*. Oxford: Oxford University Press 2010. xix+624 pp. ISBN 978-0-19-958139-9.

I thank Matthew Van Cleave for helping me with his close reading of and insights into this book.

2. In my book, *Words and Images: An Essay on the Origin of Ideas*. Oxford: Oxford University Press 2011, 59–65, I have argued that perceptions do not have propositional content. That argument does not bear on Burge's thesis because he denies that the content of perceptions is propositional.

body (Burge's example, 258), in what sense does the dog's perception represent the kind *edible body*? How do edible bodies not presently perceived come to be in any way part of what is represented? If the perception represents any kind of thing at all, why should we say that it represents the kind *edible body*, rather than the kind *greyish chunk of matter* or the kind *segment of a cow's femur*? Why should we suppose that the dog perceives the bone as *edible*? Maybe the dog perceives it as *in*edible but decides to chew on it anyway.

Burge acknowledges that this is a serious question. He writes:

> There must be some account of the connection between the perceptual states, with their representational contents, entailing veridicality conditions *as of*, and physical environmental entities that satisfy the conditions. What conditions must be in place if a perceptual state is to indicate some environmental attributes, and represent them in such a way as to specify them? What network of relations grounds explanation of the non-accidentality of the systematic connection between a perceptual state's being the kind it is by virtue of being *as of* specific environmental entities, on one hand, and the environmental entities, on the other? (85, see also 283)

In short, what makes it the case that a perception represents *a* as *F* and not *b* as *G*? Call this the *constitution question*. What Burge calls the "primary problem" of the theory of visual perception is not this, however, but rather the problem of explaining how the "immediate effects of proximal stimulation" are transformed into perceptual states with representational content (89, see 343). Call this the *transformation question*. Nonetheless, the constitution question and the transformation question are related, because a "key aspect" of the transformation question is the fact that "effects of proximal stimulation on the sensors of a perceptual system *underdetermine* both *representata* of perception and the nature of the perceptual state that represents such *representata*" (344).

So what is Burge's answer? Throughout the text, Burge tosses out a bewildering variety of hints. There are many references to biological teleology. A frog's perceptual system represents what it does "because its nature was formed through prior causal relations to the environment in the evolutionary development of the type of visual system that the individual frog has" (71; see also 81, 85, 262, 321, 346). Sometimes, though, Burge appeals to history without citing the forces of natural selection (257). In any case, a perception must have a function in enabling an animal to satisfy its biological needs, such as "mating, predating, navigating" (261; see also 94, 272f., 324, 438, 526). But he excoriates the attempt, such as one finds in the work of Ruth Millikan, to explain representation in terms of naturally selected normal functions as "quaint" (296) and "probably pointless and hopeless" (298) and as stemming from a "narrow and perverse vision of the science" (309) (though he provides no careful exposition of the view). He declares that the norms pertaining to perception are standards that perceptions must live up to in order to count as veridical (311f., 340), not norms stemming from biological function (312, 411), and they are not epistemic norms either (312).

There are many other factors that Burge sometimes seems to think of as having a place in our answer to the constitution question. We have to look for the most probable (92n) or normal (94, 98, 465) causes of perceptions. Content depends on "systematic patterns of relations to the environment" (77) or "patterns of causal interrelations, usually in evolutionary history, with attributes in the environment" (407). The transformation question, which, as we have seen, is related to the constitution question, will be answered by finding "law-like patterns of formation of representational states" (298, see also 96, 248), which are "counterparts of attributes, laws, and patterns in the distal environment" (99). Above all, our account of these transformations must account for perceptual constancies, that is, for the fact that we can perceive a single attribute across a range of different perspectives and viewing conditions (188, 274, 324, 408, 413). Moreover, our choices of representata will be confined to "biologically relevant candidates" (248, see also 256, 261). We must avoid overly "complicated" accounts of the content of perception (272f.). Sometimes Burge seems to simply dismiss the whole question on the grounds that representational contents are "primitive" (58, 310f.), although he also says that "teleology is more primitive than representation" (341).

Nowhere does Burge put all these hints together in an answer to the constitution question. However, he does say this:

> Perceptual systems have developed so that their representational states tend to correlate with the likely causal antecedent, in the systems' formative environment, of the given proximal stimulation. (345)

Burge does not say that this is his answer to the constitution question, but he offers no other answer, and this is the closest thing I can find to an answer in his book (see also 269, 397f., 400). Call this the *probability theory* of perceptual content. According to the probability theory, a perceptual state of type *S* represents *a* as *F* if and only if it picks out *a* in some sense and, in environments of the sort that formed the perceptual system (of the animal's species, through a process of natural selection), the kind of thing that is the most probable causal antecedent of states of type *S* is *F*-things.

There would be a lot to pick on in the probability theory, if that were what Burge had in mind. What typing of perceptual states are we thinking of in speaking of states of type *S*? What does it take for a perception to "pick out" one particular rather than another? The most damaging objection to the probability theory would be that it immediately succumbs to the so-called *disjunction problem*. Suppose, for the sake of argument, that what perceptual states of type *S* really represent are *F*-things. Then according to the probability theory, *F*-things are the most probable causes of perceptual states of type *S*. But now suppose that on some occasion, in the pertinent sort of environment, a non-*F* thing, namely, the unique thing that is *D*, happens to cause a perceptual state of type *S*, so that in that instance the perception is nonveridical. Then a kind of thing that is even more likely to cause states of type *S* is the kind *objects that are either F or D*. Then, by the probability theory, contrary to hypothesis, perceptions of type *S* represent not *F*-things but *F-or-D*-things (hence, the term "disjunction problem"). So the probability theory is mistaken.

Since the probability theory is the closest thing to an answer to the constitution question to be found in Burge's book, it is important that he answer the disjunction problem. He addresses it directly in chapter 8, but what he says about it is by no means an answer. He seems to dismiss it as "largely an artifact of reductive programs, detached from explanations in perceptual psychology" (322). He considers various possible interpretations of a frog's perceptions of flies, such as that they represent undetached fly parts or temporal slices of flies, and dismisses them on the grounds that "[t]he methodology of perceptual psychology excludes such alternatives" (322), but he makes no effort to say how it does that. He says, "[i]t is unlikely that empirical considerations support attributing to a frog's perceptual system a representation as of a natural kind (like <u>fly</u>)" (323) (his underlining), but he does not tell us what sort of empirical considerations he is thinking of. He says that "[t]here is some empirical support for taking the frog's perceptual system to have a perceptual attributive <u>like moving body of such and such size and shape</u>" (323), but he does not tell us what kind of shape he is thinking of. What is the shape that a frog's perception attributes to a fly? Is it *fly-shape*? But how could the frog represent that shape if it cannot, as Burge admits, represent flies as such? What kind of shape can we say a fly has as it flies around, moving its wings and legs? How could experiments on frogs zero in on a precise characterization of the kind that the frog represents? Perhaps we could answer these questions by invoking biological teleology. We could say: The shape that the frog attributes to the fly is the shape such that evolution selected a perceptual system that would respond to that shape. But Burge does not do that and his disparaging remarks about Millikan's reductionism suggest that he would not wish to answer in that manner.

Over and over Burge hammers on the theme that if we want to understand perception we need to look at what the science of perception says. The book contains broad but shallow surveys of the sensory and perceptual powers of animals and infants (421–430, and all of chapter 10). There are many footnotes containing long lists of scientific articles the contents of which are not expounded in any detail. But only 15 out of 551 pages of text are devoted to explaining how aspects of perceptual processing actually work. In those 15 pages, Burge presents four case studies of the ways in which sensory stimulation is transformed into perception. Here are the four examples: 1) The angles at which two eyes point in focusing on a surface point provide a measure of the distance of that surface (347–350). 2) As part of the process by which the lightness of a surface is perceived, gradients between light and dark are used to detect illumination edges (351–354). Sharp illumination contours are seen as reflectance edges (produced by overlapping objects of different lightnesses, for instance) rather than as luminance edges (produced by shadows, for instance). Burge does not explain the rest of the process of perceiving lightness. 3) Slanted surfaces may be detected on the basis of a regular surface texture that "shrinks" (my word, not Burge's) along one dimension of the visual field (355–359). 4) Convex bounded regions of retinal images correspond to perceptions of foreground objects, while concave bounded regions correspond to perceptions of background (359–361).

All of these examples illustrate Burge's idea that what is detected in perception is a normal cause of the perception in the environment in which the perceptual systems evolved. If one is looking cross-eyed at a regular pattern and incorrectly superimposes the image from the left eye onto the image from the right eye, one can suffer a distance illusion, but that does not normally happen. A trompe l'oeil painting can use "shrinking" texture to create the illusion of a slanted surface, but our visual systems evolved in a world without such paintings. However, despite Burge's emphasis on the importance of perceptual constancies, and even though he later refers back to these examples in connection with perceptual constancies (413), none of these examples directly illustrates the possibility of detecting the same property across a variety of viewing conditions.

Burge construes these examples as vindicating his conception of the contents of perception. He says that "a perceptual state that attributes a given distance to a particular is produced by convergence" (365), and he writes of "perceptual states attributing a given degree of lightness, or given slant" (366). This is a stretch. Instead, one should say that what these processes are doing is taking measurements. A thermometer measures temperature. One could *say* that the thermometer represents a particular, the surrounding atmosphere, as having a certain property, the property of *having a temperature of 21 degrees Celsius*, but that would obviously be a distortion of the kind of representation the thermometer produces. It would be a distortion, because it would suggest, contrary to fact, that the thermometer employs a representation of the general property *having a temperature of 21 degrees Celsius* and somehow distinguishes the region of atmosphere that surrounds it from other regions. Similarly, the eyeball convergence system measures distance to a focal point. The lightness constancy system *detects* the presence of a reflectance edge. In other words, it provides a *yes* or *no* measure of the presence of a reflectance edge. One can *say* that these systems represent particulars as belonging to kinds, but describing them that way would be misleading. We have been given no reason to think that the eyeball convergence system represents the general kind *two feet away* or that the lightness system represents the kind *edge*. So these examples do not provide support for Burge's thesis that perceptions represent particulars as belonging to kinds. I am not suggesting that perception is nothing more than the kind of measurement exhibited by a thermometer; my point is only that Burge's four examples do not illuminate the distinctive nature of perceptual representation.

Late in the book, Burge claims, in passing, that "having perceptual constancies with respect to an attribute suffices to have a capacity to represent the attribute perceptually" (466). As far as I can see, however, no sense has been given to the phrase "perceptual constancies with respect to an attribute". If a creature is capable of perceptual shape constancy, that means it can recognize one and the same particular shape across a variety of viewing conditions. If it is capable of color constancy, then that means it can recognize one and the same color (considered as a bona fide physical property) across a variety of viewing conditions. Demonstrating constancies of these kinds does nothing to locate the boundaries around the general kinds, the groupings, that, according to Burge, are attributed in perception, but that is precisely what we need to do if we want to answer the constitution question.

Burge illustrates at some length a distinction between what he considers genuine perceptual representation and what he considers to be merely non-perceptual sensory registration (421–430, 492–518). When an animal integrates signals from several directions to locate an object in space, that is an example of perception. So if a sand scorpion detects the location of a creature under the sand on the basis of vibrations reaching its eight legs, that will be perception (420, 427). But if an animal merely orients itself according to the direction from which a signal reaches it, that is not yet perception. So if a salmon finds its way back to its natal stream by serially sampling olfactory signals and calculating an average, that is not yet perception (427, 498). This seems like an important distinction, and it is plausible that, as Burge seems to think (429, 494), a capacity for perceptual constancy of some kind is necessary for what we should call perception proper. But to say that is not yet to concede that an account of perceptual constancies could be turned into a vindication of Burge's conception of the contents of perceptions.

At one point, Burge promises to explain to us what it takes for a creature to be able to perceive something as a body (450). What he actually says in the pertinent section (454–64) is that infants and animals are visually sensitive to various features of a spatial array, such as the boundary between an object and the space it occupies, and that infants and animals can track objects as they move around without expecting that those objects will retain their basic form (462f.). Burge's assumption seems to be that it is inconceivable that perception would not involve the attribution of properties or kinds at all. He even says as much in a footnote (456n). (He also says it is *a priori* that perceptions have attributive elements (535), but by that he does not mean that he cannot be wrong.) So by process of elimination, he seems to think, one can argue that the pertinent kinds must include *body*. But of course that is no answer to my doubt whether perception involves the subsumption of particulars under kinds at all. At one point, Burge even claims that, given the right background conditions, capacities for differentiating particulars from their surroundings ("segmentation") and tracking them as they move are *sufficient* for "having the perceptual attributive body" (464), but Burge provides no argument, and I cannot think of any, to show that a creature can differentiate and track a particular object only if it can perceptually subsume that particular under a general kind *body*.

Burge says many times that a perception represents a particular as belonging to a kind, and offhand it seems clear enough what that means even if it is unclear why we should believe it. But then in various ways he makes it hard to understand what he could possibly mean by that. The unclarity comes from two directions. On one side, it is not clear what Burge thinks the relation is between perceptions and what he calls *modes of presentation*. On the other side, while Burge insists that perceptions are non-conceptual and do not have propositional content, it is quite unclear what Burge thinks the difference is between perceptions and propositional thoughts.

Burge holds that perceptions can be differentiated from one another not only on the basis of what they are about and what properties they attribute but also on the basis of *mode of presentation*, which he also calls "representational content" (385, 411). But Burge tells us little about these modes of presentation, beyond the fact that they are associated with perspectives (41). He says that the attributive element in a perception is itself "inevitably from a perspective" and is only "one of many ways of perceptually attributing whatever property, relation or kind is attributed" (539). At one point, he says that the mode of presentation "depends" on the character of the proximal stimulations (388). Authors in the semantics literature usually think of the mode of presentation associated with a sentence as just another propositional content, one that represents the way in which the content communicated appears to the speaker, as opposed to the content that the speaker communicates to a hearer, but there is no hint of such a conception in Burge. In prosecuting his case against the disjunctivist contention that there is no perceptual state in common between a veridical perception and an hallucination, one might have expected him to invoke these modes of presentation. At one point, that is what he does (385), but shortly afterward he introduces something else that he calls an "attributional kind" (389), which is a product of perceptual processing (391), and says that that is the common factor between veridical and hallucinatory perception.

On the whole, it is clear that when Burge says that perceptions attribute properties and kinds, he means to speak of properties and kinds that comprise a variety of instances. But at one point, he seems to say that in characterizing a perception as, for instance, "of being cylindrical" we are not describing the content of the perception itself but are only describing in an abstract *way* a content that is necessarily more specific (74). In that case, perhaps the mode of presentation is nothing other than the property or kind attributed.

Burge is emphatic that perceptions are non-conceptual. They do not have propositional contents; their contents are not composed of concepts (36, 104, 174n). This is puzzling, because if a perception represents a particular object *o as F*, then it seems pretty clear that the perception has the propositional content *that o is F*. The difference is supposed to be that perceptions are not bound up in inferences and do not have the sort of structure that is required for representations bound up in inference (104). But Burge does say that perceptions can be "incorporated" into propositional thought (169), and perceptual attributives may be "conceptualized" (545). "The ability to *conceive* of objects as being of certain kinds, and the ability to pick out particulars in *thought*," he writes, "may make heavy use of perceptual attributives and perception of particulars" (198; see also 207, 234, 280). But Burge tells us nothing at all about the nature of this "incorporation" or "conceptualization", and so we cannot tell that it differs from the kind of inference that would make perceptions qualify as propositional thoughts.

Near the very end of the book, in just one short paragraph, Burge broaches the idea that perceptions represent in something like the way maps represent, that is, by means of structural isomorphism (540). That only makes it harder to understand in what way perceptions attribute properties. An ordinary street map may incorporate conventional signs. A red dot with a circle around it may represent a city having a population of more than 100,000. But apart from these conventional signs, maps do not attribute general properties or kinds. Presumably Burge does not think that perceptions are composed of conventional signs. On that same page, Burge tells us that "[p]erceptual attribution is never separated out from contextual singular representation that purports to refer to and single out particulars", and "[e]very occurrence of the perceptual attributive green or square or body modifies and guides a context-dependent singular element" (540). So a property is supposed to be represented by virtue of the way in which a particular is represented; but that does take us very far in understanding how properties are attributed.

So Burge has not shown that the science of perception vindicates his conception of the contents of perception. He does not even show, or even try to show, that psychologists agree with him. (In discussions of Spelke, 438ff., and Pylyshyn, 455–6n, he has to admit that some do not.) And even if it were true that they talked his way, of perceptions as representing particulars as belonging to kinds, we would not want to join in their practice without questioning whether the empirical work actually supports what they say. Throughout its brief history, the science of psychology has swung through many fads that most of us now repudiate, including Freudian psychoanalytic psychology and Skinnerian behaviorism. Judging only by the literatures in philosophy and cognitive development, which I know better than the literature in the psychology of perception, the fad *du jour* is the practice of projecting the concepts of language-using adults onto infra-linguistic humans, non-human animals and even subpersonal modules of the brain.[3]

Here are two conditions that a theory of perceptual representation must meet. First, it must identify a sense in which perceptual representations can be correct or incorrect. Second, it must explain how the character of a perceptual representation makes a difference to the sort of beliefs that may be formed on the basis of it. Burge aims for a theory that would satisfy both of those conditions; so to support my criticism, let me point out that there is another way.[4] This other way is to model perceptual representations as points in a *multi-dimensional perceptual similarity space*. The dimensions of this space correspond to dimensions along which the actual objects and scenarios perceived may vary in perceptible ways. There may be dimensions corresponding to the various aspects of color and shape, but also dimensions corresponding to perceptible variables that less readily come to mind, such as jerkiness of motion. (The motion of a squirrel is jerkier than the motion of a cat.) Corresponding to this perceptual similarity space will be an *objective quality space*, composed of dimensions of objective variation to which the dimensions of perceptual similarity space in some sense correspond.

Correctness of perceptual representation may be understood, in these terms, as *accuracy*. When an object is perceived, the result is that a *mark* is laid down

3. That this is an error is substantiated in my book, *Words and Images: An Essay on the Origin of Ideas*, *op. cit.*

4. Here I sketch the theory developed in detail in my book *Words and Images: An Essay on the Origin of Ideas*, *op. cit.*

in perceptual similarity space. A perceptual representation may be modeled as such a mark. Marks in perceptual similarity space may be mapped into points in objective quality space in either of two ways. First a mark may be mapped into the location in objective quality space that is occupied by the object sensory contact with which actually caused the mark to be recorded. To understand the second mapping, consider that there is a distinction between marks that are recorded in the *functionally normal way* and those that are recorded in some other manner. So in addition to the actual location of the cause of the mark in objective quality space, there is a potentially distinct point in objective quality space where the cause *would have been* if the mark had been recorded in the functionally normal way. The second mapping takes a mark in perceptual similarity space into the point in objectivity quality space where its cause *would have been if the mark had been recorded in the functionally normal way*. The functionally normal way of recording a mark in perceptual similarity space will be defined in terms of biological teleology. The perceptual constancies that Burge emphasizes will be aspects of this normal way of recording marks in perceptual similarity space. A perceptual representation is accurate, in one sense, to the extent that the two points it maps into in these different ways are near to one another in objective quality space.

Further, in these terms we can understand the epistemic role of perceptions. We will not get the conclusion that perceptions provide *justifications* for beliefs, and I do not believe we should expect that. But we should expect that perceptions can *guide* beliefs to the extent that credit or blame for a belief may be laid at the feet of perception. Thus, under suitable conditions, we should be warranted in saying such things as "He missed the shot because he misperceived the distance to the basket". We can have that on the approach I am proposing inasmuch as the location of a mark in perceptual similarity space may guide the perceiver's bodily motions and conceptual responses. If a perceiver thinks in response to mark *m*, "That's a dog", and thinks in response to mark *n*, "That's a cat", then if mark *o* is closer to *m* than to *n* in perceptual similarity space, the thinker will favor the thought "That's a dog" over the thought "That's a cat" in response to *o*.

Certainly this approach faces several challenges. One main challenge will be to explain in what sense a dimension of perceptual similarity space corresponds to a dimension of objective quality space, so that if a mark has a certain location on a dimension of perceptual similarity space, then we can say that it represents an object at a corresponding location on a corresponding dimension of objective quality space. We cannot assume that distance on a dimension of perceptual similarity space will be isomorphic to distance along the corresponding dimension of objective quality space, because perceptual illusions, such as the

Müller-Lyer illusion, are presumably recorded in the biologically normal way. A second important challenge will be to say how perceptual similarity space can be enhanced through learning. (Remarkably, Burge has nothing to say about how perception may be enhanced through learning.) One possibility is that whole dimensions may be added to perceptual similarity space representing similarity to certain particular objects that serve as paradigms (e.g., the family dog).

On this approach, perceptions, modeled as marks in perceptual similarity space, do, in a sense, represent the external objects and scenarios that they record, although they do not *say* anything about them. They do not classify objects as having properties or belonging to kinds or standing in relations. They do *locate* objects and scenarios on dimensions of perceptible variation. So a perception representing an object could be interpreted as "saying" that the object lay on such and such a location in objective quality space. But though a perception may "say" this, it does not "group" the object of perception together with other objects in the way Burge thinks perceptions do.

Most of the first half of Burge's book is devoted to criticizing the views of prominent philosophers of the twentieth century. Burge focuses on the sense data theorists (Russell, Broad, etc.), Strawson, Evans, Quine and Davidson. He characterizes all of these opponents as "Individual Representationalists", but I was never quite sure what Individual Representationalism was supposed to be. He clearly defines it as the view that objective representations must be accompanied by representations of the constitutive conditions for objective representation, an assumption he calls hyper-intellectualist (13). But then he denies that all Individual Representationalists are guilty of hyper-intellectualism (116f.), and he says that even behaviorism "encourages" Individual Representationalism (141f.). Some Individual Representationalists are "individualists" in that they deny that representation depends on constitutive relations to objects outside the mind, but at least Evans and Davidson are not (106, 208, 265).

In his treatment of all of these authors, there is very little by way of internal criticism. Burge criticizes the sense data theories for "conflating" modes of presentation and objects of representation (124) but makes no attempt to pinpoint this conflation in their arguments. Strawson and Evans he accuses of conflating conditions on objective representation with conditions on having a concept of objectivity (156ff., 182). He rejects Strawson's idea that representation begins with "feature placing" on the grounds that "perception is necessarily and constitutively a relation to particulars with causal powers" (166). Evans is just wrong to think that in representing an object one must be able to conceive of its differences from other things (196) and place it in allocentric spatial relations (200). Quine is just wrong to think that a thinker must be able to think quantificational thoughts in order to represent a particular object (247). Davidson is just wrong to think that content can be identified only in the context of linguistic communication (269). All of this comes before any exposition of the content of the science of perception. Nonetheless Burge's main objection, again and again, is just that these philosophers did not know what the science of perception tells us (169, 188, 264, 272, 275).

It is not clear to me that Burge's theory of perception addresses the questions these philosophers thought they were asking. Burge thinks that Strawson, Evans, Quine and Davidson were all wrong about perception. But I am not sure that their requirements on objective representation were supposed to be requirements on perception in the first place. At points, Burge himself acknowledges that they were not (234, 430). These philosophers' theories are all theories about conceptual thought. As we have seen, Burge himself distinguishes between perception and conceptual thought. So it is not clear that there is any immediate conflict between Burge's claim that perception requires no conceptual scaffold and these authors' claims that any amount of conceptual thought depends on a whole supportive scaffold of conceptual thoughts. Burge, in his endorsement of holism for conceptual thought (151, 280f.), even seems to meet them at least halfway on the nature of conceptual thought, although he later asserts, but does not argue, that Strawson et alia were wrong about conceptual thought too (544). Probably all of these authors did neglect the possibility of the kind of perceptual representation that Burge finds in animals and babies. Since all of them are concerned to explain how conceptual representation is somehow grounded in sensation or perception, they might have recognized other options if they had countenanced full-fledged perceptual representations independent of conceptual thoughts. But criticism along those lines would look quite different from Burge's criticism, which is that none of the conceptual scaffolding that they describe is necessary to reach perceptual representation.

Two points have to made about the style and tone of this book. First, it is poorly written. Burge admonishes his readers to be patient (xii, 30), but he does not reward their patience with efficient writing. I believe it is the most repetitious book I have ever read. In places it reads like a string of aphorisms (e.g., 326–341). Second, it is imperious. Over and over Burge begins a sentence with "I think" or "I believe" and then makes no effort to back up his beliefs with argument. Opponents are regularly denigrated with words like this:

> I believe that the intuitive and theoretical considerations that support anti-individualism, both about perception and about thought, are so basic that they leave no reasonable alternative. Failure to accept the view, once presented with it, tends to derive either from being distracted from the central considerations or from misunderstanding what is being claimed. (105)

For more examples of such rhetoric, see his comments on Millikan's work quoted above, and 40, 189n, 283f., 364n, 430f., 434 and 453.

I spent much of the eighties and nineties thinking about Tyler Burge's early work on social externalism.[5] In the end it took me to conclusions that Burge was unwilling to draw.[6] I still consider that work of his to be some of the most profound philosophy written in my lifetime. So I am sorry to have written such an unrelentlingly negative review of Burge's book on perception. The sections in the later chapters in which Burge reviews the empirical literature are well worth thinking about. I do not think that Burge has successfully defended, or even clearly articulated, an interpretation of what that literature shows.

References

Gauker, Christopher 2003: "Social Externalism and Linguistic Communication". In: María-José Frápolli & Esther Romero (eds.), *Meaning, Basic Self-Knowledge and Mind: Essays on Tyler Burge*. CSLI Publications, 1–33.

— 2011: *Words and Images: An Essay on the Origin of Ideas*. Oxford: Oxford University Press.

Burge, Tyler 1979: "Individualism and the Mental". In: Peter A. French, Theodore E. Uehling, Jr., & Howard K. Wettstein (eds.), *Midwest Studies in Philosophy* 4, *Studies in Metaphysics*, University of Minnesota Press, 73–121.

5. See, especially, his "Individualism and the Mental", in Peter A. French, Theodore E. Uehling, Jr., and Howard K. Wettstein (eds.), *Midwest Studies in Philosophy* 4, *Studies in Metaphysics*. University of Minnesota Press 1979, 73–121.

6. See my "Social Externalism and Linguistic Communication", in María-José Frápolli and Esther Romero (eds.), *Meaning, Basic Self-Knowledge and Mind: Essays on Tyler Burge*. CSLI Publications 2003, 1–33, and Burge's reply in the same volume, 244–250. I am misquoted on p. 249.

Grazer Philosophische Studien
85 (2012), 325–335.

ADEQUATE FORMALIZATION AND DE MORGAN'S ARGUMENT

Georg BRUN
ETH Zürich

Summary
Lampert and Baumgartner (2010) critically discuss accounts of adequate formalization focusing on my analysis in (Brun 2004). There, I investigated three types of criteria of adequacy (matching truth conditions or inferential role, corresponding syntactical surface and systematicity) and argued that they ultimately call for a procedure of formalization. Although Lampert and Baumgartner have a point about matching truth conditions, their arguments target a truncated version of my account. They ignore all aspects of systematicity which make their counter-example unconvincing.

1. *Introduction*

Logical formalization assigns formulas to ordinary language sentences or inferences, aiming at a transparent representation of a logical form suitable for use in rigorous validity proofs. In (Brun 2004), I discussed the conceptual underpinnings of this project and reconstructed criteria of adequacy guiding accepted practice of formalization. I argued that these criteria all raise problems and ultimately call for a procedure of formalization. Lampert and Baumgartner (2010; henceforth referred to by unqualified page numbers) argue that the criteria I discussed rely on incoherent distinctions between formalization and semantical analysis, and between validity in virtue of logical form and in virtue of semantical relations among non-logical expressions ("material" or "analytical" validity). Specifically, they attack my case-study of an inference traditionally attributed to De Morgan. Here is the argument with four candidates (C1)–(C4) for formalizing the conclusion (numbered (9)–(12) on 93f.):

(P) Every horse is an animal.
(C) Every head of a horse is a head of an animal.

(P1)	$\forall x(Hx \rightarrow Jx)$	*Fx*: *x* is a head of a horse
(C1)	$\forall x(Fx \rightarrow Gx)$	*Gx* *x* is a head of an animal
(C2)	$\forall x(\exists y(Hy \wedge Ixy) \rightarrow \exists y(Jy \wedge Ixy))$	*Hx*: *x* is a horse
(C3)	$\forall x \forall y(Hy \wedge Ixy \rightarrow Jy \wedge Ixy)$	*Ixy*: *x* is a head of *y*
(C4)	$\forall x(Hx \wedge \exists y Iyx \rightarrow Jx \wedge \exists y Iyx)$	*Jx*: *x* is an animal

This reply analyses the controversy over these formalizations without elaborating on deeper running differences between Lampert and Baumgartner's and my approach to theories of formalization. Sections 2 and 3 briefly sketch how my criteria are applied to (C2) and (C3), concentrating on aspects Lampert and Baumgartner ignored and on points of disagreement. Section 4 investigates their key example (C4) and section 5 their main line of argument.

2. *Truth conditions and surface rules*

This paper focuses on formalizations in classical zero- and first-order logic: Given a sentence *S* in an ordinary language **A**, a formalization Φ of *S* in a logic **L** is an ordered pair $\langle \phi, \kappa \rangle$ with a formula ϕ of **L** and a correspondence scheme $\kappa = \{\langle \alpha_1, a_1 \rangle, \ldots, \langle \alpha_n, a_n \rangle\}$ that specifies one-to-one for each non-logical symbol α_i ($1 \leq i \leq n$) in ϕ an expression a_i of **A** (augmented by auxiliary expressions for representing argument-places of predicates).[1]

In (Brun 2004, pt. III), I reconstructed three types of adequacy criteria guiding accepted practice of formalizing. Firstly, formalizations must be correct in the sense that they do not allow validity proofs for inferences which are definitely invalid by informal standards ("*i*-valid" and "*f*-valid" will abbreviate "valid according to informal standards" and "valid according to the standards of a logical formalism"). In the context of semantical formalisms, this requirement (called "TC") can be framed in terms of truth conditions as follows: A formalization $\langle \phi, \kappa \rangle$ of a sentence *S* in **L** is correct iff for every condition *c*, for every **L**-interpretation $\langle \mathsf{D}, \mathsf{I} \rangle$ corresponding to *c* and κ, $\mathsf{I}(\phi)$ matches the truth value of *S* in *c*. An **L**-interpretation corresponding to a condition *c* and a correspondence scheme $\{\langle \alpha_1, a_1 \rangle, \ldots, \langle \alpha_n, a_n \rangle\}$ is an **L**-structure $\langle \mathsf{D}, \mathsf{I} \rangle$ with a domain D and an interpretation-function I such that $\mathsf{I}(\alpha_i)$ matches the semantic value of a_i ($1 \leq i \leq n$) in c. A second type of criteria requires some syntactical correspondence of sentences and formulas. Two examples are: the logical symbols in a formalization Φ must have a counterpart in *S;* Φ's correspondence scheme must not

1. For the sake of simplicity, I assume that we formalize sentences. Inferences—i.e. sequences of sentences $\langle$premise$_1$, ..., premise$_n$, conclusion$\rangle$—are formalized as sequences of formulas using one common correspondence scheme. (Cf. Brun 2008.)

include ordinary language expressions not occurring in *S.* Thirdly, formalizing should be systematic. Ideally, adequate formalizations should be producible by a procedure of formalization.

With respect to the first two types of criteria, I basically agree with Lampert and Baumgartner that the standard formalization (C2) and its rival (C3) fare equally well (93ff.). I will therefore not discuss the problems raised by these criteria nor comment on Lampert and Baumgartner's exposition, although the latter contains some misunderstandings. I only mention why (TC) and surface rules do not decide which of the two non-equivalent formalizations (C2) and (C3) is adequate. (TC) is not distinctive enough if materially *i*-valid inferences are involved. Deciding between (C2) and (C3) would require us to settle by informal reasoning whether (C) was true even if not all horses were animals but some heads of non-animal-horses were also heads of an animal. However, we must reject the idea of taking that decision because it informally makes no sense to assess truth values with reference to semantically impossible conditions. Surface rules threaten to classify a great deal of standard formalizations as inadequate and hence cannot be taken as strict requirements but must be interpreted liberally. Either way, they fail to decide between (C2) and (C3).

The third approach to adequacy focuses on the requirement that formalizing be systematic rather than arbitrary and ad hoc (Brun 2004, chs 12.4, 13.6). This is ignored in Lampert and Baumgartner's paper (except for (HSC), cf. below), but it plays a crucial role in addressing their criticisms.[2] Section 3 briefly outlines the missing elements.

3. *Formalizing systematically*

A strategy of formalizing analogous sentences analogously is well entrenched in philosophical practice. Philosophers generally aim at finding ways of formalizing not individual sentences but sentences of a certain type. Relying on a principle of "parity of form" (Russell 1905, 483), undisputedly adequate formalizations are used as models for less clear cases. It is, for example, standard practice to formalize "All … are …"-sentences as instances of $\forall x(\phi x \rightarrow \psi x)$ even if the extension of the subject-term is empty. The problem is that this presupposes an account of parity of form.

Standard practice also uses a technique of formalizing step-by-step. While formalizing analogously is systematic in linking adequate formalizations of different sentences, the step-by-step method is systematic in relating different

2. Those omissions may be the result of their narrower notion of criteria: "criteria are no criteria if they are open for discussion" (Baumgartner/Lampert 2008, 113).

formalizations of the same sentence. In example (C), we may start with (C1), formalize "x is a head of a horse" and "x is a head of an animal" as $\exists y(Hy \wedge Ixy)$ and $\exists y(Jy \wedge Ixy)$, substitute the results for Fx and Gx in (C1), and (C2) results. Two problems are that the vague instruction of formalizing step-by-step produces inadequate results, and that it calls for explicitly formalizing not only sentences but also their parts and the way they make up the original sentence. There is, however, sound motivation for the step-by-step strategy in arguments of compositionality, which combine the two senses of "systematicity" mentioned. Such arguments call for theories capable of accounting for the productivity of ordinary languages, dealing not only with the specific examples that guided theory development but with an unlimited number of sentences and inferences.

The common theme behind surface rules and the principles of analogous and step-by-step formalization is that they all become more convincing the more we can spell out in a precise and general manner how sentences are to be formalized based on some syntactic description. If this analysis of principles guiding standard practice is correct, it reveals an ideal not typically thought to be present: formalizations should be the product of an effective procedure, an algorithm which accounts for the systematic, compositional, nature of ordinary language. Without such a procedure, the criteria discussed remain problematic and fail to provide a sufficient condition for adequate formalization (cf. Brun 2004, chs 12.4, 14).

I agree with Lampert and Baumgartner that syntactic descriptions of naive grammar are not a suitable basis for specifying satisfactory criteria and formalization procedures (the "misleading form thesis", 80, 88). But this is neither sufficient reason for ignoring considerations of systematicity nor does it follow that rigorous procedures of formalization are impossible (80), only that they must rely on a far more sophisticated analysis of ordinary language. This in turn calls for empirical investigations of language structures in relation to logical formalisms as pioneered by, for example, Montague and Davidson. This research and other contributions drawing on Chomsky's theory of language also suggest that formalisms other than first-order logic, such as generalized quantifier theory, may be more promising. That we still lack comprehensive formalization procedures does not justify dismissing this research as hopeless.

Furthermore, the argument that criteria of adequacy are conceptually prior to formalization procedures is not as conclusive as Lampert and Baumgartner assume (80f.). Of course, not just any procedure of formalizing is acceptable, but only one with results we are ready to accept as adequate. But if (TC) and surface rules do not discriminate between alternative formalizations, such as (C2) and (C3), I fail to see why we should not appeal to considerations of systematicity embodied in a procedure of formalization as additional criteria of adequacy. And if we admit that informal judgements about truth conditions come in degrees of firmness, we have reason to welcome that systematicity sometimes overwrites assessments of (TC)-correctness.

Even without satisfactory formalization procedures, we can argue about the adequacy of formalizations by pointing out that they could (not) plausibly be the product of a systematic procedure. Specifically, I suggested a criterion of hierarchical structure (called "HSC"; Brun 2004, ch. 13; cf. 90), which incorporates core aspects of the step-by-step method and compositionality: at least one of two non-equivalent formalizations of the same sentence must be inadequate if neither is more specific than the other and there is not a third adequate formalization more specific than both. A formalization $\langle \phi, \kappa \rangle$ is more specific than a formalization $\langle \psi, \kappa \rangle$ iff ϕ can be generated from ψ by substitutions $[\alpha/\beta]$ such that either (i) α is a sentence-letter occurring in ψ and β is a formula containing at least one sentential connective or a predicate-letter, or (ii) α is an n-place predicate-letter occurring in ψ and β is an open formula with n free variables containing at least one sentential connective, quantifier or predicate-letter with more than n places. (HSC) specifies a necessary, negative condition of adequacy, which can only be applied if we have two rival formalizations, but does not determine which one is inadequate. The following discussion illustrates its use.

4. *A new formalization of De Morgan's argument?*

According to (HSC), (C3) is not an adequate formalization of (C) if (C1) is adequate (cf. Brun 2004, ch. 13.6). (C1) and (C3) are not equivalent and no substitution for predicate-letters will turn (C1) into (C3), (C3) into (C1) or both into the same formula. In specifications of (C3), two $\forall$s range over the rest of the formula, but in specifications of (C1) only one $\forall$ does. (C2), on the other hand, is more specific than (C1). It results from applying $[Fx/\exists y(Hy \wedge Ixy), Gx/\exists y(Jy \wedge Ixy)]$ to (C1). Hence its adequacy is compatible with the adequacy of (C1). Formalizing step-by-step, we can develop (C2) but not (C3) from (C1).

Lampert and Baumgartner admit that within my tenets this argumentation in favour of (C2) is correct (94f.). However, they attack my discussion of De Morgan's example by introducing (C4), which is equivalent to (C3) and results from (C1) by substituting $[Fx/Hx \wedge \exists yIyx, Gx/Jx \wedge \exists yIyx]$. Their further argumentation depends on the claim that (C4) is an adequate formalization of (C) according to my account. I agree that (C4) and (C2) fare equally well with respect to (TC) and surface rules (for reasons as given for (C2) and (C3) in sect. 2). I also agree that (HSC) rules out that (C2) and (C4) are both adequate without telling us which one is inadequate. However, Lampert and Baumgartner's diagram is incorrect (95; cf. Brun 2004, 353). It places (C3) and (C4) on the same specification path, but (C3) is unrelated to (C1) and (C4). Lumping (C3) together with (C4) also leads to the false claim that "[C2] and [C4]/[C3] perform equally well on all counts" (96). If this were true, the inadequacy of (C3) would imply that (C2) and (C4) are inadequate as well.

(C4) is inadequate for other reasons. As a colloquial verbalization, we read "Every horse with a head is an animal with a head" (105). This is a different statement than (C). Apparently, the substitution used to get (C4) from (C1) turns heads of horses into horses with heads. Consider the *i*-invalid inference from (C) to (D):

(D) Every head of a horse is a horse.

Using the same correspondence scheme as above, (D) may be formalized as (D1) in analogy to (C1) and if we apply to (D1) the substitution that generated (C4), we get the inadequate (D2):

(D1) $\forall x(Fx \rightarrow Hx)$
(D2) $\forall x(Hx \wedge \exists yIyx \rightarrow Hx)$

If (D2) were adequate, it could be used to prove the *i*-invalid inference from (C) to (D). Furthermore, (D2) violates (TC) since it is a logical truth, whereas informally (D) is false. To block this result while defending (C4), Lampert and Baumgartner must either claim that (D1) is inadequate or explain why specifying Fx by $Hx \wedge \exists yIyx$ should turn an adequate formalization into an inadequate one if applied to (D1) but not if applied to (C1). Both reactions are implausible for reasons of systematicity. (C1) and (D1) perfectly illustrate the strategy of analogous formalizations. And there is no convincing reason which blocks applying the same specification to Fx in both (C1) and (D1). It seems implausible that a general procedure of formalizing could account for compositionality, yet treat (C) and (D) as fundamentally different. Advocating (C4) as adequate rather is a prototypic example of an ad hoc formalization. Against this, Lampert and Baumgartner might argue thus: the fact that (C4) is (TC)-correct, while (D2) is not, suffices to show that a systematic procedure of formalizing must treat (C) and (D) differently. This is not a viable objection to the claim that (C4) is inadequate according to my account because it incorrectly presupposes that considerations of systematicity cannot discriminate between (TC)-correct formalizations.

Finally, we must check whether the arguments against (C4) cannot be levelled against (C2) as well. Applying to (D1) the substitution that generates (C2) from (C1) yields:

(D3) $\forall x(\exists y(Hy \wedge Ixy) \rightarrow Hx)$

In contrast to (D2), (C2) and (D3) do not sanction the *i*-invalid inference from (C) to (D). (D3) can also be defended as (TC)-correct. First, we assume that there is at least one head of a horse. Then (D) is informally false since no head of a horse is a horse, and all corresponding interpretations $\langle \mathsf{D}, \mathsf{I} \rangle$ must meet two conditions: (i) there are $e_1, e_2 \in \mathsf{D}$ such that $\langle e_1, e_2 \rangle \in \mathsf{I}(I)$ and $e_2 \in \mathsf{I}(H)$; (ii) there are no $\xi, \zeta \in \mathsf{D}$ such that $\langle \xi, \zeta \rangle \in \mathsf{I}(I)$ and $\xi \in \mathsf{I}(H)$. In such interpretations, $\mathsf{I}(\exists y(Hy \wedge Ie_1y) \rightarrow He_1)$ is false and hence $\mathsf{I}(\forall x(\exists y(Hy \wedge Ixy) \rightarrow Hx))$ is false and (D3) has the same truth value as (D). Second, we assume that no heads of horses exist. Then in all corresponding interpretations there are no objects $\xi, \zeta \in \mathsf{D}$ such that $\langle \xi, \zeta \rangle \in \mathsf{I}(I)$ and $\zeta \in \mathsf{I}(H)$. Hence, $\mathsf{I}(H\zeta \wedge I\xi\zeta)$ is false for all ξ and ζ, $\mathsf{I}(\exists y(Hy \wedge I\xi y))$ is false for all ξ, and $\mathsf{I}(\forall x(\exists y(Hy \wedge Ixy) \rightarrow Hx))$ is true. This means that (D3) is (TC)-correct if we are ready to informally treat (D) as vacuously true if no heads of horses exist. This is just another instance of the standard practice which treats "All … are …"-sentences as instances of $\forall x(\phi x \rightarrow \psi x)$ and therefore as vacuously true if the subject term's extension is empty. Arguing against this practice would be of no help in defending (D2) or (C4) since they rely on the same assumption about formalizing "All … are …"-sentences.

I conclude that the claim "No tenet of Brun's theory provides any reason to give preference to [C2] or [C3]/[C4]" (96) is unfounded.

5. *Lampert and Baumgartner's main argument and "ways out"*

We are now ready to analyse Lampert and Baumgartner's main line of argument and their five "conceivable ways out". Three of them will not be discussed extensively. I agree that giving up (HSC) (99f.) as well as identifying logical form with a sequence of (non-)logical expressions in ordinary language (100ff.) do not permit a plausible account of formalization (Brun 2004, ch. 13.5.2, 267). And their (2008) "Tractarian"-programme calls for a discussion which is outside the scope of this paper. It suggests an alternative to the traditional practice of formalization and does not aim at evaluating arguments but at exhibiting their semantic structure based on the assumption that informal reasoning is immune to revision in light of logical theory.

The fourth "way out" (104f.) proposes to ignore all semantical dependencies between expressions occurring in the sentence *S* when applying (TC). This changes the nature of (TC) considerably. If we pretend that the expressions in *S* do not have the meaning they have, we no longer rely on an informal judgement about whether the given sentence S would be true under certain conditions. It is not clear to me what exactly we are being asked to do if we are to assess the truth conditions of (C) under the presumption that its predicates are semantically independent.[3] One interpretation is that ignoring all semantical dependencies between semantically dependent expressions amounts to ignoring their meaning. This boils down to applying (TC) to ordinary-language schemes instead of sentences; that is, to "Every A of a B is an A of a C" instead of (C). However, this cannot be what Lampert and Baumgartner have in mind since it is incompatible with the misleading form thesis they subscribe to and once more identifies logical forms with sequences of (non-)logical expressions. An alternative interpretation is that we are being asked to treat (C) as if it had the same meaning as some other sentence with semantically independent predicates, say "Every son of a botanist is a son of a communist". Since (TC) can be applied to this sentence without problems, we appeal to analogy and conclude that (C2) rather than (C3) is a correct formalization of (C). This reading is presumably also inacceptable to Lampert and Baumgartner who discount considerations of analogy. But it fits well into my account since it underlines that (TC) needs to be supplemented with criteria relating to the requirement of formalizing systematically.

My starting point for analysing Lampert and Baumgartner's third "way out" (102f.) and main line of argument is what I take to be the standard definition of formal validity in first-order logic (cf. Brun 2004, ch. 1.2–4, 343f.):

(L) An inference *I* is formally *i*-valid relative to first-order logic iff *I* has at least one adequate formalization in first-order logic that is *f*-valid.

Two points are crucial. First, (L) defines a technical term, "formally *i*-valid relative to first-order logic" (for short, "first-order *i*-valid"), designed to single out a subclass of *i*-valid inferences; those which are *i*-valid in virtue of their first-order logical form. (L) does not report but establishes a distinction between first-order *i*-valid inferences and inferences which are *i*-valid for some other reason, for example, in virtue of semantic dependencies between non-logical expressions. Second, (L) calls for a theory of adequate formalization. As I have argued, the criteria of adequacy discussed—all three types—are problematic and do not constitute a sufficient, but only some necessary conditions of adequacy. Consequently, we cannot expect them to yield, together with (L), a sufficient condition for first-order *i*-validity.

Lampert and Baumgartner (89, 92) use two conditionals in place of (L):

---

3. Judgements of semantic independence seem to be tricky. Why should, e.g., "*x* has four single-toed hooves" be any more semantically independent of "*x* is an animal" than "*x* is a horse" (cf. 106)?

(IFVP) If there exists at least one correct and surface faithful *f*-valid first-order formalization of an argument *S*, *S* is *formally i*-valid relative to first-order logic.

(VP) If an argument *S* is formally *i*-valid relative to first-order logic, then there exists at least one adequate *f*-valid first-order formalization of *S*.

The two conditionals do not easily recombine into a biconditional and we no longer have a definition of first-order *i*-validity. Rather, we are presented with two postulates stipulating different conditions for formalizations to be first-order *i*-valid. According to (IFVP), it suffices if there is a formalization which is correct and satisfies surface rules (and is *f*-valid in first-order logic), whereas "adequate" in (VP) requires compliance with additional criteria, such as (HSC).[4]

However, we should not, and I did not, adopt (IFVP) and (VP) instead of (L) because (IFVP) excludes important aspects of adequacy. Via (IFVP), even clearly inadequate formalizations—violating (HSC) or other considerations of systematicity—establish an inference's first-order *i*-validity. (VP) then requires that a theory of formalization provide an adequate formalization for such an inference. (L), on the other hand, excludes such a situation since it defines first-order *i*-validity in terms of *adequate* formalization. It is therefore not surprising that (IFVP) and (VP) allow for counterexamples not affecting (L). (C4) is a case in point. As I argued above, it is not an adequate formalization of (C) and thus it cannot, according to (L), establish the *i*-validity of any inference involving (C). Hence, if we adhere to (L) instead of (IFVP) and (VP), Lampert and Baumgartner's main line of argument fails (because we cannot accept their (P1) on p. 98).

Lampert and Baumgartner present two objections to replacing (IFVP) with a stronger criterion. The first is peculiar to their proposal of resorting to a principle (labelled "IFVP_{adq}") which employs not the notion of adequacy I developed but a watered-down version that adds just (HSC) to the antecedent of (IFVP) excluding all other considerations of systematicity. On this basis, they claim that (C2) and (C4) are equally adequate, yet cannot both be adequate as they violate (HSC). This brings us back to the discussion of (C4) in section 4, where my criteria of adequacy deemed (C4) inadequate and hence did better than providing "no indication whatsoever" (103) about which of the two formalizations, (C2) or (C4), is adequate. By presupposing that their (IFVP_{adq}) incorporates my adequacy criteria (102), Lampert and Baumgartner's critique misses

4. Further complications arise because the two existence-claims in (IFVP) and (VP) are independently stated yet related by an implicit reference to other formalizations in (VP) if adequacy includes (HSC).

its target. Their objection does not apply to (L) if "adequate" is used as I have explained it.

Secondly, Lampert and Baumgartner raise an "epistemic" problem relevant to any account which includes (HSC):[5] Given an adequate formalization Φ of an inference *I*, how can we ensure there is no rival formalization equally adequate in all respects except that the two formalizations violate (HSC)? If we cannot guarantee that no such formalization exists, we cannot guarantee Φ's adequacy and hence not conclusively establish *I*'s first-order *i*-validity via (L). The answer depends on what approach to formalizing one adopts, specifically on the means of determining formalizations and assessing their adequacy. If we adhere to the traditional practice, we must rely on experience and trial and error to find formalizations, which must then pass criteria of adequacy. This makes it impossible to guarantee that nobody will find new ways of formalizing which will confront us with two otherwise adequate formalizations violating (HSC). Should that happen, we may come to the conclusion that the two formalizations represent two readings of an ambiguous sentence. If this cannot be made plausible, (HSC) forces us to revise other criteria of adequate formalization. The situation is different should we rely on a procedure of formalizing that incorporates our criteria of adequacy. A formalization will then be adequate only if it is the result of applying the procedure (irrespective of whether the procedure was in fact applied). Since any acceptable procedure of formalizing needs to respect (HSC), there is simply no danger of encountering rival formalizations violating (HSC), unless we are dealing with an ambiguous sentence. Consequently, the fact that the traditional practice of formalizing has problems conclusively establishing adequacy provides reason to promote the development of procedures of formalization. As argued in section 3, they are needed in any case if the traditional practice is to be placed on solid ground.

Acknowledgements

I would like to thank Michael Baumgartner and Timm Lampert for lively debates, and Christoph Baumberger, Jaroslav Peregrin and Vladimír Svoboda for helpful comments.

5. It also affects their proposal of stipulating semantical independence, since that does not make (HSC) redundant.

References

Baumgartner, Michael & Lampert, Timm 2008: "Adequate Formalization". *Synthese* 164, 93–115.

Brun, Georg. 2004: *Die richtige Formel. Philosophische Probleme der logischen Formalisierung* (2nd ed.). Frankfurt a. M.: Ontos.

— 2008: "Formalization and the Objects of Logic". *Erkenntnis* 69, 1–30.

Lampert, Timm & Baumgartner, Michael 2010: "The Problem of Validity Proofs". *Grazer Philosophische Studien* 80, 79–109.

Russell, Bertrand 1905: "On Denoting". *Mind* 14, 479–493.

Grazer Philosophische Studien
85 (2012), 337–341.

Wolfgang KÜNNE, *Die Philosophische Logik Gottlob Freges. Ein Kommentar.* Frankfurt a. Main: Klostermann. Rote Reihe, Band 30. 2010. 840 Seiten. ISBN 978-3-465-04062-0.

Der Philosoph und Mathematiker Gottlob *Frege* (1848–1925) ist der Begründer der modernen Logik. Dies gilt in doppeltem Sinne: Frege hat die *formale* – auch „mathematisch" genannte – Logik von der Aussagen- bis hin zur Prädikatenlogik der zweiten Stufe als einen interpretierten Kalkül entwickelt, *und* er hat die Bausteine der formalen Logik philosophisch erklärt. Gegenstand des hier rezensierten Buches ist das zweite dieser beiden Unternehmen, der Autor nennt es Gottlob Freges „Philosophische Logik"; eine solche ist, wie er zu Recht bemerkt, „eine Philosophie der Logik" (20), und wenn Frege selbst die Disziplinbezeichnung „Logik" benutzt, dann meint er damit zumeist einen solchen „philosophischen Traktat" (19). So kommt es, dass sich philosophisch-logische Ausführungen in sehr vielen von Freges Schriften finden; konzentriert sind sie in den hier kommentierten Texten, insbesondere aber auch in Freges Abhandlungen *Funktion und Begriff* (1891), *Über Sinn und Bedeutung* (1892) und *Über Begriff und Gegenstand* (1892). Diese werden in Künnes Kommentar zwar nicht explizit thematisiert, aber doch an allen erforderlichen Stellen herangezogen und minutiös erörtert. Die vom Titel des Buches geweckte Erwartung, in Freges philosophische Logik *als Ganze* eingeführt und über das corpus ihrer Problemstellungen *als Ganzes* unterrichtet und aufgeklärt zu werden, wird somit uneingeschränkt erfüllt.

Insgesamt fünf Texte werden von Künne kommentiert: das *Vorwort* (1893) zu Freges Hauptwerk *Grundgesetze der Arithmetik,* drei von Frege selbst veröffentlichte und als *Logische Untersuchungen* bezeichnete Abhandlungen aus den Jahren 1918–1923 sowie ein in seinem Nachlass publiziertes Fragment unter dem Titel „Logische Allgemeinheit", das nicht vor 1923 entstanden ist und als vierte logische Untersuchung gelten kann. – Die fünf Schriften sind in Teil II. des Buches, „Texte", in historisch-kritischer Edition enthalten. Ihre Herausgaben unterscheiden sich signifikant: Während das Vorwort vom Herausgeber mit zahlreichen Anmerkungen – vornehmlich philologisch-historischer Natur – versehen ist, sind die anderen vier Texte nahezu frei von Ergänzungen. Dieser Unterschied ist eine Konsequenz von Künnes Vorgehen in Teil III., „Kommentar": Während die vier logischen Untersuchungen hier Seite für Seite erläutert werden, stellt das Kapitel 1 unter dem Titel „Das *Vorwort* im Kontext" jedenfalls in Bezug auf dessen Seiten V–XIV Einsichten aus anderen Schriften Freges bereit, und zwar Einsichten, die man braucht, um diese Seiten zu verstehen. Hier wird also nicht Absatz für Absatz kommentiert, und das macht erläuternde Anmerkungen im editorischen Teil des Buches erforderlich.

Die kritische Edition des Vorworts ist zweifellos ein Gewinn für die Fregephilologie, und auch die Wiedergaben der logischen Untersuchungen verhelfen zur genauest möglichen Kenntnis der originalen Texte Freges. Andererseits: Von dem großen Gewinn, den der Leser aus Künnes Kommentaren ziehen kann, geht nichts verloren, wenn er – zur Erleichterung des Studiums – die schon vorhandenen Herausgaben der *Untersuchungen* (H. Hermes et al., G. Patzig) neben Künnes Buch legt, um die jeweiligen Zuordnungen mühelos zu haben.

Das Buch ist gegliedert in die drei Teile: „I. Einleitung", „II. Texte" und „III. Kommentar". Hinzu treten das Vorwort und ein Anhang „Frege über metalogische Fragen (und seine Nähe zu Bolzano)". Entsprechend den edierten Texten enthält Teil III. fünf Kapitel, von denen jedes durch seine eigene Gliederung auch eine solche der behandelten Schrift vorschlägt. Letztere liegt dann der Kommentierung zugrunde. – Diese Gliederungen erleichtern die Orientierung im Text, sie sind insgesamt plausibel, Alternativen zu den gewählten Überschriften sind jedoch erwägenswert: So schiene mir der Abschnitt II.4 des Vorworts (vgl. 165) besser betitelt mit „Gesetze des Wahrseins und des Fürwahrhaltens", und Freges vorausschauender Hinweis auf möglichen „Streit [...] um mein Grundgesetz der Werthverläufe" (Vorwort VII) geht unter der Überschrift I.1 „Das Euklidische Ideal" etwas unter. Diese Alternativen sind jedoch nur solche in der überschriftsmäßigen Beleuchtung von Künnes Abschnitten. Ich stimme den Gliederungen selbst gern zu.

Das besondere Verdienst der Einleitung ist es, eine Fülle biographischer Mitteilungen zu machen, die weniger leicht sonst nur der monumentalen Biographie *Gottlob Frege. Leben – Werk – Zeit* von L. *Kreiser* entnommen werden können (vgl. 5). Hinzu tritt eine – auch systematisch orientierte – Einordnung der logischen Untersuchungen in „Freges Leben und Werk" (15). Historisch sehr aufschlussreich ist auch Künnes Darstellung von Freges Gesprächen mit Ludwig *Wittgenstein* aus den Jahren 1911–1920.

Der Anhang thematisiert Freges Überlegungen zu drei metalogischen Fragen, die man an ein Axiomensystem stellen kann: (i) die Vollständigkeitsfrage, (ii) die Frage nach seiner Widerspruchsfreiheit und (iii) die Frage, ob die einzelnen Axiome voneinander unabhängig sind, ob also das System „frei [ist] von Abundanz" (760). Künnes Ausführungen zeigen: Die jeweilige Unterscheidung zwischen den Definitionen dieser metalogischen Eigenschaften einerseits und den Beweisen für ihr (Nicht-)Vorliegen andererseits ist bei Frege angelegt, aber nicht voll entwickelt. Zu Recht lesen wir, „dass Frege die Idee eines Vollständigkeits*beweises* noch fremd ist" (761). Der Sache nach hat dies seinen Grund darin, dass diese Idee die strikte Unterscheidung zwischen der Syntax und der Semantik einer formalen Sprache verlangt, und diese Unterscheidung wurde im vollen Sinne erst von Rudolf *Carnap* auf den Weg gebracht. – Suchen wir bei Frege Antworten auf die oben genannten Fragen, so finden wir mit Künnes Hilfe Folgendes: Die *Vollständigkeit* untersucht Frege eher experimentell, indem er sich, wenn es etwa um die Logik und ein logisches Theorem geht, um einen Beweis desselben aus den Axiomen der Logik bemüht. Gelingt dies immer wieder, so hat man ein Indiz für Vollständigkeit; gelingt es nicht immer, so bleibt die Frage offen. – Die *Widerspruchsfreiheit* eines Systems folgt für Frege schon aus seiner traditionellen Auffassung von Axiomen: Diese drücken wahre Gedanken aus, und ein System (eine Menge) wahrer Gedanken kann nicht widersprüchlich sein. - Der logikgeschichtlich interessanteste Punkt ist Freges Behandlung der *Unabhängigkeits*frage. Hier knüpft Frege nämlich an David *Hilberts* Überlegungen aus dessen *Grundlagen der Geometrie* an und fragt, „ob man nicht von dem Hilbertschen Ergebnisse aus [sc.: Hilberts Beweis dafür, dass das Parallelen-Axiom der Euklidischen Geometrie von deren übrigen Axiomen unabhängig ist] zum Nachweise der Unabhängigkeit der eigentlichen Axiome gelangen kann" (762). Freges Antwort ist positiv, wie Künne mit dem von ihm „Meta" genannten Fregeschen „Grundgesetz" (764) zeigt. Dieses besagt in der Sache: Wenn es möglich ist, die Sprache, in der die Axiome $A_1, \ldots, A_n$ eines Systems und ein weiteres dazu gehöriges Axiom B formuliert sind, so umzuinterpretieren, dass dabei die Axiome $A_1, \ldots, A_n$ wahr bleiben, B aber falsch wird, dann ist das Letztere unabhängig von den Ersteren. Das Gesetz, so Künne zu Recht, „erinnert sehr an Bolzanos notwendige und hinreichende Bedingung dafür, dass zwischen ‚Sätzen an sich' *A, B, C,*... einerseits und *M, N, O,*... andererseits das Verhältnis der (wechselseitigen) Unabhängigkeit besteht" (765). – Ich ergänze: Mindestens so sehr erinnert es an den Grundgedanken der Hilbertschen Unabhängigkeitsbeweise; und mit der Akzeptanz der in diesem Gedanken enthaltenen Idee der *Uminterpretierbarkeit der deskriptiven, also nicht-logischen, Konstanten* einer formalen Sprache springt Frege weit über seinen Schatten hinaus. Er springt, um im Bilde zu bleiben, hier – aber auch nur hier – in den Schatten Hilberts.

Für die weitere Besprechung greife ich nun aus den Kapiteln 1–5 des Hauptteils einige signifikante Punkte heraus.

I) „Was ist eine Begriffsschrift [BS]?" (166ff.): Diese Frage erörtert Künne sehr umsichtig und erhellend in 1-§ 1. Die Antwort ist (vgl. 179): Eine BS ist eine formalisierte und interpretierte Schriftsprache einer „strengen Wissenschaft" wie der Arithmetik. Beispielsweise Freges „Grundgesetze der Anzahl" (GGA I, 70ff.) sind in einer arithmetischen BS formuliert. Eine *Formalisierung* haben wir hier insofern, als die Ausdrücke der

Gesetze keinerlei syntaktische, lexikalische oder kontextuelle Ambiguitäten enthalten; *interpretiert* sind diese Ausdrücke insofern, als die in ihnen vorkommenden arithmetischen Terme eine feste (arithmetische) Bedeutung haben, und zwar aufgrund der von Frege für sie gegebenen logikbasierten Definitionen. Eine BS ist eine *Begriffs-* und keine *Wort(Laut)-*schrift, weil ihre Zeichen *Begriffe* – natürlich auch: Gedanken – ausdrücken können (vgl. 173). – Eine so verstandene BS enthält stets einen Kern, der durch das Vokabular ihrer logischen Grundzeichen festgelegt ist; solche Grundzeichen sind bei Frege beispielsweise der Bedingungsstrich und „die Höhlung". Die „Formelsprache des reinen Denkens" ist ein solcher Kern, und im Titel des Buches aus dem Jahre 1879 benutzt Frege das Wort „Begriffsschrift" *auch* für diesen Kern. – Die für das Verständnis einer Wissenschaftssprache essentielle Unterscheidung zwischen logischen und deskriptiven Konstanten ist also bei Frege angelegt. – Der Untertitel zu den GGA, nämlich: „Begriffsschriftlich abgeleitet", bringt einen weiteren substantiellen Gesichtspunkt ins Spiel: Zu einer BS gehören Regeln des logischen Schließens; erst diese machen Beweise arithmetischer Theoreme aus logischen Grundgesetzen möglich. Hier stellt sich die wichtige und interessante Frage: Welche Sätze sind denn eigentlich die von Frege hergeleiteten Grundgesetze *der Arithmetik*? Sicher *nicht* sind es diejenigen in der „Zusammenstellung der Grundgesetze" des §en 47 der GGA. Künnes Antwort lautet: „Arithmetische Axiome sollen aus logischen Grundgesetzen als Theoreme abgeleitet werden" (166). In systematischer Hinsicht ist das sicherlich die richtige Antwort. Als Beschreibung dessen, was Frege tatsächlich unternimmt, halte ich sie jedoch für fraglich. Sie präsupponiert nämlich, dass Freges logizistisches Programm einer explizit axiomatisierten Arithmetik gilt, wie wir sie z. B. in der Peanoarithmetik haben. Auch in Betracht der „Beweise der Grundgesetze der Anzahl" (GGA I, 70ff.) sehe ich das jedoch nicht. Frege hatte ein Gesamtcorpus logischer Wahrheiten vor Augen, von welchem die Arithmetik als „nur weiter entwickelte Logik" (Vorwort GGA, VII) ein Teil ist. Dieses corpus wollte er – modern gesprochen – axiomatisieren, und zwar so, dass die Axiome *logische* Grundgesetze sind. Diese glaubte er dann in den bekannten sechs Sätzen aus § 47 GGA gefunden zu haben.

II) *Der Begriff* Pferd *ist kein Begriff*. „Eine echte Aporie?" (227): Sehr überzeugend unterscheidet Künne in 1-§ 2.5 zwischen der *Denotation* und der *Signifikation* eines Begriffs durch einen linguistischen Term. Damit gelingt es ihm, die oben genannte scheinbare Paradoxie aufzulösen (vgl. 233). Denn diese resultiert aus Freges Prinzip, dass gesättigte Ausdrücke nicht ungesättigte Entitäten, wie etwa Begriffe, bedeuten können. In dieser Formulierung ist das Wort „bedeuten" äquivok, es kann sowohl stehen für „denotieren" als auch für „signifizieren". Künne zeigt, dass Freges Prinzip nur bei der zweiten Lesart wahr ist, bei der ersten ist es also falsch. Der Ausdruck *der Begriff* Pferd *denotiert* – nicht *signifiziert* – den Begriff *Pferd*; er kann deshalb in dem *wahren* Satz „Der Begriff *Pferd* ist ein Begriff" als grammatisches Subjekt fungieren.

III) Unter dem Titel „Logische Gesetze als Gesetze des Wahrseins" (377, 2-§ 1) gibt Künne eine sehr erhellende Erörterung dessen, was man Freges *universalistische Wahrheitskonzeption* nennen kann (er selbst sagt „Logik-Konzeption", 380): Es gibt *ein* Universum, und jede Wahrheit, also jeder wahre Gedanke im Sinne Freges, gilt dort überall, immer und für jeden (Fregeschen) Gegenstand. Die Wissenschaft *Logik* ist dann insofern universalistisch, als dies a fortiori auf alle logischen Wahrheiten bzw. logischen Gesetze zutrifft. Diese unterscheiden sich darin nicht von empirischen Wahrheiten bzw. Gesetzen wie *Alle Menschen sind sterblich*, was Künne selbst zu Recht feststellt (380). – Mit Frege liegt es Künne nun daran, die so verstandenen logischen Gesetze als Sätze in einer *Begriffsschrift* zu formulieren. Ganz unproblematisch ist das etwa für die Reflexivität der Identitätsrelation. Wir schreiben $\forall x(x=x)$ (379), und quantifizieren damit über alle (Fregeschen) Gegenstände des *einen* Universums. – Wie aber steht es beispielsweise mit dem Gesetz der doppelten Verneinung *Duplex negatio affirmat* (vgl. 378)? Im Sinne Freges formuliert Künne dieses durch $\forall p(\neg\neg p \rightarrow p)$ und zeigt dann sehr erhellend die Unbedenklichkeit dieser Formulierung: Der Buchstabe „p" ist hier eine Variable, deren Bereich das Universum ist, und der Satz ist wahr genau dann, wenn „der Werth der Function[... $\neg\neg p \rightarrow p$] für jedes Argument [aus diesem Bereich] das Wahre ist“ (GGA I, 12), – und das ist der Fall. – Diese Deutung der logischen Gesetze als solcher des Wahrseins kann man auf den schönen Gedanken gründen, dass *die Wahrheit selbst* Gesetze erlässt und dass eben dieser Umstand diese zu *logischen* Gesetzen macht. Der Genitiv *des Wahrseins* ist dabei als ein *subjectivus* verstanden. – Eine andere Deutung ergibt dessen Lesart als eines *objectivus:* Die logischen Gesetze beziehen sich dann auf das Wahrsein als das, wovon sie gelten. Sie gelten im Reich der wahren Gedanken, z. B.: Wenn der Gedanke, dass $\neg\neg p$, dort wohnt, dann notwendigerweise auch der Gedanke, dass p. – Die Universalität ergibt sich dann daraus, dass ein logisches Gesetz für jeden wahren Gedanken gilt, unabhängig davon, auf welchen Ausschnitt des *einen* Universums dieser sich bezieht. Eine solche Deutung ergibt sich aus den – von Künne allerdings kritisierten – Überlegungen auf Seite 378f.

IV) Schemata von Aussagen über das Wahrsein und von Komposita aus solchen Aussagen: In 2-§ 2.3 thematisiert Künne die Redundanz von Wahrheitszuschreibungen. Zu Recht sieht er bei Frege die Allgemeingültigkeit folgender Schemata angelegt (vgl. 411f.):

(1) Wenn Ω(A), dann Ω(es ist wahr, dass A).
(2) Der Gedanke, dass A = der Gedanke, dass *es wahr ist, dass* A.
(3) Es ist wahr, dass A, genau dann, wenn A.

Offensichtlich macht jedes dieser Schemata Gebrauch von einer Substituierbarkeit *salva X* von *A* durch *es ist wahr, dass A,* wobei *X* eine Invariante der jeweiligen Substitutionstransformation ist. Schema (2) macht X explizit: Invariant bei dieser Gedankenkennzeichnung ist der gekennzeichnete Gedanke selbst. (2) ist ein Schema für eine bestimmte Klasse von Identitätsaussagen betreffend Gedanken. Künne hält (2) für logisch stärker („anspruchsvoller“, 412) als (3): Offensichtlich impliziert (2) das Schema (3), nicht aber umgekehrt. – Letztere Behauptung Künnes ist jedoch ergänzungsbedürftig, und die Ergänzung betrifft das Bikonditional in (3): Diese „Äquivalenzthese“ (412) besagt ja nichts über die logische Stärke des *genau dann, wenn*, diese aber variiert von Logik zu Logik. Das aussagenlogische Bikonditional etwa ist stärker als das prädikatenlogische. Nun steckt aber auch in Freges Explikat für Gedankenidentität ein Bikonditional, gelegentlich bezeichnet als *Äquipollenz* (NS, 213, vgl. auch NB, 105), und es ist kontrovers, welche logische Stärke dieses hat (vgl. hierzu auch 4-§ 11). Da gilt nun Folgendes: Wenn man für die Äquivalenz (3) ein Bikonditional wählt, das mindestens so stark ist wie die für Gedankenidentität maßgebliche Äquipollenz, dann wird auch (2) von (3) impliziert; wählt man hingegen ein schwächeres, dann besteht diese Implikationsbeziehung nicht. Deshalb muss Künnes Aussage über das logische Verhältnis zwischen (2) und (3) relativiert werden – und entsprechend auch die über die Rechtfertigungsmöglichkeiten für die Akzeptanz von (1) (vgl. 412).

V) Einige der wenigen Stellen, an denen ich Künne direkt widerspreche, ist die Anmerkung 101 (741) zum Thema *Allgemeinheit* (5-§ 6). Es geht hier um die Wiedergabe des formalen Ausdrucks

$$\forall x(x \text{ ist ein Mensch} \rightarrow x \text{ ist sterblich})$$

„in Logik-Kursen und -Lehrbüchern [...] mit dem grammatisch inkohärenten Gebilde ‚Für mindestens ein x, wenn x ein Mensch ist, dann ist x sterblich‘“. Die Wendung *Für mindestens ein x* ist hier sicherlich ein bloßes Versehen, weil es natürlich *Für jedes x* heißen muss. Kein Versehen hingegen, sondern abwegig ist Künnes Beschreibung der Funktion des ersten Vorkommnisses des Buchstaben „*x*“. Er sei nämlich im „Vorspann ein (kurioser) genereller Term“. Das ist er aber nicht. Vielmehr kürzt der Ausdruck *ein x* (bzw. *jedes x*) hier den Ausdruck *einen (jeden) Gegenstand x* ab. Dieser enthält den generellen Term *Gegenstand.* Das „*x*“ in ihm ist ein unbestimmt referierender Individuenterm, eine Individuenvariable in *kataphorischer* Rolle. In seinen weiteren Vorkommnissen spielt das „*x*“ hingegen die *anaphorische* Rolle einer Individuenvariablen. – Diese Auffassung ergibt sich im Übrigen auch aus Künnes Überlegungen auf S. 739f.

Eine letzte kritische Bemerkung: Gelegentlich störend ist der Ton, in welchem Künne sein Umfeld kommentiert. Das gilt insbesondere für seine Herabwürdigung der lesenswerten Frege-Biographie von L. *Kreiser* (vgl. z.B. 5). Diese wird dann auch nicht immer richtig wiedergegeben: So ist „Carnaps Brief an Patzig" (Anm. 46, 28), wie man bei Kreiser selbst lesen kann, ein Brief von G. *Patzig* an Kreiser.

Jedoch: Gemessen an der wissenschaftlichen Substanz von Künnes Buch sind dies Marginalia. Der Rezensent hat aus dem Rezensendum viel gelernt.

Rainer STUHLMANN-LAEISZ
Universität Bonn

Grazer Philosophische Studien
85 (2012), 342–346.

Carolyn SWANSON, *Reburial of Nonexistents: Reconsidering the Meinong-Russell Debate.* Amsterdam and New York: Rodopi 2011. 148 pp. ISBN 978-90-420-3364-1.

Gilbert Ryle expressed conventional wisdom among analytic philosophers who had been trained learning Russell's theory of definite descriptions, when in 1973 he said that Alexius Meinong's "… *Gegenstandstheorie* is dead, buried and not going to be resurrected."[1] Ryle saw the theory of descriptions as a model of the right method of philosophical analysis by replacing sentences with apparent singular terms, such as "the round square" with their Russellian paraphrases in terms of quantifiers and identity, thus supposedly avoiding the cardinal philosophical mistake of basing an account of meaning on reference. The chief moral to be drawn from Russell's theory of descriptions was that philosophical analysis should avoid ontological commitment to abstract entities proposed solely for their role in theory of meaning. Very soon after Ryle's announcement of the burial of Meinong's theory, however, it came back from the grave through the efforts of Richard Routley (later Sylvan), and then Terrence Parsons, who initiated a wave of "Neo-Meinongian" theories using novel logical devices to defend theories of "non-existent", "Meinongian," objects.[2]

Carolyn Swanson's book argues against these Neo-Meinongian views, showing that in fact a Russellian account, added to a version of "story operator" analysis of fictional discourse, provides a better account of the very evidence that the Neo-Meinongians used to motivate their accounts. Hence the "reburial" of her title. Swanson organizes her argument around Russell's two famous objections to Meinong and then a number of "paradigm facts" which were put forward as lending support to the Meinongian view. Chapters two and three present Russell's logical objections to Meinong's view. The most famous is considered first:

1) Meinong's theory infringes on the law of "contradiction" (and excluded middle.)

Thus Meinong's "round square" would seem to be both round and square, and hence not round. This is a violation of the principle of Non-Contradiction. Meinongian objects also seem to violate the Law of Excluded Middle, as the same round square has no further properties, such as being round or non-round. The violation of these logical principles correspond with two ontological features of Meinongian objects, that they can be inconsistent, having both a given property and lacking it, and indeterminate, neither having nor lacking some property.

Swanson considers a proposal to meet this objection based on a distinction between "wide" and "narrow" negation, which is to correspond to the difference between an object's lacking a property and the more general state of affairs of its not being the case that the object has the property. Intuitively the two can diverge, for it can be false that "the round square is blue" yet not the case that "is not blue" is true of the round square. Swanson's discussion argues that this proposal is inadequate when spelled out in any detail, for on a reasonable account of the distinction, one sort of negation implies the other.

Part One of the book begins a defense of the Russellian account of these apparent violations of logical principles, starting with Russell's account of "the present King of France is bald." The second of Russell's objections is that it seems that Meinongian objects are generated too easily with any properties we give them, and so infringe on commonsense knowledge. Thus the example of "The existent King of France exists". This is, however, seemingly an example of one of the very "paradigm facts" that Meinongian theories are meant to explain, that a purported Meinongian object will have the properties that characterize it, by which it is identified, thus:

2) Meinong's theory explains *characterization facts* such as "The *F* is *F*", but then has the consequence that "The existent *F* exists".

Here Swanson's main target is Terence Parsons, who invokes Meinong's distinction between "nuclear" and "extranuclear" properties to bar conclusion of 2). When the property *F* in question is one that characterizes the nature of the object it is "nuclear", rather than being one of the "extranuclear" exceptions, such as "exists." Swanson here presents a number of objections to Parsons' distinction, including the complications necessary to argue that extranuclear properties have "watered down" versions that both give a meaning to the predicate and avoid the original problem that inspired them. This notion of "watered down" properties comes from Meinong, as does the distinction between "nuclear" and "extranuclear" properties (although Meinong is said to credit his student Ernst Mally as the source of these notions.)

In Part Two Swanson turns to three kinds of "facts" or linguistic data that seem most naturally accounted for with a Meinongian theory. One kind of statement is exemplified by "The perpetual motion machine does not exist". It is suggested that even to say that there is no such thing as the perpetual motion machine we must first refer to it, and that requires some sort of being that falls short of existence. This was what Ryle saw as the source of the main error behind Meinongianism. Swanson explains how Russell's theory of descriptions was designed to represent this sort of "fact" without invoking a denotation for the description.

The second is these very cases of the "characterization" property in 2) above. Meinong proposed non-existent objects are to be distinguished from our thought of them. Hence the "round square" is an object that is round (and square), and Vulcan is a planet (albeit a non-existent planet) between Mercury and the Sun. Swanson shows how suitable choices of characterizing properties leads to outright logical contradictions, and so:

3) Meinong's theory faces the "identity problem". From the nature of a triangular square one can conclude that 3 (the number of sides of a circle) is identical with 4 (the number of sides of a square.)

That is, more generally Meinongians face:

4) The "contradiction problem". Contradictions can be derived from the natures of "the absence of blue on the round square" and "The presence of blue on the round square."

These problems were apparent to the Neo-Meinongians, but Swanson argues that their solutions, involving wide and narrow scope negations, and the nuclear-extranuclear distinction, just won't work.

Once more Swanson argues that the Russellian solution is better than any of its Neo-Meinongian alternatives. The Russellian account makes all instances of "The *F* is *G*" false if there is not exactly one *F*, when the description is "improper", but, argues Swanson, is an adequate account of "characterization" in the truth of "Any *F* is *F*", which is nevertheless true even if there are no *F*s. Swanson does not discuss the difficulty raised by the Russellian account that "The *F* is not *G*" will also be true for any case when the description is improper. She seems content with some truth in Russell's theory as an alternative to the Meinongian "truths". That the Russellian account over generates truths about "non-existents" doesn't seem to bother her.

A third class of "paradigm facts" that the Neo-Meinongian claims to be able to take at face value are reports of intentional facts as in "The round square has been a subject of thought." Meinong motivated his interest in non-existent objects by the observation that they are often the objects of intentional states. To use Meinong's example, a "perpetual motion machine" must have at least enough being to be an object of search by some or even what is claimed to not exist by others:

5) Meinongian objects can be objects of intentional states independently of their existence, merely by having being (*Sein*) and a nature or *Sosein*.

The defense of Russell in this case relies on the distinction of "scope" within intentional sentences, again with its origin in "On Denoting". It is this distinction which handles the oddity of "the present King of France is not bald", but also extends to "Joe wishes to buy a house". Here we can invoke a distinction between a desire to have some house or other from the case that there is a particular house which Joe has already selected for purchase. The claim that a use of an "operator" such as "Joe wishes it to be the case that ...", and two different scopes for descriptions, comparable to the "*de re*" and "*de dicto*" senses of "Joe believes that ..." is that basis for the third major class of "paradigm facts" that Meinongians invoke as supporting their theory.

Swanson's Part Three, her last three chapters, are devoted to an account of fictional objects. Neo-Meinongians generally claim that:

6) Fictional objects are best represented as Meinongian objects characterized by the properties that the context of their creation attributes to them.

Thus, in a standard example, "Sherlock Holmes lived at 222B Baker Street" is true because Arthur Conan Doyle located him at that address in his stories, even though it is also true that no person ever lived at that address. Swanson presents the case of fictional objects as showing the limitation of Meinongian theories of "names" for non-existent objects and instead elaborates on an alternative account borrowed from discussions of "the logic of fiction". Thus Swanson proposes that:

7) Statements about fictional accounts are best represented with "context operators" asserting that p is true in a given context w (in her notation, "C[w]: p").

The peculiar ontological features of fictional objects are removed if they are found to occur within the scope of an operator. Thus "According to the story Holmes lived a 222B Baker street" can be true, and have a consequence that "According to the story someone lived at 222B Baker street", even though "Someone lived at 222B Baker street" is not true outside the operator.

Swanson adds some details to this account using "context operators" to handle complications involving "context-crossing" familiar from the logic of fiction. London itself shows up as a "fictional character" of sorts in the Holmes stories, and is attributed properties in the story that it may not have had in reality. Swanson allows one to attribute properties to an object in a context by using a "secondary operator", which indicates that names superscripted with a context variable, and the object is to have the properties attributed to it in the context. Thus if s (Smith) in a context w (the television series *The West Wing*) admires (a relation A) the individual b (Bartlett, the president of the U.S. in the series) we express that with a "secondary operator" C^w , as follows, $[C]^w$: Asb^w. The superscripted 'w' indicates that the object b is to be taken with the properties it has in the context w. Thus Bartlett can be an admirable president of the U.S. in the context of *The West Wing*, without being a president at all in the context of "reality" at the time. (It was George W. Bush at the time.)

Swanson does not give a formal semantics for this language with context operators, but asserts that it does not involve her in Meinongian objects, either the objects in contexts, such as Sherlock Holmes, or the contexts themselves (the fictional world of the stories). Some Neo-Meinongians might feel that contexts and fictional characters in them are much like, or even literally are, possible worlds and possible objects in them. While possible objects and worlds explain the truth conditions of modal sentences as describing the properties of objects in other worlds than the actual world, so it might seem that one needs an account of the truth of sentences about what is true in different contexts and why it is that it seems that fictional objects have properties in those contexts. But Swanson has good company among "modalists" who hold that modal operators such as "possibly" and "necessarily" are just primitively true, and not in virtue of the goings on in other possible worlds. Her position would be that talk about "story operators", such as "It is true in the Conan Doyle stories that ..." are also primitive and talk about "the world of the fiction" is explicable in terms of talk about what is true according to a fiction.

Swanson concludes her book by comparing the account of these issues with a Russellian account of descriptions and names, treating both with the familiar quantificational analysis, combined with "primary" and "secondary" operators for what is true in contexts and sentences that treat objects arising from different contexts. She asserts that the Russellian solution is in the end better than the Meinongian, and so, in the end, "perhaps beingless objects should, again, be laid to rest". This is the "Reburial of Nonexistents" of her title.

I can't resist one parting quibble, but don't let it prevent anyone from carefully studying this clearly and forcefully argued book. This reviewer came to the book with more familiarity with the Neo-Meinongian "Object Theory" of Edward Zalta than Routley's *Meinong's Jungle*, or even Terry Parsons' *Nonexistent Objects*.[3] Zalta bases his theory on a distinction between two modes of predication, ordinary exemplification and what Zalta calls "encoding", based on a notion proposed by Meinong's student Ernst Mally in his 1912 essay *Grundlagen der Logik und Logistik*.[4] As developed by Zalta there is a relatively straightforward solution to the puzzle about the laws of noncontradiction and excluded middle in 1) above. Abstract (nonexistent) objects are characterized by the relation of encoding, but objects may encode only some properties, and can encode inconsistent properties. While incomplete or inconsistent with regard to the properties they encode, all abstract objects are consistent and determinate objects with regard to exemplification. The complications involving "nuclear" properties and "watered down" versions of them are avoided on this account. The "characterization facts" of 2) are similarly free of problems. An object is characterized by the properties it encodes and which it exemplifies, especially those that logically follow from properties encoded are irrelevant. No contradiction or indeed any other logical consequence as in 3) and 4) can be deduced from a property which is encoded. With regard to abstract objects as objects of intentional attitudes and with his version of points 5) through 7) above, Zalta's theory differs more substantially from Swanson's representative "Meinongians", although he gives an account of both stories and fictional objects in them as Abstract Objects.

Whether Zalta's theory preserves enough of Meinong's original doctrine to deserve the name, and in following the hints in Mally he has deviated enough to no longer deserve the title of "Meinongian" I leave to the readers of this journal, which is published by the successors of Meinong's "Graz School". Swanson has very clearly laid out the issues about which anyone following in Meinong's path must resolve. I read the book with delight, finding in one problem after another a possible motivation for the various choices made by different Neo-Meinongians over each of her issues. As well the spirited defense of the Russellian solutions, even if they are not accepted in the end, will present Meinongians with a standard that their own accounts must meet. Like Russell himself after premature reports of his death in China in the early 1920's, perhaps *Gegenstandstheorie* will survive reports of its reburial, even as its cousins, at least, still roam the earth.

Bernard LINSKY
University of Alberta

1. Gilbert Ryle, "Intentionality-Theory and the Nature of Thinking", *Revue Internationale de Philosophie* 27 (December 1973) p.255. Swanson relies primarily on Bertrand Russell's "On Denoting" from *Mind*, 1905, for the "Russellian" theory.
2. Richard (Routley) Sylvan, *Exploring Meinong's Jungle and Beyond*, Australian National University, 1979. Terence Parsons *Non Existent Objects*, Yale University Press, 1980.
3. Edward N. Zalta, *Abstract Objects*, D. Reidel, 1983, which Swanson does list in her bibliography, but only cites for an account of the nuclear/extranuclear distinction, but more importantly for the account of intentional objects, *Intensional Logic and the Metaphysics of Intentionality*, MIT Press, 1988.
4. *Zeitschrift für Philosophie und Philosophische Kritik*, 1912.

Mally, Ernst 1912: *Grundlagen der Logik und Logistik*. Ergänzungsheft zu: *Zeitschrift für Philosophie und Philosophische Kritik* 148. Leipzig: Barth.

Parsons, Terence 1980: *Non Existent Objects*. New Haven and London: Yale University Press.

Routley, Richard 1979: *Exploring Meinong's Jungle and Beyond*. Australian National University: Ridgeview.

Ryle, Gilbert 1973: "Intentionality-Theory and the Nature of Thinking". *Revue Internationale de Philosophie* 27, 255-265.

Zalta, Edward N. 1983: *Abstract Objects*. Dordrecht: Reidel.

Zalta, Edward N. 1988: *Intensional Logic and the Metaphysics of Intentionality*. Cambridge, MA: MIT Press.

Grazer Philosophische Studien
85 (2012), 347–353.

Uriah KRIEGEL, *Subjective Consciousness: A Self-Representational Theory*. New York: Oxford University Press 2009. 352 pp. ISBN 978-0199570355.

Self-representational approaches to consciousness have a long and distinguished history in the philosophy of mind. Aristotle, Brentano, and Sartre all held something like such an account of consciousness. But not until Uriah Kriegel's *Subjective Consciousness* (*SC*) has the approach received the comprehensive articulation and defense that it has long deserved. The main (but by no means only) objective of *SC* is to provide a naturalistic, self-representational account of *subjective character*—the "for-me-ness" or inward awareness of conscious states of mind. *SC* presents a well-argued and innovative account of subjective character, offering a deep and clear explanation of its relation to intransitive state consciousness and qualitative character, its likely realization in the human brain, and its contribution to the overall phenomenology (the what-it-is-likeness) for subjects undergoing conscious states. *SC* makes a welcome and important addition to the field of consciousness studies—a field that, unfortunately, has stagnated lately under its self-imposed restriction of framing the discussion about the representational nature of consciousness in terms of but two brands of representationalism: higher-order theories and first-order theories. Kriegel's new brand of representationalism is a needed shot in the arm, and it will undoubtedly advance the field in an interesting and important new direction. *SC* is a highly recommend read.

All conscious states have subjective character and qualitative character, according to Kriegel. These are the data of consciousness that need explaining. When I look out the window and I see a patch of clear blue sky, for instance, my conscious visual experiences has two important elements. One, I am aware of the experience in a way that I wouldn't be if it were unconscious or subliminal. This, as noted above, is the experience's subjective character. And two, in having the experience, the sky *looks* a particular shade of blue to me (e.g., $blue_{17}$). This is the experience's qualitative character. Both characters of my experience contribute to the overall phenomenology (what it is like) of the conscious state for me, according to Kriegel. However, the significance of their contributions is different. Subjective character makes a more fundamental contribution: It has the heavier burden of making the mental state phenomenologically conscious *at all*, whereas its qualitative character has the more refined burden of making the state the kind of phenomenally conscious state that it is (e.g., a visual experience as of $blue_{17}$ rather than as of a red_{16}).

It is because of its fundamental importance to the very existence of phenomenology that Kriegel's focus in *SC* is to explain the nature of subjective character. However, before presenting his self-representational theory of subject character, Kriegel articulates and defends a rather unique form of representationalism of qualitative character in chapter 3. On standard representationalist accounts, an experience is said to have a certain qualitative character Q (e.g., the sky looks $blue_{17}$ to me) just in case the experience represents the *observer-independent* property Q (e.g., my visual experience represents the sky as being objectively $blue_{17}$). Employing a version of Sydney Shoemaker's *argument from shifted spectra*, Kriegel shows that such standard representational accounts run into a deep problem. Such accounts, according to the argument, entail that objects in the world may come to have incompatible properties. To illustrate, consider the very real scenario of two normal human subjects (Norman and Norma) who are in no way suffering from an illusion or hallucination but nevertheless see the same patch of blue sky in slightly different ways. Norman, for example, sees the patch of sky as $blue_{17}$ while Norma sees it as $blue_{16}$ (i.e., a slightly lighter shade of blue from that which Norman sees the sky as having). Now, if Norman and Norma's respective visual experiences purport to represent the objective (observer-independent) color of the sky, as standard representational accounts of qualitative character hold, then, given that neither subject is suffering from an illusion or hallucination, their respective experiences should represent the sky's objective color correctly. But this would mean that the same patch of sky at the same moment in time is both $blue_{17}$ and $blue_{16}$, which is impossible.

Shoemaker's own suggested solution to this problem is to propose that an experience has a qualitative character Q not in virtue of it representing things as having the objective property Q but in virtue of the experience representing things as having the property of 'appearing Q to the observer.' This avoids the above problem, since there is nothing impossible about the same patch of sky *appearing* $blue_{17}$ to Norman and *appearing* $blue_{16}$ to Norma. Kriegel takes up Shoemaker's suggestion and develops his own unique account of what these appearance properties are that experience represent. According to Kriegel, these appearance properties are observer-dependent dispositional properties of objects. More specifically, they are dispositions in objects to cause certain types of neurological events in empirically specified groups of people (e.g., "normal Asian females in their twenties," 89) under certain specifiable environmental conditions. Norma's experience of seeing the patch of sky as $blue_{16}$, on this account, is taken to amount to Norma's experience representing the patch of sky as being disposed to cause a class of subjects (of which Norma is a member—e.g., normal Asian females in their twenties), under certain environment conditions (e.g., looking at a patch of clear sky at noon), to undergo a type of neurological state that realizes certain discriminatory behaviors in such subjects which are empirically indicative of their seeing $blue_{16}$ (rather than, say, $blue_{17}$). The fact that Norma's experience represents this type of observer-dependent dispositional property of the sky, according to Kriegel, is what explains why the sky looks $blue_{16}$ to her and not, say, $blue_{17}$. Kriegel recognizes, of course, that Norman too (as well as other subjects who are not "normal Asian females in their twenties") may see things as $blue_{16}$ in other conditions (e.g., perhaps when looking at a patch of clear the sky at 9 AM). So the observer-dependent property that both Norman's and Norma's respective experiences represent in these different viewing conditions, in virtue of which they both are seeing $blue_{16}$, will be a rather sophisticated *conjunctive* property, according to Kriegel. "This response-dependent property," Kriegel writes, "would have to include, then, the disposition to elicit a certain response in Norma and her group members in circumstance C [e.g., looking at the sky at noon] *and* a certain response in Norman and his group members in circumstance C* [e.g., looking at the sky at 9 AM]" (91).

There is, however, one potential shortcoming to Kriegel's innovative representational account of qualitative character. In addition to perceiving response-dependent properties, we also perceive response-independent (objective) properties (e.g., the objective color, or shape, or size of objects). However, on Kriegel's account, our perception of such objective properties does not contribute to the qualitative character of our perceptual experiences. And this may seems to some a bit odd. While looking at my rectangular kitchen table, for example, the top of the table, from where I am currently standing, looks trapezoidal to me. If I were asked to paint a realistic picture of what I am now seeing from this angle, I would paint a picture of a trapezoidal shaped table top. This is the 'appearance property' of the table top of which Shoemaker speaks and the response-dependent property that Kriegel's view proposes that my visual experience represents the table as possessing. But the table top from where I am standing also looks to me to be rectangular. If someone were to ask me what shape the table top looks to have, I would say rectangular. This is the objective, observer-independent property that my perceptual experience represents the table as having. (Palmer 1999 calls these different modes of perceptions of an object *distal* and *proximal modes of perception*, the former being the mode of perception operative in cases of perceptual constancies.) The fact that the table top *looks* rectangular to me, intuitively, is just as much a fact about the qualitative character of my perceptual experience of the table top as the fact that the table top *looks* trapezoidal to me. Kriegel's response-dependent representationalism seems equipped to explain only why the table top looks trapezoidal to me, not why it also looks rectangular to me. There appears to be a qualitative character of perceptual experiences, then, that Kriegel's account is ill equipped to explain.

In Chapter 4, Kriegel presents the heart of his self-representational theory of subjective character. Put simply, the theory holds that "conscious states are conscious in virtue of being represented by themselves" (18). Kriegel's *master argument* for his theory runs roughly as follows. PREMISE 1: Conscious mental state are conscious in virtue of their subjects being inwardly aware of them. Kriegel takes Premise 1 to be true and known to be true on the basis of "first-person knowledge of our conscious states" (19). It is in virtue of our unique first-person knowledge of our own conscious mental states, Kriegel claims, that we know firsthand (have "direct phenomenological evidence"—116) that such states of mind are always accompanied by our inward awareness of them. PREMISE 2: Inward awareness is a type of mental representation: Being inwardly aware of a mental state is just being in a state that suitably represents the mental state in question. Kriegel rejects non-representational accounts of inward awareness, such as some recent accounts (e.g., Hellie 2007) that take it to be a form of direct acquaintance (à la Russell), and offers a developed self-representational account of inward awareness in chapter 6. PREMISE 3: Inward awareness of one's own mental state must be conscious awareness. Kriegel's main justification of the truth of Premise 3 is, roughly, that if one's inward awareness of one's own conscious state were unconscious (say, like the unconscious awareness we have when undergoing subliminal perceptions), we would not have the first-person knowledge of the truth of Premise 1 that we so manifestly do. LEMMA: From Premises 1–3, Kriegel infers that mental states are conscious in virtue of their being suitably represent*ed* by a conscious mental state. PREMISE 4: The represent*ed* conscious state and the represent*ing* conscious state cannot be (or belong to) numerically distinct conscious states on pain of an infinite regress of numerically distinct represent*ing* conscious states. CONCLUSION: The represent*ed* conscious state and represent*ing* conscious state must be (or belong to) one and the same consciousness mental state.

Kriegel goes on in chapter 6 to show that the best way to develop the idea expressed in the conclusion of his master argument is on the hypothesis that all fundamental conscious mental states have conscious parts that represent the conscious mental state as a whole. However, before he does this, Kriegel puts forward and defends the *ubiquity of inner awareness* (UIA) thesis in chapter 5. The UIA thesis holds, roughly, that a subject's inward awareness of his/her conscious mental state is part of what determines what it is like for the subject to be in that state. The main challenge to the UIA thesis, Kriegel notes, is the widely held transparency thesis which effectively holds that inward awareness, which is clearly *not* one of a conscious state's first-order representational properties, cannot contribute to what it is like for a subject to be in a conscious state, for what determines what it is like to be in a conscious state is revealed to the subject only through introspection, and introspection reveals only a conscious mental state's first-order representational properties. Kriegel responds by drawing an important distinction between *peripheral* inward awareness (i.e., the kind of awareness that makes conscious states conscious) and *focal* inward awareness (i.e., introspection) and argues that while peripheral inward awareness is not revealed ("phenomenology manifest") to the subject through introspection it is revealed through peripheral inward awareness. "The awareness that constitutes the peripheral inner awareness," Kriegel writes, "and that which constitutes the general impression thereof [i.e., the phenomenological manifestation of this awareness in conscious experiences], are in fact one and the same—it is an awareness that is partly an awareness of itself" (188). And Kriegel takes this identity here to be a direct consequence of his self-representational theory of subjective character.

In chapter 6, Kriegel moves to provide a more detailed account of the underlying ontology of conscious states according to his self-representational theory, as well as a naturalistic account of the self-representational relation involved in such states. On Kriegel's theory, the basic or fundamental variety of conscious mental state—call it M—is a complex state composed of two logical (i.e., non spatial-temporal) proper parts—M* and $M^{\Diamond}$—that are themselves conscious mental states. In such a complex, the proper part M* *directly* represents proper part $M^{\Diamond}$ by virtue of $M^{\Diamond}$ bearing a causal/teleological relation to M*. The nature of this direct representation here is meant to be similar to that found to exist between (say) magnetosomes in marine bacteria and geometric North: Just as magnetosomes directly represent geometric North in these organisms by virtue of their causal/teleological relation to geometric North, mental state M* directly represents $M^{\Diamond}$ by virtue of $M^{\Diamond}$'s causal/teleological relation to M*. Also, on Kriegel's account, since mental state $M^{\Diamond}$ is a well-integrated logical proper part of the complex mental state M, M* also *indirectly* represents M as a whole by virtue of M*'s direct representation of $M^{\Diamond}$. The nature of this indirect representation here is taken to be analogous to that found in ordinary cases of indirect external perception, such as when a "visual perception may [indirectly] represent a [whole] apple by representing its front surface" (225). Figure 1 below presents a graphic illustration of what Kriegel has in mind here.

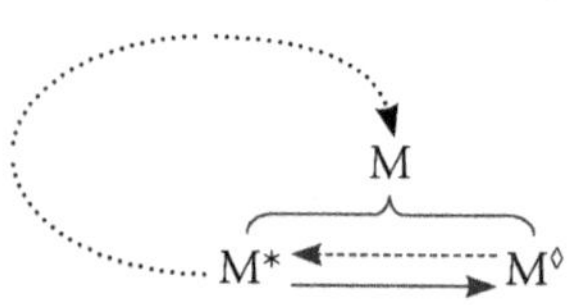

Figure 1

As way of illustration, let $M^{\Diamond}$ = Norman's conscious visual experience as of blue$_{17}$, and M* = Norman's conscious inward awareness of this experience. M would then stand for the conscious complex mental state with $M^{\Diamond}$ and M* as its logical proper parts. The dashed arrow from $M^{\Diamond}$ to M* represents the causal/teleological relation between $M^{\Diamond}$ and M* in virtue of which M* is said to directly represent $M^{\Diamond}$, which is here represented by the solid arrow from M* to $M^{\Diamond}$. And in virtue of M* directly representing $M^{\Diamond}$, *and* $M^{\Diamond}$ being an integrated logical part of M, M* is said to indirectly represent the complex whole M. The indirect representational relation here is represented by a dotted arrow from M* to M.

In chapter 7, Kriegel puts forward his cross-order integration (COI) hypothesis meant to explain how the above self-representation model of consciousness can be realized in the human brain. The COI hypothesis holds that the neural correlate of M* (that part of M which is the higher-order representational state) is to be found in the dorsolateral prefrontal cortex (dlPFC) of the brain, while the neural correlate of $M^{\Diamond}$ (that part of M which is the subject's first-order conscious experience) is to be found in the appropriate neurological center corresponding to the type of first-order conscious experience that $M^{\Diamond}$ is. So, for instance, the neural correlate of $M^{\Diamond}$ in the case of a conscious experience of the color blue$_{17}$ is likely to be found in the V4 area of the subject's occipital lobe. Finally, on the COI hypothesis, the neural correlates of these two conscious mental states is functionally integrated or bound together, and this is done by their firing rates being synchronized. The analogy here is with von der Malsburg's neural synchrony model of how first-order representations (e.g., the visual representation of blue and the representation of square) are bound together (e.g., into the visual representation of a blue square) as a result of the synchronized firing rates of the neural correlates of the two individual representations.

Finally in chapter 8, Kriegel tackles the hard problem of consciousness — that is, the problem of explaining why anything happening in the brain should constitute or produce conscious experience *at all*, and why it should constitute and produce the kind of conscious experience it actually does. After canvasing a number of responses to the hard problem, Kriegel opts to solve it by way of a deductive reductive model which involves specifying a priori the functional role of consciousness. In chapter 7, Kriegel specified the functional role of consciousness as that which "gives the subject just enough information to know how quickly and effortlessly to obtain rich and detailed information about her concurrent experience" (262). This is consciousness's functional role or the 'consciousness role.' Kriegel then offers the following functionalization of 'consciousness': Consciousness just is the actual occupant of the consciousness role. Of course, the actual occupant of the consciousness role in the human brain, according to the COI hypothesis, is the synchronization of any first-order neurological state with a dlPFC activation. And from these two points, Kriegel maintains, we have an epistemically transparent answer to the questions that constitute the hard problem.

Kriegel's articulation, defense, and development of his self-representational theory of subjective character is impressive indeed. It is without question the best self-representational account of consciousness ever offered, and for this reason alone it is required reading for philosophers and researchers in the field of consciousness studies.

The nature of consciousness is an extremely difficult and challenging topic, and so it should be of no surprise that, despite the many virtues of Kriegel's self-representational theory, it leaves some difficult questions unanswered. Here I wish to mention but two. Traditionally, self-representational theories have struggled under the weight of two major challenges. One, such theories (along with their higher-order thought cousins) have seemed to some to over-intellectualize the conditions for consciousness, making it difficult to believe (at least to any high degree of confidence) that they are true. Note that if self-representationalism were true, animals and infants that enjoy conscious experiences would be credited with being inwardly aware of their own conscious experiences. Thus, my cat who consciously hears the sound of the can opener (and rushes into the kitchen as a result), not only hears the sound of the can opener (which is occurring in the kitchen) but is inwardly aware of her own conscious experiences of the sound (which is occurring in her mind), according to self-representationalism. The problem is that we have very little if any reason at present to believe that many (if any) animals and infants have such inward awareness of their own conscious experiences. Researchers, for example, have still not found, after years of empirical study, any unambiguous evidence that even monkeys (a rather advanced species of animal) possess such inward awareness. At the moment, we simply do not know, and have no strong reasons to believe, that many (if any) animals and infants have inward awareness that, on self-representationalism, is required for consciousness. Thus, if we really were firm believers in self-representationalism, we should take a similar agnostic position regarding conscious experiences in animals and infants. But to apply an idea here from G. E. Moore's refutation of idealism, we clearly know better that many animals and infants have conscious experiences than that self-representationalism is true. Given this Moorean fact, we should instead be agnostic about the truth of self-representationalism rather than about the existence of conscious experiences in animals and infants. In chapter 5, Kriegel spends a couple of pages (194ff.) addressing a slightly different animal/infant objection against self-representationalism that is based on the assumption that researchers currently have empirical proof that animals and infants *lack* inward awareness of their own experiences. Kriegel rightly points out that at the moment researchers are not in possession of such a positive disproof of inward awareness in animals and infants. However, Kriegel's point here, though well taken, does not address the Moorean argument which purportedly shows that at present we have reason to be agnostic about the truth of self-representationalism.

The other challenge that self-representational theories have traditionally faced is that they appeal to a rather mysterious notion of representation to explain consciousness. It has seemed to many deeply perplexing how a state of mind can represent itself. Intuitively, when a mental state represents a mental state, there is, as it were, some spatial or logical distance between the represent*ed* and the represent*ing* states to allow representation to take hold. When the states are identical, they are too close, both spatially and logically, for representation to squeeze in between. Kriegel does an admirable job in demystifying the idea of self-representation in mental states. He rightly rejects a "straight representational" account "which casts the experience as wholly representing its-whole-self"—for such an account does nothing to remove the initial mystery surrounding mental self-representation—in favor of a "crooked" account according to which, as noted above, "conscious states have a part that represents directly another part, and in virtue of that represents indirectly the whole state" (215). Though laudably clear and understandable, Kriegel's own crooked account faces some challenges of its own.

First, notice that on Kriegel's crooked account, a subject's state of inward awareness (represented by M* in figure 1) is a conscious mental state that is *un*represented by a conscious state. This would seem to be inconsistent with the Lemma in Kriegel's master argument above. Kriegel responds by declaring that M* is conscious, though not in virtue of it being represented by a conscious states, but in virtue of it being a logical part of the complex conscious states M. This, however, would appear to amount to acknowledging two different kinds of consciousness, which Kriegel effectively does by describing the consciousness that belongs to M* as "sub-maximal" consciousness and that which belongs to M as "maximal" consciousness (229). (Does that mean mental state $M^{\Diamond}$ has both kinds of consciousness?). Arguably there is nothing other than Kriegel's theory that would recommend our accepting the existence of these two kinds of state consciousness. Common sense, of course, recognizes but one kind of state consciousness. Thus Kriegel's suggestion here of what makes a subject's inward awareness (M*) conscious has an an air of ad hocness about it, which is something one should avoid in one's explanatory theories.

Second, in his attempt to demystify self-representation, Kriegel appeals to the idea of indirect representation of a whole object by means of directly representing some proper part of it. The clearest and least problematic example he gives of this is that of seeing an apple (as a whole object) by seeing just the front surface of the apple. In such ordinary cases of indirect part/whole visual perception, however, there are always two distinct perceptual representations involved. There is the initial representation of the object's parts *as such* (e.g., a representation *as of* 2.5-D surfaces which is not yet a representation *as of* 3-D whole object), which then leads to a representation of the parts *as* belonging to a 3-D whole object. That there are these two distinct representational states in cases of ordinary indirect visual perception is evidenced by the fact perception of parts as belonging to wholes occurs later in visual processing (and can be prevented from forming) than perception of parts as such (see Sekuler & Palmer 1992). It would appear, then, that to properly demystify self-representation by making it a type of indirect representation akin to ordinary indirect part/whole visual perception, Kriegel's theory would need to add an extra conscious mental state to the complex M, one that played something like the role that 2.5-D surface representations play in the formation of 3-D whole object representations. The theory, then, would need to add a mental state $M^{\bullet}$ whose function is to represent $M^{\Diamond}$ *as such* (i.e., not *as* something that is a part of M but in terms of $M^{\Diamond}$'s more immediate and proximal properties, whatever they may be). M*'s function, then, would be the representation $M^{\Diamond}$ *as* part of M (the whole complex mental state). Figure 2 below gives a graphic illustration of this addition.

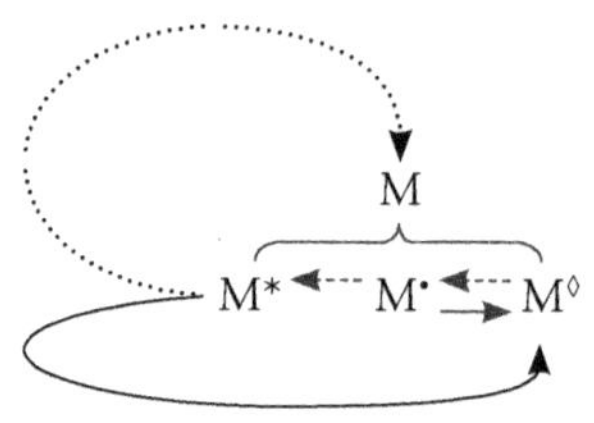

Figure 2

The solid arrow leading from M* to $M^{\Diamond}$ and the dotted arrow leading from M* to M are meant to represent collectively the idea of M* representing $M^{\Diamond}$ as a part of M; the dashed arrow from $M^{\bullet}$ to M* represents the causal process leading from one representation to the next (akin to the process in vision which leads from 2.5-D representations to 3-D representations). Now such a change to Kriegel's original self-representational model does nothing to undermine its initial clarity, and it may be needed in order to give a proper demystification of the idea of self-representation that Kriegel seeks. But this new model, with its extra logical part and new tertiary binding relation among the three parts, does not, unfortunately, fit Kriegel's COI hypothesis, as described above. On that hypothesis, recall, the brain was shown to be capable of realizing Kriegel's self-representational model by virtue of the occurrence of a unique binary relation of synchronization holding between a higher-order mental state and the first-order mental state it represents. This cannot be said here for this new tertiary self-representational model. Of course, there is no reason at this point to think that this new model cannot be realized in the human brain. Whether the brain can or cannot realize this new self-representational model is, as we say in philosophy, an empirical question.

Robert LURZ
Brooklyn College-CUNY

References

Hellie, Benj 2007: "Higher-Order Intentionality and Higher-order Acquaintance". *Philosophical Studies* 134, 289–324.

Palmer, Stephen 1999: *Vision Science*. Cambridge, MA: MIT Press.

Sekuler, Allison & Palmer, Stephen 1992: "Perception of Partly Occluded Objects: A Microgenetic Analysis". *Journal of Experimental Psychology* 121, 95–111.

GPS

Subscription Rates
There is no annual subscription rate. Each volume has its own price, varying from 60 to 90 US Dollars. Individual subscribers get a 50% discount.

Information for Contributors

GPS publishes articles on philosophical problems in every area, especially articles related to the analytic tradition. Each year at least two volumes are published, some of them as special issues with invited papers. Reviews are accepted only by invitation.

Manuscripts in German or English should be submitted electronically as an e-mail attachment, either in MS Word or in rtf format, prepared for anonymous reviewing (i.e. without the author's name and affiliation), together with an English abstract of 60–100 words. Footnotes should be kept to a minimum, and references should be incorporated into the text in (author, date, page) form. An alphabetical list of references should follow the text.

A submitted paper will normally be sent to a referee.

Authors are responsible for correcting proofs. Corrections that deviate from the text of the accepted manuscript can be tolerated only in exceptional cases. Authors will receive a free electronic offprint of their article.

Manuscripts should be sent to the following address:
martina.fuerst@uni-graz.at

Email Addresses
Editors
johannes.brandl@sbg.ac.at
david.1@nd.edu
maria.reicher-marek@rwth-aachen.de
stubenberg.1@nd.edu
Managing Editor
martina.fuerst@uni-graz.at

GPS online
Further information about GPS (back volumes, electronic publication, etc.) is available at the publisher's web site: http://www.rodopi.nl, section "Series & Journals".

www.ingramcontent.com/pod-product-compliance
Lightning Source LLC
Chambersburg PA
CBHW071415230926
51846CB00001B/99
9789042035584